Sustainable Development Goals Series

The **Sustainable Development Goals Series** is Springer Nature's inaugural cross-imprint book series that addresses and supports the United Nations' seventeen Sustainable Development Goals. The series fosters comprehensive research focused on these global targets and endeavours to address some of society's greatest grand challenges. The SDGs are inherently multidisciplinary, and they bring people working across different fields together and working towards a common goal. In this spirit, the Sustainable Development Goals series is the first at Springer Nature to publish books under both the Springer and Palgrave Macmillan imprints, bringing the strengths of our imprints together.

The Sustainable Development Goals Series is organized into eighteen subseries: one subseries based around each of the seventeen respective Sustainable Development Goals, and an eighteenth subseries, "Connecting the Goals", which serves as a home for volumes addressing multiple goals or studying the SDGs as a whole. Each subseries is guided by an expert Subseries Advisor with years or decades of experience studying and addressing core components of their respective Goal.

The SDG Series has a remit as broad as the SDGs themselves, and contributions are welcome from scientists, academics, policymakers, and researchers working in fields related to any of the seventeen goals. If you are interested in contributing a monograph or curated volume to the series, please contact the Publishers: Zachary Romano [Springer; zachary.romano@springer.com] and Rachael Ballard [Palgrave Macmillan; rachael.ballard@palgrave.com].

Carolina Rojas Quezada
Editor

Urban Wetlands in Latin America

Protection, Conservation, Innovation, Restoration, and Community for Sustainable and Water Sensitive Cities

Springer

Editor
Carolina Rojas Quezada
Instituto de Estudios Urbanos y Territoriales
Centro de Desarrollo Urbano Sustentable CEDEUS
Instituto Milenio SECOS
Santiago, Chile

ISSN 2523-3084 ISSN 2523-3092 (electronic)
Sustainable Development Goals Series
ISBN 978-3-031-69589-6 ISBN 978-3-031-69590-2 (eBook)
https://doi.org/10.1007/978-3-031-69590-2

This Springer imprint is published by the registered company Springer Nature Switzerland AG
The registered company address is: Gewerbestrasse 11, 6330 Cham, Switzerland

Preface

In the ever-evolving narrative of urbanization, the intrinsic value of urban wetlands is often obscured beneath the shadows of filled and degradation. Urban wetlands are very valuable blue-green infrastructure spaces for human settlements and people, yet they are disappearing as a result of urbanization. The international science literature has tried to document the global destruction, mapping and the multiple ecosystem services provided by these ecosystems, such as water and air cleansing, flood mitigation and recreation, among others (e.g. Barbier et al., 2011; Barbier, 2019; Davidson, Fluent-Chouinsrd & Finlayson, 2018). We want to draw attention to the fact that the loss of urban wetlands affects the sustainability of cities and future generations. This phenomenon of disappearing is worldwide but is particularly intense in the latest decade in Latin America (Darrah et al., 2019; Rojas et al., 2019). For that, this book, entitled *Urban Wetlands in Latin America: Protection, Conservation, Innovation, Restoration, and Community for Sustainable and Water Sensitive Cities*, emerges as a comprehensive exploration, seeking to redefine the discourse surrounding urban wetlands in the Latin American context. While much of the existing literature tends to underscore the decline of these vital ecosystems and their ecosytems services, our approach is decidedly holistic. Through a lens that emphasizes success stories, community engagement, and applied research, the intention is not merely to diagnose the challenges but to present actionable solutions that can pave the way for sustainable and water-sensitive cities in Latin America.

This book is about the contribution of urban wetlands in livable cities. Therefore and given the weakness of public policies implemented in the region for their protection, it is essential to highlight good practices, mechanisms and strategies aimed at Protection, Conservation, Innovation, Restoration, and work with the Community, all of this hand in hand with the 17 Sustainable Development Goals.

In Latin America, the urban wetlands are also directly related to current crises such as climate change and climate justice, the water scarcity and the loss of biodiversity, so also with this book we want to take a step and, without ignoring the effects of urbanization, we want to promote the transition from the paradigms of sustainable cities to water-sensitive cities, that is, cities that manage their water bodies in a sustainable and resilient manner and promote the use of nature-based solutions.

The opening chapters focus on protection and conservation. These chapters, authored by esteemed scholars and practitioners, lay the foundation for a comprehensive exploration of the challenges, regulations, law and transformative potential of urban wetlands in the context of 17 Sustainable Development Goals.

Urban Wetland Losses and Land-Use Conservation Challenges in Three Latin American Cities
Authored by Juanita Aldana from Universidad del Norte, Colombia, and colleagues from Chile and Perú, this chapter serves as a poignant introduction to the overarching theme. Aldana delves into the stark reality of urban wetland losses, emphasizing the challenges posed by land-use changes in three urban wetlands of coastal Latin American cities (Aconcagua in Chile; Ciénaga de Mallorquín in Colombia; and Pantanos de Villa in Perú). Through a meticulous examination, the chapter sets the stage for a deeper understanding of the urgent need for conservation efforts to safeguard these critical ecosystems.

Regulation and Protection of Urban Wetlands: A Comparative Analysis in Chile, Colombia, and Peru
Paola Moschella is the corresponding author from Pontificia Universidad Católica de Perú and colleagues from Colombia and Chile. This chapter conducts a comparative analysis of the regulatory frameworks and protective measures surrounding urban wetlands in Chile, Colombia and Peru. By juxtaposing these approaches, Moschella offers insights into the varied strategies employed across Latin American nations, fostering a nuanced understanding of the legal landscape governing urban wetlands.

Urban Wetlands for Sustainable Cities in Mexico
Authored by Luis Zambrano and Ina Falfán from Mexico, this chapter unfolds as an in-depth examination of the hydrologically diverse landscape of urban wetlands in fostering sustainability within Mexican cities. The authors provide valuable insights into the intricate relationship between urban development and wetland preservation, shedding light on Mexico's unique challenges and potential solutions. Despite historical depletion, Mexico preserves significant wetlands, notably 'Xochimilco and San Gregorio' in Mexico City and the 'Interdunary Lagoons System' in Veracruz, along with four non-Ramsar sites. This chapter serves as a critical lens through which readers can appreciate the significance of integrated planning for sustainable urban growth.

Urban Wetlands Protection Law in Chile. A Successful Tool for Urban Planning
Carolina Rojas, representing Pontificia Universidad Católica de Chile, takes center stage in this chapter. Here, the focus turns to the Chilean experience, exploring the efficacy of urban wetlands protection law as instrumental tools for urban planning. Rojas unfolds a narrative of success, demonstrating how legal frameworks can act as catalysts for the preservation of urban wetlands. This chapter becomes a compelling case study, offering valuable lessons for other regions grappling with similar challenges.

Urban Wetlands in Latin America as Support to 17 SDG: A Guideline to Sustainable Cities to SDG 11

Continuing the exploration, this chapter, authored by Carolina Rojas from Pontificia Universidad Católica de Chile and colleagues from Colombia and Perú, broadens the lens to align urban wetlands with the global agenda of Sustainable Development Goals (SDGs). This chapter serves as a guideline, illustrating how urban wetlands can be pivotal in supporting SDG 11—Sustainable Cities and Communities. Rojas navigates through the intersections of urban wetlands and the broader framework of sustainable development, emphasizing their role as essential contributors to a resilient urban future.

As these initial chapters unfold, the reader is invited on a journey through the complexities and opportunities presented by urban wetlands in Latin America. Each chapter, authored by experts in the field, contributes a unique perspective, laying the groundwork for a deeper understanding of the multifaceted challenges and innovative solutions that define the intricate relationship between urbanization and wetland conservation in the region.

Second part is dedicated to: Innovation and Restoration, the narrative shifts towards innovation and restoration, unveiling a mosaic of chapters that illuminate the transformative potential of urban wetlands in Latin American cities. Authored by distinguished experts, each chapter unfolds as a testament to the dynamic interplay between human ingenuity, environmental stewardship, and the resilience of urban ecosystems.

Exploring the Behavior of Llanquihue's Urban Wetlands: Lessons Learnt and Perspectives for Water-Sensitive Urban Design

Initiating this exploration, Camila Teutsch, the corresponding author, and colleagues from Chile, navigate the intricacies of Llanquihue's urban wetlands in Chile. This chapter not only serves as a comprehensive analysis of their behavior but also extracts invaluable lessons for informing water-sensitive urban design. Teutsch's work is a beacon for understanding the dynamic relationship between urban development and the preservation of wetland ecosystems.

The Use of Urban Wetlands as an Integrating Element for Wastewater and Stormwater Solutions in an Ecosystem-Based Disaster Risk Reduction Approach: A Case Study in Rio de Janeiro, Brazil

Antonio Beleño, the corresponding author and colleagues from Brazil, takes the stage in this chapter with a groundbreaking exploration in Rio de Janeiro, Brazil. This chapter unfolds as a case study, demonstrating the pivotal role of urban wetlands in integrating wastewater and stormwater solutions within an ecosystem-based disaster risk reduction approach. Beleño's work becomes a blueprint for leveraging urban wetlands to address pressing environmental challenges.

Urban Wetlands as Resilient Landscape Infrastructure—The Case of Llanquihue Green Infrastructure Plan, Chile

Osvaldo Moreno, the corresponding author of this chapter, illuminates the innovative Llanquihue Green Infrastructure Plan in Chile. This chapter unveils urban wetlands as resilient landscape infrastructure, showcasing how strategic planning can transform these spaces into pillars of sustainability. Moreno's insights provide a roadmap for cities seeking to harness the ecological resilience of wetlands in their urban planning endeavors.

Restoration of the Rocuant Andalién Coastal Wetland (Metropolitan Concepción, Chile): A Transdisciplinary Proposal

In this chapter, corresponding author Octavio Rojas and colleagues from Chile propel us into the realm of restoration with a transdisciplinary proposal for the Rocuant Andalién Coastal Wetland in Chile. Rojas' work epitomizes the collaborative approach needed to rejuvenate degraded wetlands, offering a model that intertwines ecological restoration with the broader fabric of societal and scientific contributions.

Quebrada Park: Social and Ecological Restoration in Puerto Varas, Chile

Camila Teutsch returns as the corresponding author in this chapter, guiding us through the nuanced landscape of social and ecological restoration in Quebrada Parque, Puerto Varas. This chapter becomes a canvas for understanding how restoration efforts can be intricately intertwined with the well-being of communities, showcasing the harmonious integration of natural and human systems.

The "Re-naturalization" of Green Public Spaces: A Multi-scale Landscape Architecture Proposal for the Integration of Natural Systems in Urban Contexts. The Case of Bosque Londres, Valdivia, Chile

Concluding this section, Paulina Ibieta, Camila Wirsching, and Angela Mimica present this chapter, unveiling a multi-scale landscape proposal for the integration of natural systems in urban contexts. Through the case study of Bosque Londres in Valdivia, this chapter becomes a testament to the transformative power of integrating urban nature, paving the way for harmonious coexistence between urban development and ecological preservation.

As we embark on this journey through innovation and restoration, these chapters collectively echo a resounding call for creative solutions, collaborative approaches, and a renewed commitment to the preservation and rejuvenation of urban wetlands in Latin America.

The Final Section Is Dedicated to engaged Community for Sustainable and Water Sensitive Cities

In the third and captivating section of Urban Wetlands in Latin America, our focus turns to the heart of change—engaged communities. The chapters in this section, penned by insightful authors, illuminate the intricate dance between wetlands, environmental education, and the collective action of communities. As we embark on this section, we discover stories of re-existence, citizenship, co-creation, and revaluation, underscoring the indispensable role communities play in shaping the destiny of urban wetlands.

Wetlands in Resistance: Critical Environmental Education and Community Participation in the Lower Paraná Delta

Authored by Luciano Iribarren and colleagues from Argentina, this chapter is a compelling exploration of wetlands in conflict within the lower Paraná Delta. Iribarren intricately weaves together narratives of critical environmental education and community participation, unveiling a powerful tale of re-existence. This chapter serves as a testament to the resilience of communities in the face of environmental challenges, demonstrating how engagement becomes a catalyst for change.

The Role of Citizenship in the Protection of Wetlands. Comparative Analysis of Three Urban Wetlands in Chile, Colombia, and Peru

Caroline Stamm and colleagues from Chile, Colombia and Perú, the corresponding author of this chapter, take us on a comparative journey across urban wetlands in Chile, Colombia and Peru. This chapter unfolds as an insightful analysis of the role of citizenship in wetlands protection. Stamm's work becomes a valuable resource for understanding the nuanced dynamics of community engagement, offering comparative insights that transcend national boundaries.

Observing the Wetlands of the Huasco River, Chile—Co-creation of a Space for the Dissemination of the Environmental and Cultural Values of the Landscape

In this chapter, Ximena Arizaga and Osvaldo Moreno from Chile lead us to the wetlands of the Huasco River (North of Chile), where co-creation takes center stage. This chapter is an exploration of a space meticulously crafted for the dissemination of environmental and cultural values. Arizaga's work stands as a beacon, showcasing the potential of co-creation in fostering a deep connection between communities and their surrounding landscapes.

Mapocho Wetlands Bioroutes—Itineraries Proposal and Participatory Records

Sandra Iturriaga, the corresponding author of this chapter, guides us through the Biorutas Wetlands of the Mapocho River. This chapter unfolds as an opportunity for revaluation, focusing on itineraries and communities. Iturriaga's work becomes a lens through which we witness the transformative potential of community engagement, where the revaluation of urban wetlands becomes intertwined with the daily lives of those who inhabit them.

As we navigate the chapters in this section, we find ourselves immersed in the stories of communities actively shaping the destiny of urban wetlands. These narratives exemplify the power of community engagement not just as a protective force but as a driving catalyst for innovation, restoration, and the holistic well-being of both people and the ecosystems they call home.

As you see in the book you will find contributions in the following fields.

1. A Holistic Approach to Urban Wetlands:

The prevailing discourse on urban wetlands frequently fixates on their diminishing extent and the consequential loss of ecosystem services. This book, however, takes a pioneering step towards a more positive and constructive narrative. It delves into the success stories and best practices surrounding urban wetlands, accentuating their positive impact on city sustainability and the well-being of future generations. By highlighting mechanisms and strategies for Protection, Conservation, Innovation, Restoration, and Community involvement, the book aims to instigate a paradigm shift, steering away from the prevalent narrative of loss towards a vision of thriving urban wetlands.

2. Addressing Gaps in Global Research:

Global urban wetland research often grapples with a dearth of evidence on critical dimensions such as community engagement, political dynamics influencing real estate development, alignment with the 17 Sustainable Development Goals (SDGs), successful restoration endeavors and the integration of green-blue infrastructure and nature-based solutions. Recognizing these gaps, *Urban Wetlands in Latin America* endeavors to be a catalyst for change. By delving into these underrepresented topics, the book seeks to provide a comprehensive understanding that goes beyond the mere quantification of loss, aiming to be a beacon for informed decision-making and sustainable urban development.

3. Emphasizing Blue-Green Infrastructure:

In an era where urban policies often prioritize the ‘green’ aspects of city planning, this book directs attention towards the often-overlooked 'blue' spaces—urban wetlands. By doing so, it advocates for a balanced and integrated approach to urban development. Urban wetlands, traditionally sidelined, are brought to the forefront as crucial components of blue-green infrastructure. This emphasis not only underscores their ecological significance but also demonstrates their potential in enhancing the quality of life in cities. Through a meticulous exploration of blue spaces, the book aspires to inspire urban planners, policymakers and communities to manage these areas for recreation, water management and overall urban well-being.

4. Filling the Gap in Latin American Literature:

A notable void exists in the literature addressing urban wetland studies in Latin American cities, and this book positions itself as a pioneering contribution. The absence of relevant literature in Springer Nature catalogs amplifies the significance of this endeavor. By addressing this gap, the book not only becomes a unique addition to scholarly discourse but also establishes itself as an essential resource for those who have expressed interest in publications focusing on urban ecosystems and green infrastructure. Through its meticulous examination of Latin American urban wetlands, the book seeks to be a cornerstone for future research, policy development and community engagement in the region.

This inspired book, bringing together academics and practitioners dedicated to the research and realization of projects rescuing the value of urban wetlands in Latin American cities, offers comprehensive coverage. It combines the need to protect and conserve urban wetlands from the pressures of urban growth, while also highlighting successful experiences of restauration and regulation and applied research with the communities. The series serves as a cornerstone for future research, policy development, practitioners and community engagement in the Latin American region.

Santiago, Chile Carolina Rojas Quezada

References

Barbier, E. B. (2019). The value of coastal wetland ecosystem services. In *Coastal Wetlands* (pp. 947–964). Elsevier. https://doi.org/10.1016/b978-0-444-63893-9.00027-7

Barbier, E. B., Hacker, S. D., Kennedy, C., Koch, E. W., Stier, A. C., & Silliman, B. R. (2011). The value of estuarine and coastal ecosystem services. *Ecological Monographs, 81*(2), 169–193. https://doi.org/10.1890/10-1510.1

Darrah, S. E., Shennan-Farpón, Y., Loh, J., Davidson, N. C., Finlayson, C. M., Gardner, R. C., & Walpole, M. J. (2019). Improvements to the Wetland Extent Trends (WET) index as a tool for monitoring natural and human-made wetlands. *Ecological Indicators, 99*, 294–298. https://doi.org/10.1016/j.ecolind.2018.12.032

Davidson, N. C., Fluet-Chouinard, E., & Finlayson, C. M. (2018). Global extent and distribution of wetlands: Trends and issues. *Marine and Freshwater Research, 69*(4), 620. https://doi.org/10.1071/mf17019

Rojas, C., Munizaga, J., Rojas, O., Martínez, C., & Pino, J. (2019). Urban development versus wetland loss in a coastal Latin American city: Lessons for sustainable land use planning. *Land Use Policy, 80*, 47–56. https://doi.org/10.1016/j.landusepol.2018.09.036

Contents

Part I

Protection and Conservation

Urban Wetland Losses and Land-Use Conservation Challenges in Three Latin American Cities

Juanita Aldana-Domínguez,
Carolina Rojas Quezada, Juan Munizaga,
Paola Moschella, Caroline Stamm,
and Carolina Martínez

Contents

Abstract

Wetlands are valuable and threatened ecosystems throughout the world. They are essential for the maintenance of life and ecological balance, in addition to providing a larger number of benefits to human societies. Wetlands reduce the risk of flooding by regulating the flow of water, being essential to reduce the impacts of climate change, especially along the coasts and coastal cities of the world. Urban growth poses a risk to wetlands in and around urban areas. In this work, we quantify the loss trends of wetlands in urban areas in three Latin American countries between the years 2002 and 2019. We

J. Aldana-Domínguez (✉)
Department of Chemistry and Biology, Universidad del Norte Colombia, Barranquilla, Colombia
e-mail: ajuanita@uninorte.edu.co

C. Rojas Quezada
Instituto de Estudios Urbanos y Territoriales, Centro de Desarrollo Urbano Sustentable CEDEUS, Instituto Milenio SECOS, Santiago, Chile
e-mail: carolina.rojas@uc.cl

J. Munizaga
EULA Center, Universidad de Concepción, Concepción, Chile
e-mail: juanmunizaga@udec.cl

P. Moschella
Department of Humanities—Geography and Environment Section, Pontificia Universidad Católica del Perú, Lima, Perú
e-mail: pmoschella@pucp.edu.pe

C. Stamm
Institute of Urban Studies, Pontificia Universidad Católica de Chile, Santiago, Chile
e-mail: castamm@uc.cl

C. Martínez
Department of Geography, Pontificia Universidad Católica de Chile, Santiago, Chile
e-mail: camartinezr@uc.cl

C. Rojas Quezada (ed.), *Urban Wetlands in Latin America*, Sustainable Development Goals Series,
https://doi.org/10.1007/978-3-031-69590-2_1

selected as case studies the wetlands of Aconcagua in Chile, Ciénaga de Mallorquín in Colombia, and Pantanos de Villa in Peru. Through multitemporal analysis of Quickbird satellite imagery collections, we classify land covers at 2.5-m resolution and identify areas covered by wetland. We quantify the losses and gains of the wetland areas and the areas surrounding them. We found that the three wetlands analyzed lost area, confirming the worldwide trend of loss. The Ciénaga de Mallorquín wetland was the one that decreased the most due to coastal dynamics and urbanization. The Pantanos de Villa wetland reduced its area due to the increase in urban areas that grew filling the wetland under the pressure of informal settlements and beach condominiums. Lastly, the Aconcagua wetland was reduced due to the increase in grasslands together with the increase in bare soil, also affecting the beaches and dunes. In all three case studies, urban wetlands decreased due to urban growth. For this reason, we call attention to the urgency of improving urban planning to ensure the maintenance of these key ecosystems to ensure more sustainable and resilient cities, as proposed by Sustainable Development Goal 11.

Keywords

Wetland loss · Land-use changes · Urban wetland · Latin America · Colombia · Chile · Peru

1 Introduction

Wetlands are aquatic ecosystems formally defined by the Ramsar Convention (or global treaty for the conservation of wetlands) as: "areas of marsh, fen, peatland or water, whether natural or artificial, permanent or temporary, with water that is static or flowing, fresh, brackish or salt, including areas of marine water, the depth of which at low tide does not exceed 6 m" (Ramsar 1971). Worldwide, wetlands cover an approximate area of 12.1×10^6 km^2, of which 54% is permanently inundated and 46% is temporarily inundated (Davidson et al. 2018).

Wetlands are degrading and disappearing even though they are fundamental ecosystems for ecological balance, biodiversity maintenance, mitigation and adaptation of climate change, and for sustaining human well-being (Maltby and Acreman 2011; McInnes 2014; Ramsar 2019). It is estimated that since 1900, the world has lost more than 50% of these ecosystems, and the increase in urbanization has been identified as one of the main causes for this loss (Boyer and Polasky 2004; Faulkner 2004; Bishop et al. 2006). In fact, more than 55% of the world's population lives in cities, and it is expected that by 2050, this figure will reach 68, accentuating the problems associated with urbanization (United Nations 2018). As urban growth increases, wetland area decreases. Unfortunately, Latin America leads this worldwide tendency, reporting a loss of 59% of wetlands over the last decades (1970–2015) (Darrah et al. 2019). This loss is combined with the fact that it is one of the poorest and most economically unequal regions on the planet (Cepal 2019). The rate of wetland loss is positively associated with poverty and negatively with socio-ecological well-being parameters such as human development, environmental performance, social progress, and economic status (Let and Pal 2023), highlighting the great challenge that Latin America has to conserve its wetlands and, at the same time, generate better socioeconomic conditions for its population.

Wetlands are particularly relevant for cities because: (1) most large cities are located on the coast, (2) world urban growth has been concentrated in low-elevation coastal zones, and (3) the assistance of wetlands when facing frequent urban disasters has not been properly recognized. According to United Nations (2014), 233 of the world's cities are located in zones that are at high risk for flooding, affecting approximately 663 million people. These urban areas require precaution and more resilient infrastructure. Ramsar highlights the role of wetlands in cities by decreasing the impact of urban flooding caused by strong rainfall, as well as providing a buffer from swells and tsunamis, making cities more resilient (Ramsar 2018). The importance of wetlands could be even greater when facing extreme events, such as the flooding that occurred in Phoenix, USA in 2014 (Kim et al. 2017). In fact, in the northeastern USA, wetlands helped save $625 million dollars in direct damage from the floods caused by Hurricane Sandy in 2012 (Narayan et al. 2017). Wetlands are also essential to combat urban heat islands, as they have a cooling capacity, even 42.3 times higher than that of green spaces (Xue et al. 2019). They also act as carbon sequestering systems, coming to be considered as the world's biotic carbon hotspots (Temmink et al. 2022). However, Latin America is dealing with a loss of wetlands, despite being one of the regions most exposed to flooding (United Nations 2011). Evidently, this tendency also increases the region's vulnerability to climate change (Seto et al. 2011; Hallegatte et al. 2013).

We are facing an alarming scenario, where sustained urban growth is the generalized trend in Latin American cities (United Nations 2018) which in many cases occurs at the expense of natural spaces (Rojas et al. 2013; Aldana-Domínguez et al. 2019). Facing pressure to expand, cities have converted wetlands to developable land by legal and illegal landfills as is the case in Chile and Argentina (Pintos and Sgroi 2012; Rojas 2018). Legal clearing is protected by urban laws that do not incorporate factors such as ecological connectivity, or geomorphological processes (Rojas et al. 2019). As for the illegal clearing, in addition to the environmental impact, high-risk situations are generated, caused by ground instability. This progressive growth of Latin American cities has altered and fragmented wetlands, whether it be for housing or transportation infrastructure, creating coastal zones with degraded and devalued ecosystems that are marginalized from the territory.

Due to this situation in Latin America, the study of urban wetlands is highly relevant, since it can contribute to expanding our base of knowledge, supporting conservation initiatives and generating inputs to improve urban planning. Urban wetlands have been little studied, to the point that the Ramsar Convention recognizes this debt, admitting that urban wetlands have been forgotten (Hettiarachchi et al. 2015). In response to this, these authorities have highlighted the relevance of urban wetlands, and in 2018, they declared them as key sites for making cities healthy and habitable (Ramsar 2018). On the other hand, the lack of information on the boundaries of urban wetlands has further complicated their recognition in urban planning. They are, in fact, a unique ecosystem, very challenging to physically define with remote sensing techniques and geographic information systems, especially when they coexist in heterogeneous landscapes and are located in coastal cities (Adam et al. 2014; Gibril et al. 2020). Furthermore, they do not have a uniform vegetation cover, are highly dynamic and their spectral reflectance is easily disrupted. Additionally, land use has further increased their variability, causing modifications in their vegetation and water levels (Gallant 2015).

The objective of this study is to carry out a spatial and multitemporal analysis with remote sensing, to determine temporal changes in wetland surface area in three Latin American cities: Concón (Chile), Barranquilla (Colombia), and Lima (Peru). It seeks to analyze changes in land use that have led to loss and gain in urban wetlands areas during the period (2002–2019) when the greatest loss of these ecosystems has been reported (Darrah et al. 2019). These data will allow us to detect the main spatial dynamics related to the reduction and alteration of these ecosystems, thereby taking a first step toward the recognition of the status of urban wetlands. This study is part of the "Urban Wetlands in Latin America: a solution for sustainable cities SDG 11" Project (2019–2021).

1.1 Overview of Urban Wetlands in Latin America

Although Latin American countries have joined the Ramsar Convention, and represent 11% of the world's wetlands, the region leads in wetland loss, reaching 59% (Darrah et al. 2019). Specifically, the region has observed losses of wetland surface in cities in Chile, Peru, Colombia, Argentina, and Brazil. Decreases are registered on the Brazilian coast (de Sousa et al. 2011; Wittmann et al. 2015), the Andes and Caribbean regions of Colombia (Patino and Estupinan-Suarez 2016), and the Luján River basin in Argentina (Pintos and Sgroi 2012). However, figures of each Latin American country's contribution and the distribution of most of the Latin American wetlands are still unknown, which is worrisome (Mitsch and Gosselink 2015). In addition to the negative effects associated with the loss of wetlands on human well-being, the ecosystem services that these ecosystems provide to society are diminished (IPBES 2018).

Coastal cities in Latin America are vulnerable to the effects of climate change, which together with the socio-natural risks like flooding means that more people are exposed to danger. Chile, Colombia, and Peru have been chosen as example countries with wetlands that are subject to urbanization and vulnerability to climate change. The three countries have experienced significant urban growth and project further increases for 2030. Additionally, 80% of their population lives in cities, which is representative of the environmental conflicts caused by urbanization and conflicts with wetlands located in cities.

Chile

In Chile, more than 80% of the population lives in cities. There are more than 5 million ha of wetlands, but only 8.7% is protected. None of these protected wetlands are in urban areas. The pressure to build has led to the elimination of wetlands in cities spanning from north to south, especially in the southern-central zone, which happens to have a high recurrence of waterlogging and flooding (Rojas et al. 2014). The Aconcagua-Concón wetland is under pressure from industrial activity and from urban growth. There is evidence of an expansion in the real estate market for second homes (Hidalgo et al. 2016; Martínez et al. 2020), and a concentration of economic activities (ports, tourism, and services). In addition, since 2015 this coastal zone has been seriously affected by extreme events associated with climate change, including swells, meteotsunamis, coastal erosion, and tsunamis like Japan 2011 and Illapel, Chile (Martínez et al. 2011, 2018; Carvajal et al. 2017); all of which have caused considerable damage to the infrastructure and connectivity (Winckler et al. 2017). Additionally, the areas adjacent to the wetland (the Aconcagua River estuary) have been categorized as a low-quality landscape because of the high degree of anthropic intervention, which has led to a further loss of naturalness on the coastal landscape (Rangel-Buitrago et al. 2018).

Colombia

In Colombia, wetlands make up more than 30 million ha, which is equivalent to 26% of its territory. Wetlands have historically been associated with the development of human cultures (Jaramillo et al. 2015). The main cause of wetland transformation has been the change in land use to pastures for raising cattle, agriculture, and deforestation, and to a lesser extent, urbanization (Patino and Estupinan-Suarez 2016). These anthropic changes have led to the transformation of 24% of the country's wetlands. It is estimated that by 2025, the main factors of change in wetlands will be, in first place, the expansion of ranching, and in second place, urban and transportation development (Ricaurte et al. 2017). The Magdalena-Cauca basin (i.e., where the Metropolitan Area of Barranquilla is located) is expected to suffer the greatest changes in land use, with negative effects on wetlands. Barranquilla is a coastal city located at the mouth of the Magdalena River in the Caribbean Sea. This estuary creates a vast area of wetlands that have been greatly altered by anthropic activities throughout history and is highly vulnerable to the effects of global change (Rodríguez 2015; Aldana-Domínguez et al. 2018). The transformation of these ecosystems has had a negative impact on the ecosystem services, mainly on the regulation services (Aldana-Domínguez et al. 2019), which means that new guidelines that allow for recovering and conserving the urban ecosystems must be a priority.

Peru

In Peru, the expansion of cities is one of the main causes of wetland degradation (MINAM—Ministerio del Ambiente 2015), especially in the coastal and desert strip where wetlands have a special value as nuclei of biodiversity and freshwater reserves. The wetland ecosystems cover a total surface of 6.9 million ha, which is equivalent to 5.4% of the territory (MINAM—Ministerio del Ambiente 2019). Among them, the coastal wetlands only occupy 0.04%, and are especially vulnerable to growing real estate pressure and the impact of urbanization on the aquifer. For example, the expansion of the city of Lima between 1990 and 2013 led to the loss of 203 ha of wetlands (Moschella 2018). Studies on the main wetlands in Lima show the degradation and loss of ecosystem services caused by anthropic pressure (Moschella 2012; Aponte and Cano 2013; Pulido and Bermúdez 2018). Although there is a legal framework for protecting wetlands, there are weaknesses in the instruments of protection and an adequate regulation for the application of norms is lacking (Ramirez and Aponte 2018). Further studies are also required to understand the hydrology of these ecosystems and contribute to their conservation (Rodríguez 2017). In this sense, the National Wetland Strategy (MINAM—Ministerio del Ambiente 2015) has identified a lack of studies

Fig. 1 Location of the studied urban coastal wetlands

on the valuing and management of wetlands as well as a weak participation for the conservation of these ecosystems.

2 Methodology

2.1 Study Areas

The following urban coastal wetlands were chosen for this study: (1) Aconcagua (also known as the Concón wetland) in the city of Concón, Chile (32°54′59″–71°30′22″W), (2) la Ciénaga de Mallorquín in Barranquilla, Colombia (11°2′42″N–74°50′42″W), and (3) Pantanos de Villa in Lima, Peru (12°12′35″S–76°59′25″W). They are all located in coastal areas and are subject to the different pressures of changes (Fig. 1).

2.2 Data Processing

Free satellite images from the last two decades (2002–2019) were selected from the collection available from the Quickbird satellite on Google Earth Pro. Then, a buffer of approximately 500 m from the perimeter of each wetland was defined. The images were georeferenced and classified by photointerpretation at a scale of 1:2.000 in ArcGIS Pro.

2.2.1 Preparation and Selection of Satellite Images for Land Cover Classification

Satellite images were selected from Google Earth Pro, according to the following criteria:

- Time: Search for the collections that represent changes over the last 20 years.
- Seasonality: Preference for images with similar seasonal conditions (winter–summer).

These criteria defined the studied time horizon from 2002 to 2019, and the 2.5-m resolution Quickbird sensor (see Table 1). A priori these images are useful for the detection of smaller surfaces such as urban wetlands. The images were georeferenced by ArcGIS Pro by taking control points in KML format on Google Earth Pro. A total of 18 control points were taken in each image for each wetland, which were complemented with points on the ArcGIS Pro basemap and fieldwork in March 2020. Once georeferencing was completed, points with an error greater than 1 m were eliminated (Table 1).

Table 1 Selected images per wetland study area

Wetland study area	Collection year	Image source	Error (RMS)
Aconcagua	2004–2019	Google Earth Pro—ArcGIS Pro	0.45–0.32
Ciénaga de Mallorquín	2002–2019	Google Earth Pro—ArcGIS Pro	0.69–0.62
Pantanos de Villa	2002–2019	Google Earth Pro	0.98–0.73

2.2.2 Classification and Land-Use Covers

Classification was done through a combination of photo interpretation and image tracing of the Quickbird images on ArcGIS Pro and the calculation of the Normalized Difference Vegetation Index (NDVI) that allowed for observing the different types of vegetation and separating the water bodies from bare soils. The recognition of the categories was based on the definitions and reference images shown in Table 2 that allowed for distinguishing natural areas from artificial areas, and using the time series of imagery, to observe trends from 2002/2004 to 2019. Surrounding vegetation that is not part of the wetland was not classified in detail. For the Ciénaga de Mallorquín, two types of wetlands were identified following the Colombian wetland classification system (Ricaurte et al. 2019): permanent wetland (permanently flooded), and semi-permanent wetland (periodically flooded; mainly covered by mangrove forests in this area).

2.2.3 Validation of Land Use and Covers

The validation was done through the capturing of 840 points of control, whose verification was

Table 2 Land-use covers, definitions, and reference signatures

Land covers	Land Use	Description	Referential Photograph
Artificial and productive areas	Transportation Networks	Paved, gravel or dirt roads and tracks.	
	Urban Areas	Populated sectors, plazas, shopping centers and disperse urbanization.	
	Urban green areas	Sectors with artificial vegetation for recreation and relaxation.	
	Industrial	Industrial sectors, factories and supply zones.	
	Agriculture	Agriculture and fodder zones.	
	Plantations	Wood production zonas with plantations	
	Bare soils	Land devoid of or with very scarce vegetation, with no productive use.	
Natural	Water Bodies	Water bodies that do not belong to the wetland category, i.e., ocean.	
	Other types of vegetation	Other types of vegetation, such as scrubs, bushy areas and native woods.	
	Wetlands	Permanent wetland: Wet areas, intertidal zones, tide marshes, swamps and estuaries.	
		Semipermanent wetland: Zones covered in mangroves, specific case for la Ciénaga de Mallorquín.	
	Grasslands	Natural areas covered by low-lying vegetation, temporary or permanent meadows.	
	Dunes	Accumulations of sand stabilized by vegetation.	
	Beaches	Beach areas, without vegetation and dominated by wave action.	

Authors' elaboration, incorporating a subset of images from Google Earth Pro, 2002-2004.

optimized in GIS and with fieldwork conducted in March of 2020. Then, the veracity of the interpreted classification was evaluated with the Kappa Statistic Index.

3 Results

The results of the land use and cover typing for the three wetland areas are shown in Figs. 2, 4, and 6; they generated Kappa Index values of >85% of precision. Therefore, the maps allow for making a statistically valid interpretation. The distribution and changes in land use are reported in Tables 3, 4, and 5, where values from the initial year, gains, losses, and net changes can be observed in the land-use categories identified for each wetland.

A sum of 268 ha of decrease was observed totalizing the three studied wetlands (Table 3), of which 175 correspond to loss (Losses–Gains). The main causes were coastal erosion and the increase in artificial and productive covers such as urbanization and grasslands, among others. Urbanization increased by a total of 162 ha considering the three studied cases.

Of the studied wetlands, La Ciénaga de Mallorquín (Colombia) has suffered the greatest surface loss, followed by Aconcagua (Chile) and Pantanos de Villa (Perú). Additionally, among the three wetlands, Pantanos de Villa is under the most pressure from urbanization in its surroundings (i.e., it is virtually surrounded by urban development with the exception of the "Green Area").

3.1 Aconcagua Wetland Area (Concón, Chile)

In the case of the Aconcagua wetland study area, during the period of analysis (2004–2019) the main interactions and exchanges were between the wetland, grassland, bare soils, beaches, and urban and industrial areas (Fig. 2). First, the category experiencing the greatest gain was grasslands, with 37 ha of net change, closely followed by industrial area with 31 ha and then by urbanization with a net gain of 14 ha (Table 4).

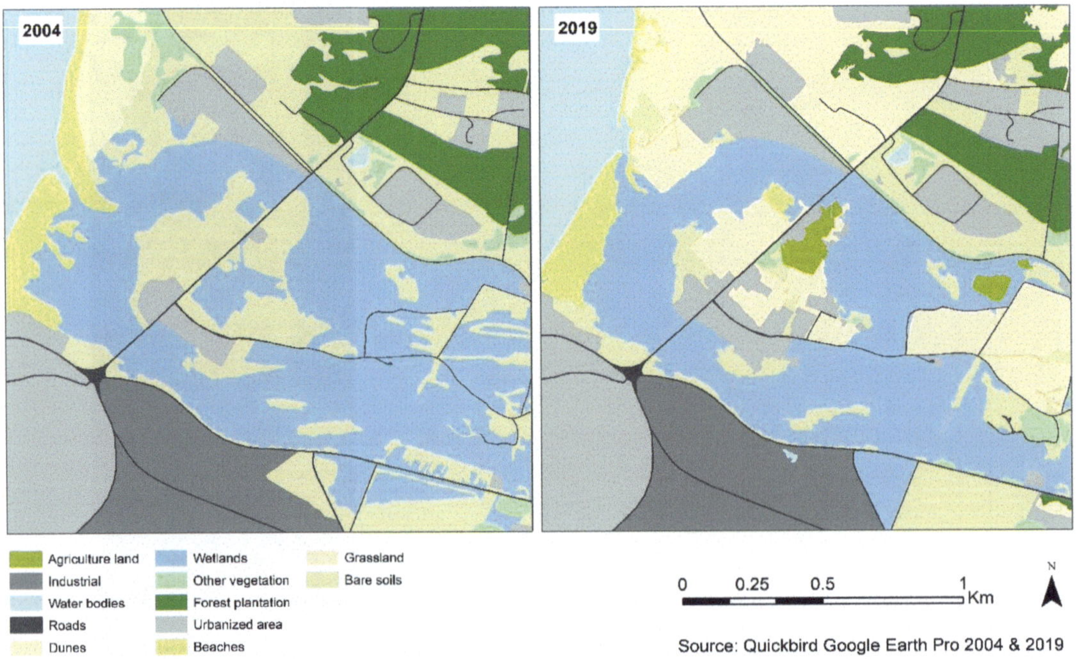

Fig. 2 Distribution of the land uses and cover types detected in the studied wetlands and surrounding areas for the Aconcagua study area (Concón, Chile) from 2004 to 2019

Table 3 Extension in the initial year 2002 (La Ciénaga de Mallorquin and Pantanos de Villa) and 2004 (Aconcagua) and in the final year 2019, total gains and losses in the area of the three wetlands

Wetlands	Extension of the wetland year 2002/2004 (ha)	Extension of the wetland year 2019 (ha)	Total losses (ha)	Total gains (ha)
Aconcagua	116.4	93.2	33	10
La Ciénaga de Mallorquín	873.3	701.2	224	81
Pantanos de Villa	301.2	293	11	2
Total	1290.9	1087.4	268	93

Table 4 Land cover changes in the Aconcagua study area from 2004 to 2019, the analyzed image included the wetland and its surrounding areas

	Aconcagua between 2004 and 2019		
Land use	Losses (ha)	Gains (ha)	Net change (ha)
Water bodies	0	9	8
Roads	0	0	0
Dunes	−5	0	−5
Wetlands	−33	10	−23
Other vegetation	−6	4	−2
Plantation forest	−9	1	−7
Beaches	−10	5	−5
Bare soils	−64	12	−52
Urban	−1	15	14
Agriculture	0	3	3
Grasslands	−29	66	37
Industrial	0	31	31

Note: any difference in net change totals is due to computer round-off

The Aconcagua wetland experienced a net loss of 23 ha, but with a total loss of 33 ha from 2004 to 2019, mainly caused by the conversion to grasslands (17 ha), followed by bare soils (8 ha), beaches (3 ha), and finishing with less pressure of urban zones (2 ha) (Fig. 3). The losses were somewhat compensated for by a gain of 10 ha from bare soils (9 ha) and beaches (1 ha).

Besides the loss of wetlands, bare soils experienced a loss of 64 ha and gain of 12 ha for a net loss of 52 ha. This loss accounted for some of the increase in grasslands surrounding the wetland. Some of the new grasslands also came from beaches. A similar situation has occurred with the dunes—conversion of dunes to grassland; they experienced a net loss of 5 ha (Table 4).

Table 5 Land cover changes in Ciénaga de Mallorquín study area from 2002 to 2019, the analyzed image included the wetland and its surrounding areas

	Ciénaga de Mallorquín between 2002 and 2019		
Land use	Losses (ha)	Gains (ha)	Net change (ha)
Water bodies	−14	128	114
Roads	0	7	6
Other vegetation	−77	18	−59
Beaches	−20	36	15
Bare soils	−82	39	−42
Urban	−1	70	70
Grasslands	0	17	17
Permanent wetland	−175	3	−172
Semi-permanent wetland	−49	78	29
Industrial	−6	28	22

Note: any difference in net change totals is due to computer round-off

3.2 La Ciénaga de Mallorquín Wetland Area (Barranquilla, Colombia)

Both permanent and semi-permanent wetlands at La Ciénaga de Mallorquín changed during the study period (Table 5, Figs. 4 and 5). The main wetland change has been the loss of permanent wetland surface (net loss of 172 ha; Fig. 5a). This loss was mainly due to coastal erosion resulting from the displacement of the sand bar that separates the wetland from the Caribbean Sea. This caused an increase in the water bodies coverage (94 ha; Fig. 5a). The water bodies are represented by the ocean and the river. The wetlands have connections to both but are separated by a sand bar and a spur. The permanent wetland has also been reduced as a result of an increase in the semi-permanent wetland (52 ha), probably because of the natural formation of mangrove areas on the western side of the Ciénaga de Mallorquín, and a recent mangrove reforestation

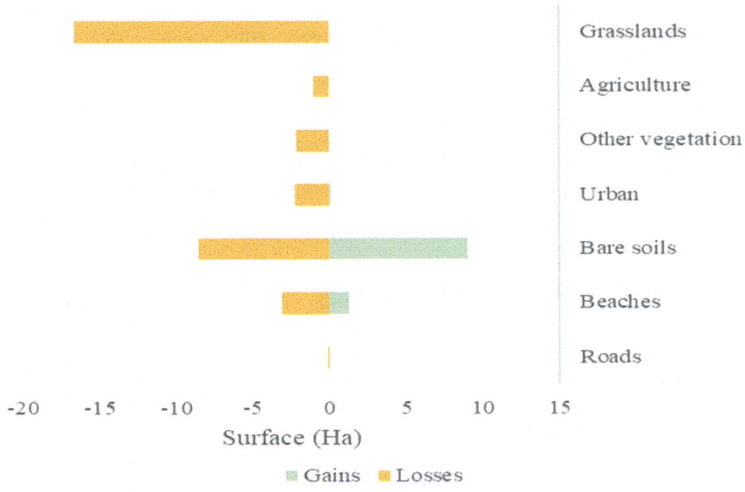

Fig. 3 Gains and losses in the Aconcagua Wetland (Concón, Chile) from 2004 to 2019

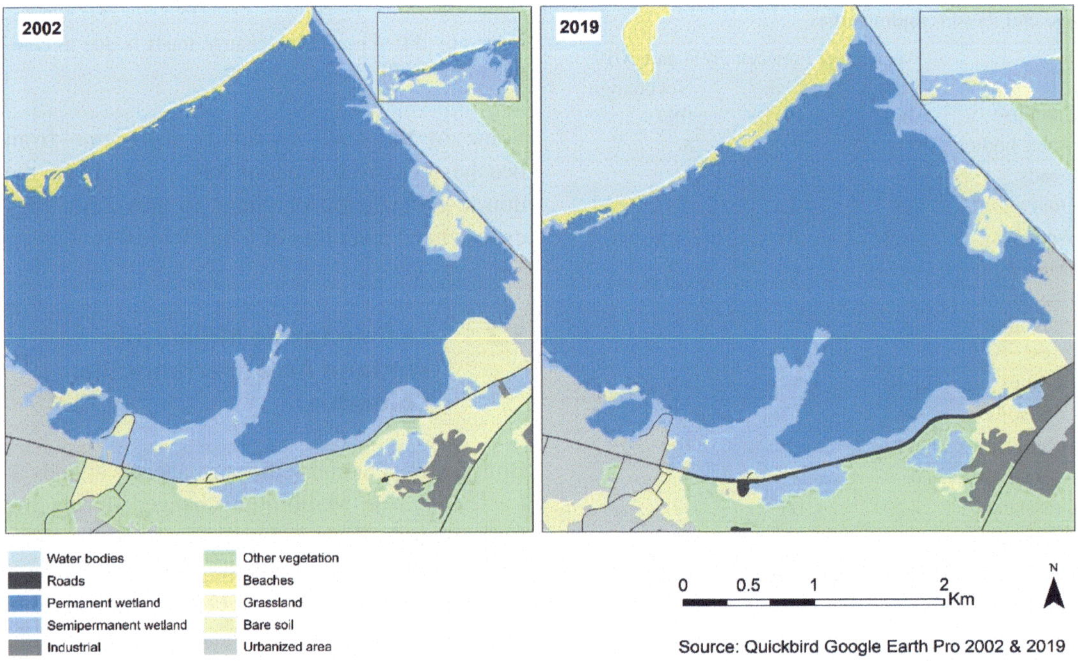

Fig. 4 Distribution of the land uses and cover types detected in the studied wetland and surrounding area for the Ciénaga de Mallorquín study area (Barranquilla, Colombia) from 2002 to 2019

effort. The formation of beaches and conversion to urban development were also responsible for the loss of 22 ha and 6.7 ha, respectively, of permanent wetland.

On the other hand, the semi-permanent wetland gained 52 ha at the expense of permanent wetland and 25 ha from bare soils, while it lost some surface area to water bodies (12 ha), grasslands (10 ha), and urban zones (8 ha; Fig. 5b). Another significant change in the area surrounding the wetland was the loss of other vegetation (59 ha; Table 5), which includes the last few fragments of tropical dry forest surrounding the city of Barranquilla. For the entire study area, as expected, urbanization has increased (70 ha; Table 5), affecting both types of wetlands

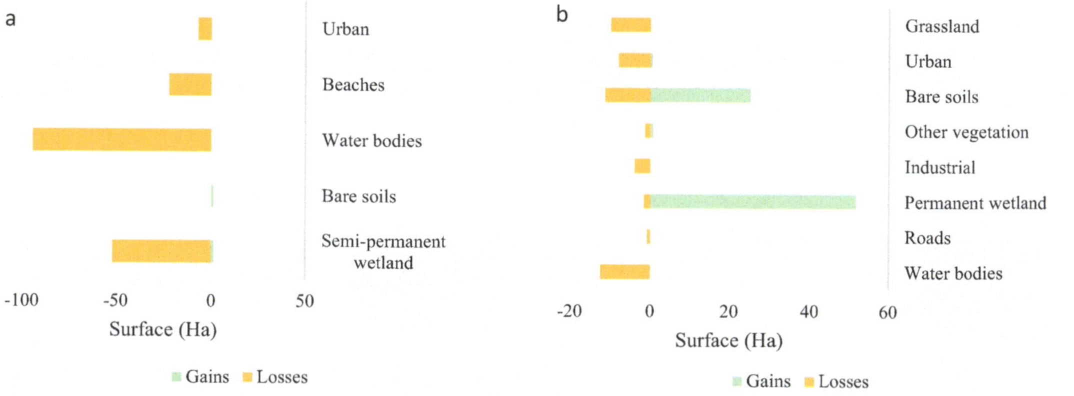

Fig. 5 Gains and losses in (**a**) the permanent wetland and (**b**) the semi-permanent wetland at the Ciénaga de Mallorquín (Barranquilla, Colombia) from 2002 to 2019. The conversion toward water bodies accounted for most of the permanent wetland loss. The figures also show changes in wetland type (permanent v. semi-permanent) (permanent and semi-permanent wetland) and the tropical dry forest.

Table 6 Land cover changes in the Pantanos de Villa study area from 2002 to 2019, the analyzed image included the wetland and its surrounding areas

	Pantanos between 2002 and 2019		
Land use	Losses (ha)	Gains (ha)	Net change (ha)
Water bodies	0	9	9
Roads	−1	5	4
Wetlands	−11	2	−8
Beaches	−11	2	−9
Bare soils	−76	7	−70
Urban	−8	86	78
Agriculture	−1	0	−1
Green areas	−3	0	−3

Note: Any difference in net change totals is due to computer round-off

3.3 Pantanos de Villa Study Area (Lima, Peru)

For the Pantanos de Villa study area, the greatest change from 2002 to 2019 was the increase of urban areas (86 ha; Table 6 and Fig. 6), which mostly happened on bare soils, and to a lesser extent on wetlands and beaches. Wetland coverage declined by 11 ha: 6 ha due to urbanization, 2 ha lost to bare soils, and the remainder to beaches and roads. Wetland gained only 2 ha, with the most significant corresponding to wetland expansion over an unbeaten path (Fig. 7).

4 Discussion

Our study shows an important reduction in the surface area of most of the studied urban coastal wetlands, mainly caused by three factors: coastal erosion/creation of water bodies, expansion of grassland, and growth of urban areas surrounding them. The land-use cover changes observed over the last two decades themselves confirm the general trend of wetland loss in the Latin American region reported by Darrah et al. (2019).

The Ciénaga de Mallorquín wetland has suffered the greatest loss, despite being included in the Ramsar Site “Estuarine system of the Magdalena River Ciénaga Grande de Santa Marta”. The main changes are due to coastal erosion caused by coastal dynamics and changes in the water body and vegetation. Coastal erosion is a complex phenomenon and, in this area, it is related to oceanographic processes (i.e., currents and waves generated by the wind and tides) (Rivillas-Ospina et al. 2017) and the anthropogenic impacts originated by the construction of infrastructures to maintain the Barranquilla port and the navigable channel (Villalón 2017;

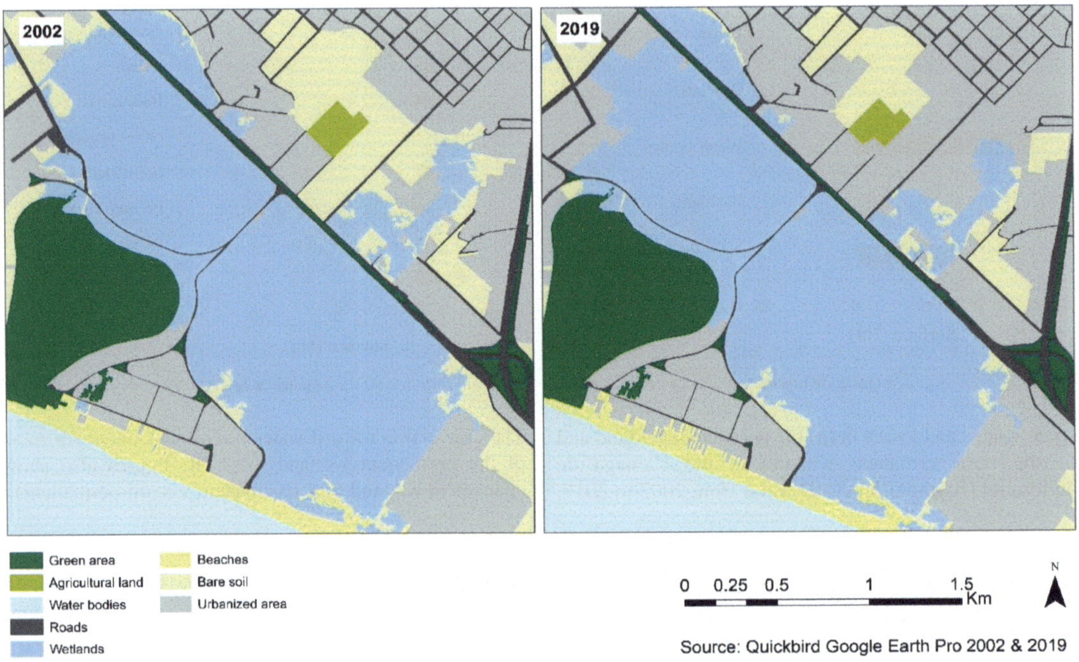

Fig. 6 Distribution of the land uses and cover types detected in the studied wetland and surrounding area for the Pantanos de Villa wetland (Lima, Peru) from 2002 to 2019. The larger "Green Area" is a private golf course

Fig. 7 Gains and losses in the Pantanos de Villa Wetland (Lima, Perú) from 2002 to 2019. There were small changes in the wetland area, mainly a few hectares of losses

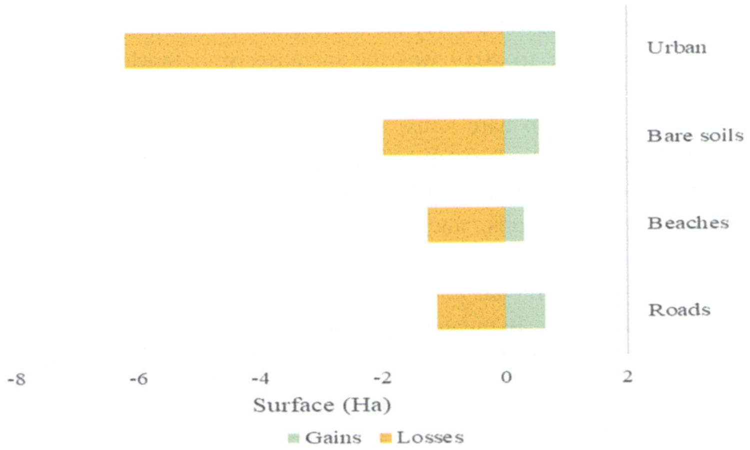

Aldana-Domínguez et al. 2018). The sand bar that maintains the wetland and its physical-chemical conditions separated from the Caribbean Sea undergoes a retraction process that has been documented by Rivillas-Ospina et al. (2017) who observed an erosion rate of 0.14 m/year from 1973 to 2016.

The spreading of grasslands could be interpreted as pre-construction activity, where grassland increases over bare soils and agricultural areas mostly in the case of Aconcagua. The Aconcagua wetland area, although located in an area under a great deal of historic pressure from industrial activity (Martínez et al. 2020), is currently being altered by an increase of grasslands and bare soils especially when the dunes started to increase with low vegetation due to natural conditions as a precipitation, which is also affecting the losses in beaches. The Aconcagua wetland is less affected by urban growth and agriculture areas.

Clearly, urbanization is affecting these three study areas. The Ciénaga de Mallorquin is the most urbanized wetland studied, losing 15 ha due to urban expansion. Two sectors of urban growth are recognized: the eastern side of the wetland where unplanned settlements built by displaced people from other Colombian regions, and even from other countries, come to Barranquilla seeking better life opportunities (Aldana-Domínguez et al. 2018). And on the western side of the wetland, in addition to the informal settlements, there are the planned social interest urban developments to fulfill the housing deficit of the population with lower monetary resources. Urbanization has also caused the destruction of the tropical dry forest, one of the most threatened ecosystems and at risk of collapse in Colombia (Etter et al. 2017), which is a key ecosystem for the supply of ecosystem services in Barranquilla (Aldana-Domínguez et al. 2019). Despite the fact that Barranquilla has stood out for promoting the concept of BiodiverCity (Aldana-Domínguez et al. 2022) to include biodiversity in urban planning, and the construction of an eco-park in the Ciénaga de Mallorquín is currently being completed, reality shows that the ecosystems around this wetland continue to disappear due to indiscriminate urban growth (Brieva 2022).

In a similar way, in Pantanos de Villa the urban areas have increased at a faster rate, although our results show that the variation in the wetland's extension seems limited, it is worth mentioning that this wetland has been converted to urban land for several decades before the period of this analysis (Moschella 2018). While most of the new urban areas are registered here as conversions from bare soils without use, these bare soils are actually cleared wetland (i.e., they were mainly filled to dry the land for future plotting and building). Currently, the wetland is practically restricted to the area that is protected and recognized as a Ramsar site. Since its borders are mostly developed, it is most likely that its expanse will not vary. Additionally, it is worth mentioning that there are significant differences between the degree of urban planning and its impact on the wetland's functioning. To the northeast of the wetland, irregular low-income settlements and filled wetlands can be found. Here, recent land-use changes affect the main springs that supply the wetland and the water channels that connect them, putting the quality and quantity of surface flow entering the wetland at risk. Specifically, there is an increase of dwellers that use the spring water as they lack piped drinking water. Also, a near and unstable landfilling might block the channels. For instance, a public sports field obstructs the water flow. Meanwhile, the southwest end presents mostly beach urbanization through construction of high-income condominiums, which have drained the wetlands and have also reduced the ecosystem's connection with the coast. The urbanization of the Aconcagua wetland is less than the growth of the grasslands in the study period. This is consistent with the changes in the coastal zone that have deteriorated the wetlands and dunes of Concón (Castro and Aliaga 2010; Martínez et al. 2020).

Latin America faces a great challenge in advancing toward urban sustainability according to the 17 Sustainable Development Goals established by the 2030 Agenda (United Nations 2015) where Goal 11 has a target to make cities and human settlements inclusive, safe, resilient, and sustainable. The region also needs to improve urban planning and management regarding the identification, definition, and norms for permitted uses of urban wetlands and their surroundings. Urban wetlands should be "recognized spaces" in the city—natural habitats important for the well-being of the city residents and kept safe, resilient, and sustainable facing the challenges of climate change. Despite the global trend of wetland loss and the trend for Latin America described in the results, all is not lost. Wetlands are ecosystems that can be restored using the best scientific knowledge (Zedler 2000); they can be managed involving different stakeholders such as universities, citizens, and children, as in the case of Pantanos de Villa; and they can adapt to the effects of climate change if urban planning and coastal management leave more accommodation space, that is, the vertical and lateral space available for fine sediments to accumulate and be colonized by wetland vegetation (Schuerch et al. 2018). So, if we want to continue enjoying the ecosystem services provided by wetlands, it is necessary to make decisive progress in the effective management of these ecosystems and their conservation and restoration must be among humanity's main priorities (Albert et al. 2021).

5 Conclusion

The analysis of Quickbird images from the period of 2002 to 2019 has allowed for the creation of valid maps that show land-use cover, including wetlands, during two time periods. Consequently, it has permitted spatial and temporal tracking of wetland trends in selected urban wetlands in Latin America and provided information on land cover changes over the past 20 years. Our study shows a significant reduction in the surface area of all studied coastal urban wetlands, confirming the global trend of wetlands loss. The main factors responsible for this trend are: coastal erosion, expansion of grasslands, and growth of urban areas.

In the three case studies, we show that urbanization continues to impact urban wetlands, as well as, the ecosystems that surround them such as beaches, grassland, and tropical dry forest. The Ciénaga de Mallorquín wetland was the one that changed the most, with a loss of 143 ha due to coastal dynamics and urbanization. In contrast, the Aconcagua wetland lost 23 ha due to the increase in grasslands along with the increase in bare soil. The Pantanos de Villa wetland lost the least amount of wetland surface, reducing by 9 ha during the study period due to the increase in urban areas.

Latin American cities have great challenges to conserve their wetlands and, at the same time, improve the socio-economic conditions for their growing urban population. For this reason, cities must transform their relationship with urban wetlands and place them at the center of their efforts to achieve more inclusive, safe, resilient, and sustainable cities and human settlements as established by SDG 11.

Acknowledgments This study is part of the research project "Urban Wetlands in Latin America: a solution for sustainable cities SDG 11" (2019–2020). It was financed by the CODS—Latin American Sustainable Development Goals Center, whose general objective is to strengthen interdisciplinary research and cooperation on urban wetlands, recognizing that they are key ecosystems fundamental for fostering more sustainable and resilient cities (i.e., by 2030, as mandated in the SDG 11 for Latin American countries), and therefore strengthening their connection with public policy.

References

Adam E, Mutanga O, Odindi J, Abdel-Rahman EM (2014) Land-use/cover classification in a heterogeneous coastal landscape using RapidEye imagery: evaluating the performance of random forest and support vector machines classifiers. Int J Remote Sens 35:3440–3458. https://doi.org/10.1080/01431161.2014.903435

Albert JS, Destouni G, Duke-Sylvester SM, et al (2021) Scientists' warning to humanity on the freshwater biodiversity crisis. Ambio 50:85–94. https://doi.org/10.1007/s13280-020-01318-8

Aldana-Domínguez J, Montes C, González JA (2018) Understanding the Past to Envision a Sustainable Future: A Social–Ecological History of the Barranquilla Metropolitan Area (Colombia). Sustainability 2018, Vol 10, Page 2247 10:2247. https://doi.org/10.3390/SU10072247

Aldana-Domínguez J, Palomo I, Arellana J, Gómez de la Rosa C (2022) Unpacking the complexity of nature's contributions to human well-being: lessons to transform the Barranquilla Metropolitan Area into a BiodiverCity. Ecosystems and People 18:430–446. https://doi.org/10.1080/26395916.2022.2097477

Aldana-Domínguez J, Palomo I, Gutiérrez-Angonese J, et al (2019) Assessing the effects of past and future land cover changes in ecosystem services, disservices and biodiversity: A case study in Barranquilla Metropolitan Area (BMA), Colombia. Ecosyst Serv 37:100915. https://doi.org/10.1016/j.ecoser.2019.100915

Aponte H, Cano A (2013) Estudio florístico comparativo de seis humedales de la costa de Lima (Perú): actualización y nuevos retos para su conservación. Revista Latinoamericana de Conservación 3:15–17

Bishop MJ, Powers SP, Porter HJ, Peterson CH (2006) Benthic biological effects of seasonal hypoxia in a eutrophic estuary predate rapid coastal development. Estuar Coast Shelf Sci 70:415–422. https://doi.org/10.1016/j.ecss.2006.06.031

Boyer T, Polasky S (2004) Valuing urban wetlands: A review of non-market valuation studies. Wetlands 24:744–755

Brieva H (2022) El ecocidio de Ciudad Mallorquín. https://www.elheraldo.co/columnas-de-opinion/

horacio-brieva/el-ecocidio-de-ciudad-mallorquin-columna-de-horacio-brieva-951225
Carvajal M, Contreras-López M, Winckler P, Sepúlveda I (2017) Meteotsunamis Occurring Along the Southwest Coast of South America During an Intense Storm. Pure Appl Geophys 174:3313–3323. https://doi.org/10.1007/s00024-017-1584-0
Castro CP, Aliaga C (2010) Evaluación de la pérdida de suelo, asociada al proceso de expansión urbana y reconversión productiva: Caso: comunas de Los Andes, Quillota y Concón, valle del Aconcagua. Revista de Geografia Norte Grande 45:41–49
Cepal (2019) Informe de avance cuatrienal sobre el progreso y los desafíos regionales de la Agenda 2030 para el Desarrollo Sostenible en América Latina y el Caribe. Online https://www.cepal.org/es/publicaciones/44551-informe-avance-cuatrienal-progreso-desafios-regionales-la-agenda-2030-desarrollo
Darrah SE, Shennan-Farpón Y, Loh J, et al (2019) Improvements to the Wetland Extent Trends (WET) index as a tool for monitoring natural and human-made wetlands. Ecol Indic 99:294–298. https://doi.org/10.1016/j.ecolind.2018.12.032
Davidson NC, Fluet-Chouinard E, Finlayson CM (2018) Global extent and distribution of wetlands: trends and issues. Marine and Freshwater Research. Mar Freshw Res 69:620–627
de Sousa PT, Fernandez Piedade MT, Candotti E (2011) Brazil's forest code puts wetlands at risk. Nature 478:458. https://doi.org/10.1038/478458b
Etter A, Andrade A, Saavedra K, et al (2017) Estado de los ecosistemas colombianos: una aplicación de la metodología de la Lista Roja de Ecosistemas ver. 2.0. Bogotá, Colombia
Faulkner S (2004) Urbanization impacts on the structure and function of forested wetlands. Urban Ecosyst 7:89–106. https://doi.org/10.1023/b:ueco.0000036269.56249.66
Gallant AL (2015) The challenges of remote monitoring of wetlands. Remote Sens (Basel) 7:10938–10950
Gibril MBA, Kalantar B, Al-Ruzouq R, et al (2020) Mapping heterogeneous urban landscapes from the fusion of digital surface model and unmanned aerial vehicle-based images using adaptive multiscale image segmentation and classification. Remote Sens (Basel) 12:1081. https://doi.org/10.3390/rs12071081
Hallegatte S, Green C, Nicholls RJ, Corfee-Morlot J (2013) Future flood losses in major coastal cities. Nat Clim Chang 3:802–806. https://doi.org/10.1038/nclimate1979
Hettiarachchi M, Morrison TH, McAlpine C (2015) Forty-three years of ramsar and urban wetlands. Global Environmental Change 32:57–66. https://doi.org/10.1016/j.gloenvcha.2015.02.009
Hidalgo R, Arenas F, Santana D (2016) ¿Utópolis o distópolis?: Producción inmobiliaria y metropolización en el litoral central de Chile (1992-2012). Eure 42:27–54. https://doi.org/10.4067/S0250-71612016000200002
IPBES (2018) Summary for policymakers of the regional assessment report on biodiversity and ecosystem services for the Americas of the Intergovernmental Science-Policy Platform on Biodiversity and Ecosystem Services. J. Rice, C.S. Seixas, M.E. Zaccagnini, M. Bedoya-Gaitán, N. Valderrama, C.B. Anderson, M.T.K. Arroyo, M. Bustamante, J. Cavender-Bares, A. Díaz-de-León, S. Fennessy, J. R. García Marquez, K. Garcia, E.H. Helmer, B. Herrera, B. Klatt, J.P. Ometo, V. Rodrig, Bonn, Germany
Jaramillo U, Cortés-Duque J, Flórez C (2015) Colombia Anfibia, un país de humedales. Volumen 1. Instituto Alexander von Humboldt, Bogotá, Colombia
Kim Y, Eisenberg DA, Bondank EN, et al (2017) Fail-safe and safe-to-fail adaptation: decision-making for urban flooding under climate change. Clim Change 145:397–412. https://doi.org/10.1007/s10584-017-2090-1
Let M, Pal S (2023) Socio-ecological well-being perspectives of wetland loss scenario: A review. J Environ Manage 326:116692
Maltby E, Acreman MC (2011) Ecosystem services of wetlands: pathfinder for a new paradigm. Hydrological Sciences Journal 56:1341–1356. https://doi.org/10.1080/02626667.2011.631014
Martínez C, Contreras-López M, Winckler P, et al (2018) Coastal erosion in central Chile: A new hazard? Ocean Coast Manag 156:141–155. https://doi.org/10.1016/j.ocecoaman.2017.07.011
Martínez C, López P, Rojas C, et al (2020) A sustainability index for anthropized and urbanized coasts: The case of Concón Bay, central Chile. Applied Geography 116:102166. https://doi.org/10.1016/j.apgeog.2020.102166
Martínez C, Quezada M, Rubio P (2011) Historical changes in the shoreline and littoral processes on a headland bay beach in central Chile. Geomorphology 135:80–96. https://doi.org/10.1016/j.geomorph.2011.07.027
McInnes RJ (2014) Recognising wetland ecosystem services within urban case studies. Mar Freshw Res 65:575–588. https://doi.org/10.1071/MF13006
MINAM—Ministerio del Ambiente P (2019) Mapa Nacional de Ecosistemas del Perú. In: https://sinia.minam.gob.pe/mapas/mapa-nacional-ecosistemas-peru
MINAM—Ministerio del Ambiente Perú (2015) Estrategia Nacional de Humedales. Aprobada por Decreto Supremo N° 004-2015-MINAM.
Mitsch W, Gosselink J (2015) Wetlands
Moschella P (2018) Peri-urbanization and land management sustainability in Peruvian cities. Https://tel.archives-ouvertes.fr/tel-02144701, Université de Strasbourg
Moschella P (2012) Variación y protección de humedales costeros frente a procesos de urbanización: casos Ventanilla y Puerto Viejo. Pontificia Universidad Católica del Perú
Narayan S, Beck MW, Wilson P, et al (2017) The Value of Coastal Wetlands for Flood Damage Reduction in the

Northeastern USA. Sci Rep 7:9463–9462. https://doi.org/10.1038/s41598-017-09269-z
Patino JE, Estupinan-Suarez LM (2016) Hotspots of Wetland Area Loss in Colombia. Wetlands 36:935–943. https://doi.org/10.1007/s13157-016-0806-z
Pintos PA, Sgroi AM (2012) Efectos del urbanismo privado en humedales de la cuenca baja del río Luján, provincia de Buenos Aires, Argentina. Estudio de la megaurbanización San Sebastián. AUGMDOMUS 4:25–48
Pulido VM, Bermúdez L (2018) Estado actual de la conservación de los hábitats de los Pantanos de Villa, Lima, Perú. Arnaldoa 25:679–702. https://doi.org/10.22497/arnaldoa.252.25219
Ramirez DW, Aponte H (2018) Por qué los Humedales de Puerto Viejo perdieron su protección legal: analizando los motivos. Rev Peru Biol 25:49–54. https://doi.org/10.15381/rpb.v25i1.14349
Ramsar (1971) The Ramsar Convention. Ramsar Center, In: Rue Mauverney Gland, Switzerland CH 1196 28:
Ramsar (2019) Humedales: Una solución natural al cambio climático. In: http://www.worldwetlandsday.org
Ramsar (2018) Wetlands: essential for a sustainable urban future. Fact Sheet 10. In: https://www.ramsar.org/sites/default/files/urban_wetlands_en.pdf
Rangel-Buitrago N, Contreras-López M, Martínez C, Williams A (2018) Can coastal scenery be managed? The Valparaíso region, Chile as a case study. Ocean Coast Manag 163:383–400. https://doi.org/10.1016/j.ocecoaman.2018.07.016
Ricaurte LF, Olaya-Rodríguez MH, Cepeda-Valencia J, et al (2017) Future impacts of drivers of change on wetland ecosystem services in Colombia. Global Environmental Change 44:158–169. https://doi.org/10.1016/j.gloenvcha.2017.04.001
Ricaurte LF, Patiño JE, Zambrano DFR, et al (2019) A Classification System for Colombian Wetlands: an Essential Step Forward in Open Environmental Policy-Making. Wetlands 39:971–990. https://doi.org/10.1007/s13157-019-01149-8
Rivillas-Ospina GD, Ruiz-Martinez G, Silva R, et al (2017) Physical and morphological changes to wetlands induced by coastal structures. In: Finkl CW, Makowski C (eds) Coastal Wetlands: Alteration and Remediation. Springer, Cham, Switzerland, pp 275–315
Rodríguez M (2017) Variación de humedales costeros e irrigaciones agrícolas: el caso de la Albúfera de Medio Mundo y el área agrícola de Huaura. Pontificia Universidad Católica del Perú
Rodríguez M (ed). (2015) ¿Para dónde va el Río Magdalena? Riesgos sociales, ambientales y económicos del proyecto de navegabilidad. Foro Nacional Ambiental, Bogota, Colombia
Rojas C (2018) Desafíos en la Planificación Territorial: Humedales Urbanos una oportunidad de gestión y participación para ciudades más sustentables y resilientes. La Vía Medio Ambiental Desafíos y Proyecciones para un Chile Futuro 191–201
Rojas C, Munizaga J, Rojas O, et al (2019) Urban development versus wetland loss in a coastal Latin American city: Lessons for sustainable land use planning. Land use policy 80:47–56. https://doi.org/10.1016/j.landusepol.2018.09.036
Rojas C, Pino J, Basnou C, Vivanco M (2013) Assessing land-use and -cover changes in relation to geographic factors and urban planning in the metropolitan area of Concepción (Chile). Implications for biodiversity conservation. Applied Geography 39:93–103. https://doi.org/10.1016/j.apgeog.2012.12.007
Rojas O, Mardones M, Arumí JL, Aguayo M (2014) A review of fluvial flooding in Chile, period 1574-2012: Causes, effects and recurrence geographical cos. Revista de Geografia Norte Grande 57:177–192
Schuerch M, Spencer T, Temmerman S, et al (2018) Future response of global coastal wetlands to sea-level rise. Nature 561:231–234. https://doi.org/10.1038/s41586-018-0476-5
Seto KC, Fragkias M, Güneralp B, Reilly MK (2011) A meta-analysis of global urban land expansion. PLoS One 6:e23777. https://doi.org/10.1371/journal.pone.0023777
Temmink RJM, Lamers LPM, Angelini C, et al (2022) Recovering wetland biogeomorphic feedbacks to restore the world's biotic carbon hotspots. Science (1979) 376:eabn1479.
United Nations (2018) World Urbanization Prospects: the 2018 revision
United Nations (2014) World urbanization prospects:The 2014 Revision, Highlights. New York, NY, USA
United Nations (2011) World urbanization prospects:The 2011 Revision, Highlights. https://www.un.org/en/development/desa/publications/world-urbanization-prospects-the-2011-revision.html
United Nations (2015) The 17 goals. https://sdgs.un.org/goals. Accessed 1 Nov 2022
Villalón J (2017) Barranquilla y el deterioro de su entorno natural. Huellas 100:4–19
Winckler P, Contreras-López M, Campos-Caba R, et al (2017) El temporal del 8 de agosto de 2015 en la bahia de Valparaíso. Lat Am J Aquat Res 45:622–648
Wittmann F, Householder E, Lopes A, et al (2015) Implementation of the Ramsar Convention on South American wetlands: an update. Research and Reports in Biodiversity Studies 4:47–58. https://doi.org/10.2147/rrbs.s64502
Xue Z, Hou G, Zhang Z, et al (2019) Quantifying the cooling-effects of urban and peri-urban wetlands using remote sensing data: Case study of cities of Northeast China. Landsc Urban Plan 182:92–100. https://doi.org/10.1016/j.landurbplan.2018.10.015
Zedler JB (2000) Progress in wetland restoration ecology. Trends Ecol Evol 15:402–407. https://doi.org/10.1016/S0169-5347(00)01959-5

Regulation and Protection of Urban Wetlands: A Comparative Analysis in Chile, Colombia, and Peru

Paola Moschella, Carolina Rojas Quezada, Juanita Aldana-Dominguez, Caroline Stamm, Carlos Javier Velásquez, and Ingrid Capera

Contents

Abstract

Latin American countries have a set of legal instruments and public policies that contribute to the protection of urban wetlands against the pressure of city expansion that threatens their extent and quality. With the aim of discussing the role of public policies and regulations in the protection of urban wetlands, the situation in Chile, Colombia, and Peru is analyzed in a comparative manner. These countries are examples of diverse regulations and wetland biodiversity in urban environments. Legal instruments for environmental protec-

P. Moschella (✉)
Department of Humanities—Geography and Environment Section, Pontifical Catholic University of Peru, Lima, Peru
e-mail: pmoschella@pucp.edu.pe

C. Rojas Quezada
Instituto de Estudios Urbanos y Territoriales, Centro de Desarrollo Urbano Sustentable CEDEUS, Instituto Milenio SECOS, Santiago, Chile
e-mail: carolina.rojas@uc.cl

J. Aldana-Dominguez
Department of Chemistry and Biology, University of the North, Barranquilla, Colombia
e-mail: ajuanita@uninorte.edu.co

C. Stamm
Institute of Urban Studies, Pontifical Catholic University of Chile, Santiago, Chile
e-mail: castamm@uc.cl

C. J. Velásquez · I. Capera
Department of Law, University of the North, Barranquilla, Colombia
e-mail: cvelasquez@uninorte.edu.co; icapera@uninorte.edu.co

C. Rojas Quezada (ed.), *Urban Wetlands in Latin America*, Sustainable Development Goals Series, https://doi.org/10.1007/978-3-031-69590-2_2

tion and conservation, spatial planning, and urban planning relevant to the management of urban wetlands are analyzed. In addition, advances in regulation and protection of urban wetlands are evaluated and contrasted based on national-level indicators. The lack of effective protection measures is demonstrated. The three countries have adhered to the Ramsar Convention to protect their wetlands and have established Ramsar sites; however, their existence does not stop the pressures and threats to the wetlands. It is identified that Peru has a larger surface area of Ramsar sites and that Colombia has the most extensive Ramsar urban wetland. It is highlighted that Chile has a recent urban wetlands law, while in Colombia and Peru the category of urban wetland does not exist in legislation. Meanwhile, in the protected natural areas, urban wetlands practically have no presence. Among the main weaknesses are the disarticulation between sectors and planning scales; in addition to outdated planning instruments regarding current challenges such as climate change, the water crisis, and the loss of biodiversity.

Keywords

Public policies · Regulation · Protection

1 Introduction

Each Latin American country has a set of legal instruments and public policies or has adhered to global initiatives to protect their wetlands, such as the Convention on Wetlands of International Importance, known as the Ramsar Convention. Latin America is the second region with the largest surface area of wetlands on the List of Wetlands of International Importance, with almost 60 million hectares, implying the commitment of the countries for their conservation (Ramsar Convention on Wetlands [RCW] 2018). However, it is also the region that leads the recent loss of wetlands worldwide between 1970 and 2015 (Darrah et al. 2019).

The causes of wetland loss in Latin America are diverse, and among them is urban growth, which has exerted strong pressure on wetlands causing their reduction and loss of habitats (see Chapter "Urban Wetland Losses and Land-Use Conservation Challenges in Three Latin American Cities" and Rojas et al. 2020). In addition, urban wetlands present particularities, such as a smaller extent and limited biodiversity due to the effects of urbanization in their environment, or a high relevance at the local level rather than at the national or international level. Therefore, they demand special treatment in wetland conservation policies.

Research on policies for urban wetlands is limited. Likewise, the efficiency of national and international policies on wetlands of all types may not adapt well to the situation of urban wetlands due to ecological differences and urban dynamics in the context of globalization (Hettiarachchi et al. 2014).

Therefore, it is necessary to evaluate the legal framework of each country for the conservation of urban wetlands, as well as to demonstrate the effect of the regulatory framework on the conservation of wetlands and urban wetlands. To this end, the objective is to discuss the role of public policies and regulations in the protection of urban wetlands and to comparatively analyze the situation in Chile, Colombia, and Peru as examples of diverse regulations and also as emblematic countries in wetland biodiversity in urban environments.

This study is part of the project "Urban Wetlands in Latin America: a solution for sustainable cities SDG 11" with the objective of "Strengthening research and interdisciplinary cooperation in urban wetlands as key and fundamental ecosystems to enhance more sustainable and resilient cities by 2030, as is the mandate of SGD 11 in Latin American countries and thus strengthen their connection with public policy". As part of the project, an interdisciplinary analysis was carried out at the local and national level in environmental, regulatory, and public policy aspects.

Table 1 Legal and public policy instruments prioritized for analysis in the workshop

Category of instruments	Chile	Colombia	Peru
International agreement	Convention on wetlands of international importance especially as waterfowl habitat (Ramsar)		
Environmental conservation	National Wetland Protection Plan (2018–2022) Law of National Monuments (Nature Sanctuaries) Urban Wetlands Law	Basin Management and Planning Schemes Management Plan for Protected Areas of the National System of Protected Areas National Policy on Inland Wetlands	Basin Water Resources Management Plans Master Plan for Environmental Management and Handling of the Protected Natural Area National Wetlands Strategy
Urban and spatial planning	Urban Wetlands Law Regional Territorial Planning Plan Regulatory Plans (municipal, inter-municipal, metropolitan)	Planning process for physical-territorial and environmental development (PEMOT-POD-POT) Urban Environmental Management Policy	Territorial Planning Plan (POT) Territorial Conditioning Plan (PAT) Urban Development Plan (PDU)

2 Method of Analysis

The study included the comparative analysis of the legal framework for the protection of urban wetlands in Chile, Colombia, and Peru through three components: the review of normative documents and public policy, the calculation of indicators on the state of protection of wetlands, and the participatory evaluation of the legal framework. The methods used in these components are detailed below.

First, an inventory of legal and public policy instruments relevant to the management of urban wetlands was developed for each country. These instruments were classified into three categories: international agreement, environmental conservation, and urban and spatial planning. In total, 60 rules, plans, and public policy documents from the three countries were reviewed.

Subsequently, indicators were calculated from the analysis of documents and spatial analysis to demonstrate the effect of the regulatory framework on the conservation of wetlands and urban wetlands.

Finally, progress in regulation and protection of urban wetlands was evaluated in a participatory workshop in October 2020 with 15 public policy specialists from the three countries. The workshop included the joint development of a SWOT analysis by country to identify the strengths, opportunities, weaknesses, and threats of the prioritized instruments in three categories (see Table 1). Likewise, the following questions were answered: What stands out from the legal framework in the three countries? What public policies (existing or not) are more valuable to implement transversally in Latin America, in order to avoid the loss of urban wetlands?

In the following sections, the results are presented and organized into three categories of instruments: international agreement, environmental conservation, and urban and spatial planning.

3 Ramsar, the International Framework for Wetland Conservation

Among international policies and treaties, wetland conservation is mentioned within the Sustainable Development Goals of the 2030 Agenda and in the Sendai Framework for Disaster Risk Reduction. However, the only international treaty focused on wetlands is the Convention on Wetlands of International Importance especially as Waterfowl Habitat, known as the Ramsar Convention. This convention was adopted in the city of Ramsar (Iran) in 1971 for national action and international cooperation for the conservation and rational use of wetlands.

Currently, all Latin American countries are part of the Convention. In most of these countries, the Convention came into force in the 1990s; however, the first Latin American country where it came into force was Chile, in 1981.

This Convention provides the main instrument for the protection of the most important wetlands in each country, through the designation as a "Ramsar site" for inclusion in the List of Wetlands of International Importance. In addition, among the three countries analyzed, only in Peru there is a specific norm that establishes the national process for the designation of Ramsar sites, highlighting the need for the commitment of local or regional authorities with the support of local actors to ensure the sustainability of the proposed area, including resources for its management.

In terms of advances in wetland protection under this mechanism, Chile has 16 Ramsar sites; Peru has 13; and Colombia has 12. Under this figure, in Peru 57% of the country's wetland surface is protected, in Colombia, 7.5% and in Chile, 0.33% (see Table 2).

It should be noted that the Ramsar Convention does not specifically mention urban wetlands. However, at the 10th Conference of the Parties, Resolution X.27 on Wetlands and urbanization was approved, where the concepts of urban wetland and peri-urban wetland are defined. In addition, this resolution recognizes the importance of urban wetlands and their high risk of degradation; and encourages the designation of urban Ramsar sites.

Thus, among the countries analyzed, Colombia and Peru have an urban wetland designated as a Ramsar site. In Peru, the urban Ramsar wetland is Pantanos de Villa in the city of Lima, designated in 1997 with an area of 263 ha. In Colombia, it is the complex of urban wetlands of the capital district of Bogotá designated in 2018. and composed of 11 ecological parks with an area of 667 ha. Both cases are strategic refuges for dozens of migratory bird species.

The results of the workshop with specialists show that the Ramsar Convention has direct and indirect positive effects on wetland conservation. Directly, the designation of a Ramsar site carries legal implications in some countries, such as the restriction of infrastructure projects in Chile or the prohibition of certain activities in Colombia, which allows for reducing impacts on the ecosystem. Even though there may not always be a legal implication, the international commitment that is assumed lends legitimacy to conservation efforts and even motivates environmental defenders. At the same time, the Convention helps to highlight the importance of wetland conservation. In addition, specialists from Peru and Colombia agree in highlighting that the Convention provides a framework that promotes citizen participation and of different

Table 2 Protected wetlands as Ramsar sites in Chile, Colombia, and Peru

Indicator	Chile	Colombia	Peru
Year of entry into force of the Ramsar Convention	1981	1998	1992
Number of Ramsar sites	16	12	13
Total area of Ramsar sites (ha)	363,927	1,991,536	6,784,041
Area of wetlands in Ramsar sites (ha)	14,898	1,518,983	3,892,501
Percentage of total national wetlands protected in Ramsar sites (%)	0.33	7.49	57.38
Number of urban Ramsar sites	0	1	1
Total area of urban Ramsar sites (ha)	0	667	263

actors, favoring the construction of governance for wetland conservation.

The main limitation of the Convention is the limited impact on the effective protection of urban wetlands. In the three countries analyzed, less than 5% of Ramsar sites are urban wetlands and these cover less than 0.02% of the area of wetlands in Ramsar sites. This is because urban wetlands are hardly prioritized at the national level given the existence of other wetlands of greater extent and species richness.

Furthermore, in all three countries, weaknesses related to public management limit the potential for conservation within the framework of the Ramsar Convention. In Chile, the designation is expected to be requested by a third party and does not imply funds to ensure its management. In Colombia, there are problems with inter-institutional coordination, non-compliance with state commitments, and institutional weakness at the local and environmental sector level. Despite their recognition and importance, the appropriate measures are not taken to comply with their provisions. Similarly, in Peru, the lack of coordination between regional and local management, weak local environmental public policy, and the lack of coordination between wetland protection mechanisms were identified as weaknesses.

Thus, the Ramsar Convention is considered an opportunity to form an international network for conservation and offers a series of tools to improve conservation measures that transcend the local level, making it necessary to make greater efforts to produce tangible conservation results from its application.

4 Policies and Instruments for Urban Wetland Conservation

4.1 Main Policies and Instruments for Urban Wetland Conservation in Selected Countries

The adoption of policies, strategies, or national plans that favor wetland conservation is a recommendation within the framework of the Ramsar Convention. In 1987, only 4 countries of the Ramsar Convention reported having a strategy or plan (RCW 2010), and the number has progressively increased to 73 countries in 2018 with a national wetland policy or equivalent instrument (RCW 2018). Likewise, it is recommended that the national wetland policy consider the following issues: setting conservation objectives, strengthening coordination between government agencies, creating conservation incentives, promoting good management, generating information, raising awareness, and promoting local participation (RCW 2010). Chile, Colombia, and Peru have instruments of this type, under different names, which are presented below.

Chile's National Wetland Protection Plan (2018–2022) was announced in 2018 by the President of the Republic as an initiative of the Government of Chile, through the Ministry of the Environment (MMA), which seeks to contribute to stopping the deterioration of wetlands and preserving their rich biodiversity. It is also a response to the strong civil demand for the protection of these ecosystems. The plan is a commitment made by the Government to declare 40 wetlands in a relatively short period as "nature sanctuaries" and to a lesser extent as "national parks", which contribute to the representativeness of ecosystems with low protection at the national level (MMA 2018).

The National Policy for Inland Wetlands of Colombia 2002, developed by the Ministry of the Environment and the National Environmental Council, aims to promote the conservation and rational use of Colombia's inland wetlands in order to obtain ecological, economic, and sociocultural benefits as an integral part of the country's development. As specific objectives, this policy establishes: integrating the country's wetlands into the processes of land-use and spatial planning; promoting their conservation, rational use, and rehabilitation according to their ecological and socioeconomic characteristics, and promoting and strengthening awareness and sensitization processes at the national, regional, and local level, in relation to their conservation and rational use. The wetland policy states that the declaration of inland wetlands should be framed within the planning processes for development and reiterates the importance of municipal land management for the safeguarding of these ecosystems.

Peru's National Wetland Strategy was established in 2015 and was developed by the Ministry of the Environment with the aim of promoting the conservation and sustainable use of wetlands through four strategic axes: reducing their vulnerability; strengthening the regulatory framework and capacities for management; strengthening participatory management, and promoting traditional knowledge and techniques in wetland management. In this sense, provisions have been approved for the multisectoral management of wetlands that specify the different competent authorities and their specific functions in wetland management. It is worth highlighting the establishment of the Sustainable Management Plan in prioritized wetlands that local governments must develop for wetlands.

It is also important to note that Chile has the recent and unprecedented Urban Wetlands Law No. 21.202 of 2020. This new instrument arises from a high environmental awareness for the protection of wetlands and a series of conflicts in the country for the protection of wetlands in cities, either due to the effects of increasingly frequent extreme events, such as tsunamis and floods, or due to issues of infrastructure and equipment location on and around wetlands. The law modifies other legal bodies such as the Environmental Bases Law and the General Law of Urbanism and Construction, regulating the occupation of ecosystems in urbanized or partially urbanized environments according to the urban limit established by a spatial planning instrument. For the first time, the concept of "urban wetland" is introduced, which is defined as "those extensions of marshes, swamps and peatlands, or surfaces covered with water, whether they are of natural or artificial regime, permanent or temporary, stagnant or flowing, fresh, brackish or salty, including extensions of sea water, whose depth at low tide does not exceed six meters and which are totally or partially within the urban limit". In addition, the law empowers local municipalities to declare their wetlands, considering that they previously had no tools to protect them and stop real estate advancement. Through its regulations, the criteria for delimiting wetlands are established, as well as the development of ordinances and new governance including citizen participation. Finally, a clear territorial impact is that every "urban wetland" declared under this law by the Ministry of the Environment must be included in the spatial planning instruments of its respective city as a "natural value protection area".

Within the framework of these policies, the three countries have developed relatively updated inventories of wetland surface area, with the most recent being Chile's in 2020. The inventories show that Colombia has the largest extent of wetlands, with more than 20 million hectares, representing 19.2% of its continental surface; Peru has 7.7 million hectares of wetlands representing 27% of its continental area; and Chile has 5.6 million hectares constituting 7.4%. Although it is important to consider that the scale of official information gathering from national inventories may omit small urban wetlands. Currently, only Chile takes into account the particularity of urban wetlands in its inventory (see Table 3).

Additionally, all three countries have national systems of natural or wilderness protected areas, where portions of territory are geographically delimited with the purpose of conserving and ensuring biological diversity and associated values of cultural, landscape, and scientific interest. These areas can contain and conserve wetland ecosystems under different conservation categories.

In this way, Chile has the National System of State Protected Wilderness Areas (SNASPE); this has recognized 393,000 ha, only 8.7% of the wetland area. SNASPE areas are not necessarily Ramsar. Thanks to the National Wetland Protection Plan (2018–2022), protected areas in wetlands have increased in the last 5 years. In Colombia, there is the National System of Protected Areas (SINAP) that protects 9.3% of the territory covered by wetlands with 1.8 million hectares. In Peru, there is the National System of Natural Areas Protected by the State, managing

Table 3 Environmental policy indicators and wetland conservation

Indicator	Chile	Colombia	Peru
Wetland policy or law	Urban Wetlands Law (2020) National Wetland Protection Plan of Chile (2018–2022)	National Policy for Inland Wetlands (2002)	National Wetland Strategy (2015)
The policy or law defines urban wetland	Yes	No	No
Year of the Wetland Inventory	2020	2015	2019
Scale of the Wetland Inventory	1:250,000	1:100,000	1:100,000
Total wetland area	5,600,000 ha	20,252,500 ha	7,780,831 ha
Wetland area protected in natural or wilderness protected areas	393,870 ha	1,883,482 ha	633,066 ha
Percentage of wetlands in natural or wilderness protected areas	8.7%	9.3%	8%

to cover 8% of the wetlands with 633,000 ha (see Table 3).

Likewise, there are other mechanisms for the protection of wetlands in their condition as a water resource. In Peru, there are marginal strips, which consist of the land adjacent to the bodies of water that constitute public domain goods for protection, the primary use of water, free transit, fishing, surveillance roads or other services, and where the use for human settlement purposes or others that affect them is prohibited. Similarly in Colombia, the water rounds are the zones or strips of public domain up to 30 m wide from the maximum flood level adjacent to the bodies of water, which aim to allow the normal functioning of the hydrological, geomorphological, and ecosystem dynamics of these bodies of water. In Chile, the zones or strips adjacent to the wetlands can be public or private property, where there may be use restrictions according to the zoning established in each case.

Similarly, wetlands can also be protected through water resource planning instruments. Thus, in Colombia, the Basin Hydrographic Management and Planning Plans (POMCA) carry out the tasks of delimiting the water round and are articulated with management plans for priority wetlands and their delimitation; for this, they have technical guides that guide their formulation. In Peru, the Basin Water Resources Management Plan can consider the conservation of wetlands of interest, although no specific guidelines have been established for this.

4.2 Evaluation of Environmental Conservation Policies and Instruments

The main normative instrument to highlight is the Urban Wetlands Law of Chile, which addresses the specific needs of wetland conservation in urban environments. Also, specialists in the workshop recognized this law as an excellent tool for decentralization, as many decisions depend on regional governments, favoring a process of declaring urban wetlands more expedient than the declaration of a sanctuary that must be approved by the national government. It is also an opportunity to carry out joint work between the Ministry of the Environment and the Ministry of Housing and Urban Planning, to coordinate joint actions in relation to the management of urban wetlands as safe and sustainable park projects and green areas. In addition, the application of the Urban Wetlands Law in the evaluation of projects by the Environmental Impact Assessment System is recognized as a strength, since prior to this law its evaluation was by project typology and not by ecosystem location.

Regarding national wetland policies, in Colombia and Peru, specialists consider them to be outdated. In Colombia, it is identified that the policy does not consider the challenges and the current scenario, which is why it is in a process of review and updating. Meanwhile, in Peru, it is considered that the Strategy is little known which limits its potential. In addition, the national wetland inventory requires an adaptation of the methodology to better identify the different types of wetlands, such as the case of small coastal wetlands that are unprotected or seasonal wetlands. It would be very useful to consider small wetlands in a grouped manner, as is the case with the Bogotá Wetland Complex. Likewise, it is important to recognize and protect the springs that supply the wetlands.

On the other hand, in Chile, the increase in declarations of wetlands as Nature Sanctuaries, reaching 17 sanctuaries by the year 2020, is highlighted as a strength; thus, the representativeness of protected wetlands at the national level has been increasing. The weaknesses lie in the fact that the sanctuaries are not under environmental guardianship and the lack of a Biodiversity and Protected Areas Service, which has not yet been approved by the national Congress. This weakness impacts the lack of financial resources for the management and monitoring of wetlands.

Meanwhile, in Colombia, the variety and articulation of instruments that contribute to the conservation of wetlands, from environmental conservation and water resource management, are highlighted as strengths. In addition, the technical guides that guide the preparation of these. It is highlighted that when there is a management plan, it implies that there is knowledge at a sufficient level of detail to adopt effective measures to protect the ecosystem. Unfortunately, it must be emphasized that, in some cases, all these instruments seem symbolic, since they end up depending on the action of the responsible environmental authority in each case, which does not always formulate the plans as indicated in the regulations and their harmonization with instruments from other sectors is not achieved. Another identified weakness is that greater control of the authorities that are part of the National Environmental System is required.

In Peru, it is considered that the regulatory framework generates a fragmented vision of wetlands, where the Ministry of the Environment considers it as an ecosystem, while the National Water Authority can see it as a water resource. In addition, from the management of water resources, there is a lack of political will to protect wetlands. The lack of environmental control and supervision, and the low valuation of degraded wetlands and of the services that wetlands provide at the local level, are also identified as weaknesses. Among the opportunities, the increasingly active citizen participation that is key in urban areas stands out; and the decentralization of public policies that can favor the conservation of wetlands of local importance.

5 Protection of Urban Wetlands from Urban and Spatial Planning

5.1 Main Instruments for Urban and Spatial Planning in Selected Countries

Each country has processes and planning instruments for urbanism and land management, taking into account the different scales. In Chile, the main spatial planning instruments are the Communal Regulatory Plans (PRC), the Intercommunal Regulatory Plans (PRI), and the Metropolitan Regulatory Plans (PRM). It is important to note that, in Chile, these instruments plan only the urban area leaving out the rural areas. There are also the Regional Territorial Planning Plans (PROT), which cover the entire territory at a regional scale. In Peru, local governments are in charge of Planning the organization of space, land use, and the protection and conservation of the environment are key. The main tools on urban planning are the Urban Development Plans (PDU) or Metropolitan in urban areas and the Territorial Conditioning Plans (PAT) at the provincial level. Additionally, recently, some regional governments have started developing spatial planning plans. In Colombia, Territorial Planning Plans (POT) are formulated at the municipal level, Metropolitan Strategic Territorial Planning Plans (PEMOT), and Departmental Planning Plans (POD).

Regarding urban and spatial planning aspects, urbanization is highly concentrated in specific areas of vast territories; in all three cases, it does not cover 1% of the country. Chile is the country with the highest local urban regulation; it has the most coverage, with 65% (Ministerio de Vivienda y Urbanismo [MINVU] 2020), which shows that planning has focused more on urban aspects than environmental ones in its territory. In Colombia, although all municipalities have a spatial planning instrument, only 17.4% of the POT or equivalents are updated (Ministerio de Vivienda, Ciudad y Territorio [Minvivienda] 2018), a low percentage for adequate planning of cities that have urban wetlands. Similarly, in Peru, only 15.1% of municipalities have a PDU (Instituto Nacional de Estadística e Informática [INEI] 2019) (see Table 4).

In relation to the incorporation of wetlands in spatial planning instruments, in Chile, only from the Urban Wetlands Law, recognized wetlands must be considered in all territorial planning instruments. Before this law, they could be planned as green areas or flood areas, where lands that were not buildable due to their special nature and location had to be indicated, or be classified as a municipal natural reserve. Likewise, urban wetlands recognized at each spatial scale must be included as areas of natural value protection, in order to establish the conditions under which permits for urbanizations or constructions that are developed in them should be granted. In Colombia, wetlands must be incorporated into the various spatial planning plans. In this incorporation, the terms and conditions of an action in the territory are defined and established, constituting themselves as superior norms of mandatory compliance, for the elaboration and adoption of the plans. The above, with the purpose of protecting and conserving areas that require it due to their special nature, as well as articulating the vision of what is necessary to ensure is conserved from an environmental point of view, with spatial and urban planning processes. In Peru, wetlands can be considered in all planning instruments and there must be correspondence between the different levels of planning. Thus, in the PAT, wetlands can be defined as an environmental protection zone. Likewise, the PDU must incorporate the protection and conservation areas that were identified in the PAT or the Economic Ecological Zoning (ZEE), under the conservation land-use category.

Regarding the permitted land uses in wetland areas, it is noted that in Chile they are defined depending on the place where they are located, as follows: "every territorial planning instrument must include existing urban wetlands, and recognized by the Ministry of Environment within the framework of the new Urban Wetlands Law, at each territorial scale as a Natural Value Protection

Table 4 Indicators linked to urban planning

Indicator	Chile	Colombia	Peru
Urban surface	289,781 ha	369,878 ha	345,800 ha
Percentage of urban surface in the country's surface	0.36%	0.36%	0.26%
Percentage of updated urban planning and spatial planning instruments at the local level	65.87% of communes with a PRC	17.4% of municipalities with POT/PROT/EOT	15.1% of municipalities with a PDU
Number of mentions of the word "Wetland" in the National Urban Policy or Plan	2 times in National Urban Development Policy 2014	9 times in the Urban Environmental Management Policy	0 times in the National Housing and Urban Planning Policy

Sources: Instituto Geográfico Agustín Codazzi (2015), Center for International Earth Science Information Network & Columbia University (2013), Instituto Nacional de Estadísticas & MINVU (2018), Minvivienda (2018), MINVU (2020), and INEI (2019)

Area" (art. 60, General Law of Urbanism and Constructions). In some cases, when in rural sectors the wetland has been defined as a limited management protection zone, the Ministry of Agronomy will define the land use, and, if it is within the SNASPE, it will be defined as one of the categories defined by law by categorizing through Supreme Decree. In Colombia, the land uses of the watersheds are defined by the Watershed Management and Planning Plans (POMCA), which are formulated by the competent environmental authority in the jurisdiction of the wetland. Subsequently, the different spatial planning instruments at the different scales (POT, PEMOT, and POD), must incorporate the land uses defined by the environmental instrument as a superior norm and structural content of the plans (environmental determinant). In this way, the importance of the environmental vision is recognized at the urban level, which is reflected in the instruments formulated by the environmental authorities that must, in any case, prevail over everything that opposes it and that has not been considered under the category of determinant. It is worth noting that when there is some land with a declared environmental vocation, such as a wetland, the land is classified as protection land. In the case of Peru, if the wetland is located in a Protected Natural Area, the permitted uses are established in the Master Plan of the area. In the case of wetlands located in an urban area, the use of the land is defined through the Urban Development Plan, formulated by the local government, from a predominantly urbanistic vision. Likewise, a wetland can be delimited by the National Water Authority or considered as a protection zone in the PAT.

Finally, as an indicator of the relevance of wetlands for the National Urban Policy, in Colombia the word wetland or wetlands is mentioned nine times in the Urban Environmental Management Policy, in Peru it is not mentioned in the National Housing and Urbanism Policy to 2030, and in Chile it is mentioned twice in its National Urban Development Policy 2014, which constitutes an indicator of the scant importance that wetlands have in Chilean policies that guide urban planning (see Table 4).

5.2 Evaluation of Urban and Spatial Planning Instruments

Chile

Specialists in the workshop recognized that the main strength is the change that the Urban Wetlands Law introduces in spatial planning, due to the recognition of decreed urban wetlands as areas of natural value. It is even mentioned that this impact is more significant than that which will have in the Environmental Impact Assessment System, which still does not clarify if it will consider all wetlands within the urban limit or only those declared are highlighted. The role of local governments in new governance for urban wetlands is also emphasized, including the participation of citizen organizations.

As weaknesses, we have that planning instruments have been losing over the years their attribution for the zoning of areas of natural value, especially under urban pressure, changing this condition in some plans. In addition, risk studies are not necessarily binding and are not reflected in the zoning of spatial planning instruments due to decisions of the municipal council, being insufficient to prevent building in urban wetlands (e.g., in the case of risk studies after the earthquake and tsunami of 2010 in the Biobío region).

On the other hand, regarding Law 21.200, it is mentioned that, given the number of urban wetlands present in the country, requests for recognition of urban wetlands can be very numerous, which can overload the level of work associated with the declaration of urban wetlands. In the same way, it can also constitute a technical and financial difficulty for municipalities to carry out files. In addition, due to a lack of resources, there may be discoordination or technical ignorance to delimit wetlands between the Regional Environment Secretariats (decentralized bodies of the Ministry in the regions) and the municipalities due to a lack of human resources to assume the regulation of the Urban Wetlands Law. In this sense, the Ministry of the Environment is developing a delimitation guide for municipalities.

The opportunities related to spatial planning instruments lie in the possibility of implementing real landscape planning actions, covering geographical areas such as watersheds. In turn, synergies are seen with the new Regional Territorial Planning Plans with binding character; which can give broad environmental guidelines for the recognition of the value of wetlands and condition their use.

In relation to property, the law also opens the opportunity to recognize wetlands as a national good for public use, although we know that they are not just bodies of water due to property rights. Taking advantage of this instance to change the vision of the owners regarding wetlands, for example, through the real right of conservation, implies changes or giving tax incentives so that owners bet on conservation. Likewise, it is highlighted that the same real estate companies, when making projects with a view to the wetland or a preserved landscape, can pay taxes to acquire resources for conservation.

Colombia

In relation to the spatial planning process, it is emphasized that there has been an evolution and clarification in the planning processes and instruments and, therefore, an improvement in their formulation, especially in environmental aspects.

In virtue of the above, in the second generation POT in municipalities and districts, environmental and risk determinants have been required with greater rigor; for their part, in the PEMOT and POD they have been explicitly included with details related to technical aspects, cartographic representations, and strategic visions. The rural dimension is also required with rigor in the POT, as well as its relationship with urban activities and suburban dynamics, which favors the protection of wetlands. In addition, it is considered that the greater effectiveness of the instrument for the conservation of wetlands is related to the participation of numerous actors and the inclusion as an environmental determinant in spatial plans.

Similarly, the weaknesses do not lie as such in the instruments, but in the application by the competent authorities and in the interaction between those who enjoy and exercise concurrent competencies in the urban-rural and regional-environmental-risk. In many regions of the country, the capacities, skills, and knowledge of municipal mayors are limited, so the proper application of the instrument is uncertain. The above is because there is an imbalance in the appropriation and management of economic resources in the municipalities, which even impacts the construction of sufficiently detailed cartography in aspects such as risk management. In virtue of this, it is necessary to continue limiting or defining in greater detail each of the elements that make up each instrument, so that it is not left to the interpretation of the authorities.

It is also appropriate to establish criteria to achieve the much-desired coherence, agreement, and harmonization among the various planning instruments for development (social, economic, environmental, territorial), as they occur on multiple scales, for which it is becoming difficult to achieve this purpose. In relation to the harmonization of efforts, the threat lies in the undeniable institutional weakness in territorial entities to understand and act harmoniously in relation to the protection of wetlands. What is commonly observed is the antagonistic action at different levels and scales on shared resources that should be managed integrally.

Peru

As a strength of urban planning instruments, the possibility of preserving urban wetlands and regulating their areas of influence as special regulation zones is pointed out; although this land-use category is imprecise for the management that urban wetlands demand, so some municipalities establish their own categories. As these are plans regulated by the Ministry of Housing, Construction, and Sanitation, their regulation and the technical teams that develop them have a predominantly sectorial approach and there is weak articulation with environmental protection strategies. This is observed, for example, in the lack of political will to change zoning for the protection of wetlands.

In addition, it is considered that urban planning instruments are not adequately developed or executed. Likewise, there is a lack of articulation of institutions to apply the instruments, as well as supervision from local governments to ensure compliance with the rules.

Regarding spatial planning, it is pointed out that it is not yet a public policy due to recent changes and lack of direction at the national government level, so there is a significant weakness of instruments for this purpose. In addition, the main actors in territorial management have a weak valuation of wetlands compared to other ecosystems or land uses, hindering their proper incorporation into these instruments.

6 Conclusions

While the Latin American countries analyzed have a varied and broad body of environmental, urban, and spatial public policy regulations that directly or indirectly consider wetlands, these have not been sufficient to ensure their conservation in Latin American cities and stop the threats.

It is noteworthy that Chile has a recent law on urban wetlands, while in Colombia and Peru there is no category of urban wetland in legislation. In addition, in the three countries, there is little flexibility of the conservation frameworks to regulate a highly dynamic ecosystem.

The lack of effective protection measures is demonstrated. The three countries have adhered to the Ramsar Convention to protect their wetlands and have established Ramsar sites; however, their existence does not stop the pressures and threats to the wetlands.

It is identified that Peru has a larger surface area of Ramsar sites and that Colombia has the most extensive Ramsar urban wetland. The designation of a Ramsar site is considered as a complementary conservation strategy, requiring other strategies such as incorporation into national protected area systems.

Meanwhile, in the natural protected areas of the countries, urban wetlands practically have no presence. Among the main weaknesses is the disarticulation between sectors and planning scales; in addition to the lack of updating of existing territorial instruments regarding current challenges such as climate change, the water crisis, and the loss of biodiversity.

7 Recommendations

In normative aspects, the Chilean Urban Wetlands Law is an excellent example that can be replicated in other countries, in case the environmental policy and the growth of cities have normative gaps to integrate a natural ecosystem into spatial planning. Of course, the effectiveness of this law is yet to be evaluated and its progress can be followed on a portal of the Chilean Ministry of the Environment.

Given the real estate pressure of urban development, the inclusion of the concept of urban wetland in the regulations of countries is necessary, as these ecosystems are endangered and present challenges and opportunities different from wetlands in rural environments.

It is important to integrate environmental and spatial planning, as shown by the case of Colombia and the recent Urban Wetlands Law in Chile, prioritizing environmental and sustainability criteria.

Spatial planning should be improved with land use standards that are compatible with the rational use of wetlands and sustainable urban development. That would strengthen housing policy and urban land management without reducing the surface of wetlands due to informal and unsafe constructions.

It is also necessary to include and improve in the spatial planning instruments the characterization of threats and vulnerability in the definition of risk areas, especially of flooding in wetland areas, as well as the inclusion of specific studies of the ecosystem services they provide.

Compliance with existing standards should be promoted and better supervised, including the commitments acquired in the Ramsar Convention on internationally important wetlands.

The conservation of Ramsar wetlands or with other protection figures demands the increase of public funds for the realization of inventories, diagnoses, management plans, and interdisciplinary research.

Acknowledgments We thank the Center for Sustainable Development Goals for Latin America and the Caribbean (CODS) for the funds assigned to this research. In addition, we thank the support of Lesley Calero and Jasson Fernández from the Pontifical Catholic University of Chile in the calculation of indicators.

References

Center for International Earth Science Information Network & Columbia University. (2013). Low Elevation Coastal Zone (LECZ) Urban-Rural Population and Land Area Estimates, Version 2. Palisades, New York: NASA Socioeconomic Data and Applications Center (SEDAC). https://doi.org/10.7927/H4MW2F2J.

Darrah, S. E., Shennan-Farpón, Y., Loh, J., Davidson, N. C., Finlayson, C. M., ... y Walpole, M. J. (2019). Improvements to the Wetland Extent Trends (WET) index as a tool for monitoring natural and human made wetlands. *Ecological Indicators*, 99, 294–298. https://doi.org/10.1016/j.ecolind.2018.12.032

Hettiarachchi, McAlpine, C. & Morrison, T. H. (2014). Governing the urban wetlands: a multiple case-study of policy, institutions and reference points. *Environmental Conservation*, 41 (3): 276–289 https://doi.org/10.1017/S0376892913000519

Instituto Nacional de Estadísticas & Ministerio de Vivienda y Urbanismo—MINVU, Chile. (2018). Metodología para medir el Crecimiento Urbano de las Ciudades de Chile. https://geoarchivos.ine.cl/File/geo/metodologia-para-medir-el-crecimiento-urbano-de-las-ciudades-de-chile.pdf

Instituto Nacional de Estadística e Informática, Perú. (2019). Planificación Municipal. En Perú: Indicadores de Gestión Municipal 2019. https://www.inei.gob.pe/media/MenuRecursivo/publicaciones_digitales/Est/Lib1720/index.html

Instituto Geográfico Agustín Codazzi. (2015). Estadísticas generales del país. https://geoportal.igac.gov.co/contenido/datos-abiertos-igac

Ministerio del Medio Ambiente (MMA), Gobierno de Chile. (2018). Plan Nacional de Protección de Humedales 2018–2022. https://mma.gob.cl/wpcontent/uploads/2018/11/Plan_humedales_Baja_confrase_VERSION-DEFINITIVA.pdf

Ministerio de Vivienda y Urbanismo, Chile. (2020). Seguimiento IPT. http://seguimientoipt.minvu.cl/

Ministerio de Vivienda, Ciudad y Territorio—Minvivienda, Colombia. (2018). 20 años de ordenamiento territorial en Colombia: experiencias, desafíos y herramientas para los actores territoriales. Bogotá: Universidad Nacional de Colombia, Instituto de Estudios Urbanos, Ministerio de Vivienda, Ciudad y Territorio.

Ramsar Convention on Wetlands. (2010). National Wetland Policies: Developing and implementing National Wetland Policies. Ramsar handbooks for the wise use of wetlands, 4th edition, vol. 2. ramsar.org/sites/default/files/documents/library/hbk4-02.pdf

Ramsar Convention on Wetlands. (2018). Global Wetland Outlook: State of the World's Wetlands and their Services to People. ramsar.org/sites/default/files/documents/library/gwo_e.pdf

Rojas, C., Aldana-Domínguez, J., Munizaga, J., Moschella, P., Martínez, C. y Stamm, C. (2020). Urban wetland trends in three Latin American cities during the latest decades (2002–2019): Concón (Chile), Barranquilla (Colombia), and Lima (Peru). *Wetland Science & Practice*, 37(4), 283–293.

Urban Wetlands for Sustainable Cities in Mexico

Ina Falfán and Luis Zambrano

Contents

Abstract

As a hydrologically diverse country, Mexico has thousands of natural and artificial wetlands throughout its territory. Of this large number of wetlands, 144 are included in the Ramsar List, and four of them correspond to wetlands embedded in urban areas. For centuries, there has been a severe depletion of aquatic systems in Mexico; however, there are still many relevant wetlands in urban areas, some of which are currently protected by Mexican laws. Here we focused on continental and coastal lacustrine and estuarine systems for the sustainability of cities. We outlined the importance as cases of conservation in Mexico of two Ramsar sites, the 'Xochimilco and San Gregorio' wetland in Mexico City and the 'Interdunary Lagoons System in the city of Veracruz', and four non-Ramsar sites, 'La Laguna de las Ilusiones' in Villahermosa; 'Laguna La Escondida' in Reynosa; 'El Estero El Salado' in Puerto Vallarta; and the artificial wetland 'El Palote' dam, in León. These urban wetlands support populations of endemic, vulnerable, endangered, and/or critically endangered species of flora and fauna, and, at the same time, they are essential spaces for leisure and recreational activities. However, even with their conservation status, these systems are under intense urban pressure, being necessary to develop better restoration and management strategies for land and biodiversity. The long-term conservation of urban wetlands in Mexico and the ecosystem services they provide to citizens, as well as improving the resilience and sustainability of cities, still require the combined efforts of people and government.

Keywords

Aquatic ecosystems · Urban lakes · Urban lagoons · Urban water bodies · Wetlands benefits · Natural protected areas

I. Falfán · L. Zambrano (✉)
Laboratorio de Restauración Ecológica, Instituto de Biología, Universidad Nacional Autónoma de México, Ciudad de México, Mexico
e-mail: ina.falfan@st.ib.unam.mx; zambrano@ib.unam.mx

C. Rojas Quezada (ed.), *Urban Wetlands in Latin America*, Sustainable Development Goals Series,
https://doi.org/10.1007/978-3-031-69590-2_3

1 Introduction

Worldwide, wetlands are part of urban systems mainly due to two different processes. On the one hand, for centuries, humans have settled near or around wetlands and water bodies to survive and thrive, as these provided food, water, and other raw materials (e.g., fuel, fiber) and a way to dispose of people's waste. Over time, wetlands were embedded by the growing urban centers and often modified so that people could continue to use them for washing clothes, as rainwater storage, as garbage pits and sewers, and as communication and transportation routes (Ghermandi et al. 2010; Molina-Prieto and Rubio 2016; Stefanakis et al. 2014). On the other hand, once the city was/is fully established, the creation of wetlands to provide a wide range of services to people has become a relatively common practice (Stefanakis et al. 2014; White et al. 2020).

Urban wetlands include a wide range of permanent and temporary natural and artificial systems within cities, such as coastal zones, lakes, lagoons, rivers, streams, ponds, estuaries, marshes, tidal flats, peatlands, mangroves, canals, drains, reservoirs, ponds, and stormwater treatment areas (Alikhani et al. 2021; Forman 2014; Haase 2015). However, while natural wetlands are declining globally, artificial ones (generally smaller in size) have increased at the same scale (MEA 2005; Davidson 2014; Darrah et al. 2019; Birch et al. 2022), a trend that is replicated in Latin America (Darrah et al. 2019). Approximately 62% of 6,968,452 ha of natural wetlands in Mexico have been lost for several reasons (Landgrave and Moreno 2012; Moreno 2008). However, within a particular historical logic of desiccating natural wetlands and constructing artificial ones is that, for example, more than 50% of 1834 lakes, lagoons, and other confined water bodies in the 145 most populated Mexican cities are of artificial origin nationwide (Falfán and Zambrano 2023).

Wetlands in urban areas are valuable because of the benefits they provide to people, wildlife, and cities. These benefits, derived from their structure and functions, include fish and fiber supply, water supply, water purification, climate regulation, flood regulation, coastal protection, recreation, and tourism (Alikhani et al. 2021; Ampatzidis and Kershaw 2020; Haase 2015; MEA 2005). Through interaction with natural and artificial wetlands, people benefit their physical, mental, and emotional health and quality of life (Johansson et al. 2019; White et al. 2020). Urban wetlands provide habitat and refuge for aquatic, semiaquatic, and terrestrial wildlife groups (native and exotic) such as birds (including migratory), amphibians, reptiles, fishes, and mammals (Almazán-Núñez and Hinterholzer-Rodríguez 2010; Ramírez-Jiménez et al. 2017; Rueda et al. 2017). Within an appropriate management and planning framework, urban wetlands can help maintain the connectivity of ecosystems fragmented by urbanization, thus providing wildlife corridors (Dallimer et al. 2012). Through urban wetlands system, the individuals of the different species can move from one place to another in search of suitable conditions for their survival, thus contributing to the conservation of species diversity in the fragments and the wetlands themselves (Lee et al. 2022; Yang et al. 2022).

While urban wetlands can contribute to the resilience and sustainability of cities by providing services, they are also vulnerable spaces. Namely, when they cause annoyances or disservices, such as the presence of mosquitoes, bad smells, or flooding, and are therefore perceived as wastelands, the tendency is to disappear/desiccate them to convert them to other urban land uses (Forman 2014; Ibrahim et al. 2012; Poó 2012). Furthermore, they are the first that receive the negative effects of urbanization, urban lifestyle, and urban metabolism, such as polluted discharge water from industries and households, loss of area due to garbage, and the presence of exotic and invasive species (Birch and McCaskie 1999; Mancini et al. 2012; Schueler and Simpson 2001). However, in the current global climate change, urban wetlands, along with green spaces, can be considered part of nature-based solutions aiming to mitigate its adverse effects on cities (Cui et al. 2021; Haase 2017; MEA 2005).

Mexico is ecologically, climatically, geomorphologically, and hydrologically diverse, which has favored the presence of thousands of natural and artificial wetlands throughout its territory, including those in urban areas (Falfán and Zambrano 2023; Ortiz-Arrona et al. 2022; Thornhill et al. 2022). Of these thousands, 144 are Ramsar-listed (CW 2023a), and only four correspond to urban wetlands, namely, Ramsar-listed wetlands in urban areas (CW 2023b). However, no Mexican city is accredited as a Ramsar Wetland City, which is an accreditation, different from the Ramsar List, granted in 2018 and 2022 to cities located near and dependent on wetlands of international importance to highlight and strengthen their conservation, wise use, and sustainable socioeconomic benefits for local people (CW 2023c). Mexican public policies and legal frameworks have favored the establishment of some urban wetlands as natural protected areas (NPAs), mainly of state attribution, recognizing them for their ecological, ecosystem, and conservation value, as well as places for tourism and recreation (GEJ 2020; Palacios and Castillón 2013). NPAs constitute an environmental policy instrument with a broader legal definition for the conservation of biodiversity (CONANP 2023). Here we describe some study cases of protection and conservation of urban wetlands, outlining their processes, why they are essential for conservation, and how these different types of Mexican wetlands favor the sustainability and resilience of the cities where they are located.

2 Methods

We focused on continental and coastal lacustrine and estuarine systems for the sustainability of cities. We reviewed and outlined the importance, as cases of conservation in Mexico, of two Ramsar sites, the 'Xochimilco and San Gregorio' wetland in Mexico City and the 'Interdunary Lagoons System in the city of Veracruz', and four non-Ramsar sites, two lagoons; 'Las Ilusiones' in Villahermosa and 'La Escondida' in Reynosa, one estuary; 'El Salado' in Puerto Vallarta, and the artificial wetland; 'El Palote' dam in León (Figs. 1 and 2). All these urban wetlands are declared NPAs with management plans (Table 1).

Fig. 1 Location of described Mexican urban wetlands

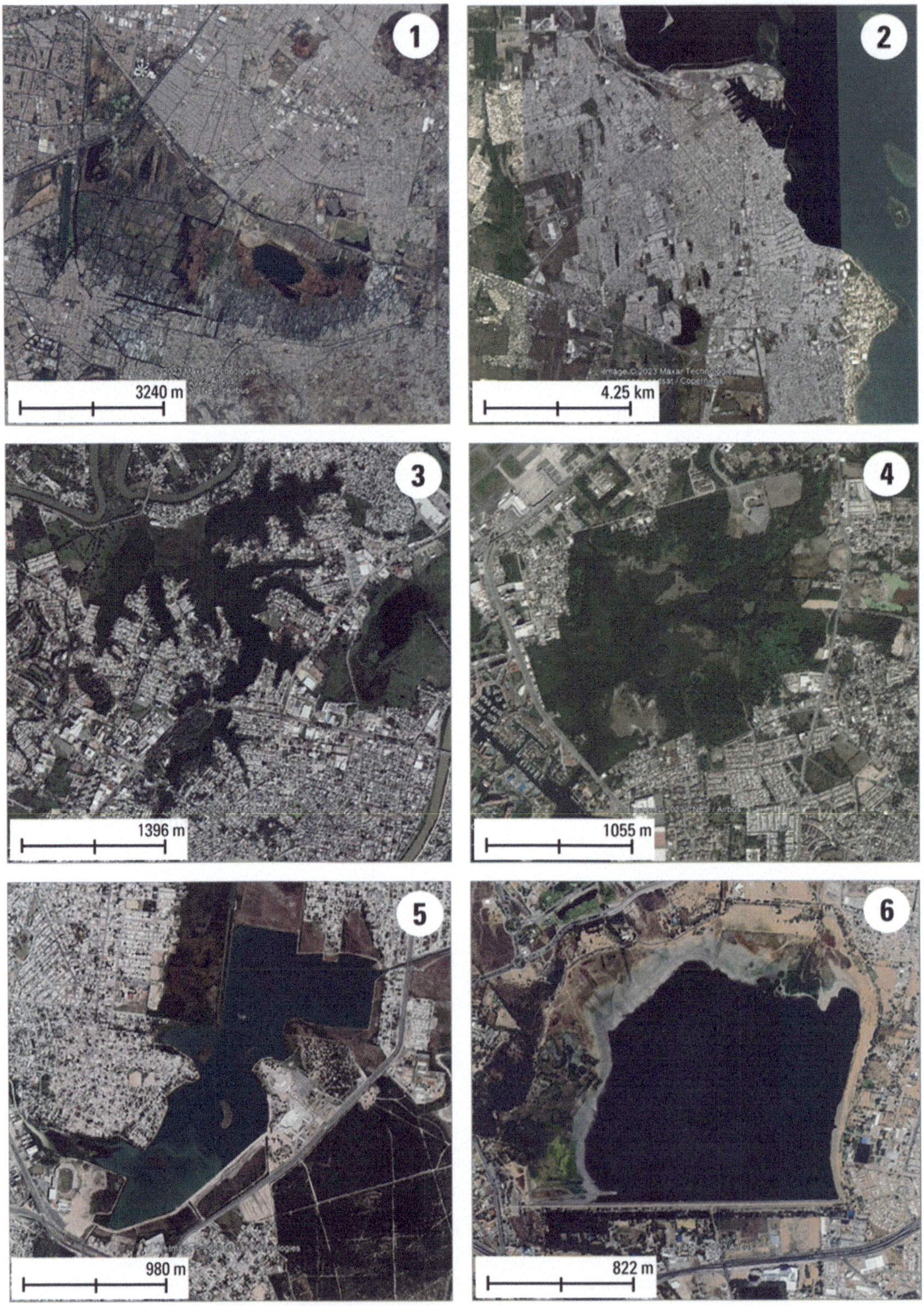

Fig. 2 Views in Google Earth of described Mexican urban wetlands. **1**: Xochimilco and San Gregorio wetland, date: 04/2023, @ 2023 Maxar Technologies, @ 2023 Airbus, @ 2023 CNES/Airbus. **2**: Interdunary Lagoons System of Veracruz, date: 02/2021, @ 2023 Maxar Technologies, Landsat/Copernicus. **3**: Las Ilusiones lagoon, date 04/2022, @ 2023 Maxar Technologies. **4**: El Salado estuary, date: 11/2020, @ 2023 CNES/Airbus. **5**: La Escondida lagoon, date 05/2022, @ Maxar Technologies. **6**: El Palote dam, date: 04/2023, @ 2023 Airbus

Table 1 Mexican urban wetlands information related to their condition as NPAs

Urban wetland (Year declared as state NPA)	NPA category	NPA conservation subprograms	Services provided by the urban wetland
1. Xochimilco and San Gregorio wetland (1992)	Zone subject to ecological conservation	Protection Hydraulic management Ecological restoration Productive use Regulation of recreational, sports, and government activities Research, monitoring, and evaluation Community participation Institutional coordination Administration	Habitat for endemic, native, and endangered species Preservation of remnant ecosystems and wetlands Preservation of an ancient system of production and cultural heritage Water storage and filtration, aquifer recharge and discharge, flood prevention Water for agricultural and urban use Carbon sequestration, biomass export, and climate regulation Tourism, recreation, sports
2. Interdunary Lagoons System in the city of Veracruz (2016)	Multifunctional biological corridor	Protection Management Restoration Knowledge Culture Administration	Preservation of remnant ecosystems Habitat for endemic, native, and endangered species Groundwater recharge Sediment retention, microclimate regulation Water filtration and storage, flood protection Recreation, educational values, and social interaction Aesthetic values and higher capital gain of neighborhoods
3. Las Ilusiones lagoon (1995)	Ecological reserve	Protection Management Restoration Knowledge Culture Administration	Habitat for endemic, native, and endangered species Carbon sequestration, water flow regulation, and microclimate regulation City's identity Aesthetic values Tourism, culture, and recreation Educational values
4. El Salado estuary (2000)	Ecological conservation zone	Protection and surveillance Interpretation and environmental education Research, monitoring, and scientific cooperation Institutional cooperation, extension, and collaboration Natural resources management Construction and maintenance	Preservation of remnant and critical ecosystems Habitat for endemic, native, and endangered species Water storage and filtration, flood prevention Protection against natural hazards and infrastructure failures Carbon sequestration Educational values and recreation Tourism
5. La Escondida lagoon (1997)	Urban Park	Protection Knowledge Management Restoration Culture Administration	Preservation of remnant ecosystems Habitat for endemic, native, and endangered species Water flow regulation and water storage Water for urban use Recreation, sports, and tourism Educational values
6. El Palote dam (2000)	Ecological park	Natural resources management Environmental education and public use Research and monitoring Administration and surveillance	Habitat for endemic, native, and endangered species Water flow regulation and water retention Water for urban use Recreation Educational values

A general panorama of the paths and ways of conservation of urban wetlands in Mexico and how these spaces contribute to sustainable Mexican cities in this descriptive study is based on a literature review and synthesis of the information found. We use information from publicly available sources such as hemerographical, institutional, and tourism websites, NPA decrees, and management programs.

3 Results

3.1 Ramsar Sites

Of the thousands of natural and artificial wetlands in Mexico, to 2023, 144 are included in the Ramsar List (CW 2023a). We address two of the four Ramsar-listed urban wetlands: The Xochimilco and San Gregorio Atlapulco Ejidos Lake System in Mexico City and the Interdunary Lagoons System in the city of Veracruz (CW 2023b).

3.1.1 Xochimilco and San Gregorio Wetland in Mexico City

The Xochimilco and San Gregorio wetland in Mexico City has been a Ramsar site since 2004 covering 2657 ha (CW 2023a). It has been a UNESCO Cultural and Natural World Heritage since 1987 (UNESCO 2023) and an FAO Globally Important Agricultural Heritage System since 2017 (FAO 2017). It is also an NPA by the local government since 1992 with the category of Zone Subject to Ecological Conservation (GDF 2006). All these recognitions for this unique wetland are because of its natural and social history. The broadest objective of the wetland as an NPA is to conserve the natural resources of the ecosystem and the cultural landscape through the integral management of these same resources and the revaluation of pre-Hispanic production techniques (GDF 2006).

Xochimilco and San Gregorio represent a 2% remnant of the system of five ancient lakes over which Mexico City and other settlements are established (Zambrano et al. 2020). Present-day Mexico City was founded centuries ago by indigenous people who settled on a peninsula in the central part of the lakes, using the southern part for food production in the chinampas (Poó 2012; Zambrano et al. 2020). Chinampas is a production system that is still practiced in this wetland nowadays. Producers cultivate ornate flowers and vegetables for subsistence and local sale (González 2019). Some of them are transitioning to more agroecological forms of production that include the rehabilitation of native fauna and flora in Xochimilco and San Gregorio (Figueroa et al. 2022; LRE-UNAM 2023).

The chinampas system increased the number of habitats and biodiversity in Xochimilco, being home to native and endemic, terrestrial, riparian, aquatic, and subaquatic species of flora (halophytic vegetation, tulars) and fauna (fishes, reptiles, amphibians, birds, mammals), several of which are in some category of risk according to official Mexican standards (GDF 2006; Ramsar 2004a; Zambrano et al. 2020). The most emblematic and charismatic species in this system is the axolotl salamander *Ambystoma mexicanum*, a species threatened with extinction due to the loss and degradation of its habitat, as well as the introduction of invasive non-native species that pose a direct threat to the axolotl population (Zambrano and Rojas 2021; Zambrano et al. 2020). Because of its biological, ecological, and conservation values, the axolotl is used as a flagship species for the restoration of the site (Zambrano et al. 2020).

This NPA is also an important tourist place for sports, recreation, and leisure activities, visited annually mainly by local tourists, with the positive and negative impacts of this productive activity in the wetland system. One of the proposals for tourism in this area is the transition to conservation tourism and gastronomic tourism, aiming to act with ecological awareness and conserve the integrity of the wetland (Vázquez-Medina et al. 2023). At the same time, Xochimilco and San Gregorio, a complex water system of ~40 km^2 of artificial channels, small lagoons, and temporary wetlands transitioning between rural and urban areas (Contreras et al. 2009), provides ecosystem services such as carbon sequestration, flood prevention, aquifer recharge and discharge, biomass export, and climate regulation (Castelán et al. 2015; GDF 2006; Zambrano and Rojas 2021).

However, the system is far from ideal, being polluted and eutrophicated because of anthropic activities such as domestic wastewater, agrochemicals use, and solid waste (GDF 2006), so it is necessary to understand this aquatic environment as a socioecological system (Zambrano and Rojas 2021). People live in and actively participate in this natural yet highly transformed system, the reason why are essential for the protection, restoration, and long-term conservation of Xochimilco and its biological and cultural diversity for their own benefit and that of Mexico City in general (Zambrano and Rojas 2021; Zambrano et al. 2020).

3.1.2 The Interdunary Lagoons System in the City of Veracruz

The Interdunary Lagoons System in the city of Veracruz was designated as a Ramsar site on 2 February 2005, with 141 ha (CW 2023a). It is part of a larger NPA: The Multifunctional Biological Corridor Archipelago of Interdunary Lagoons in the Urban Area of the Municipalities of Veracruz and La Antigua, Ver., of state attribution, decreed in November 2016 (GEV 2016). While the Archipelago is composed of 33 urban, semiurban, and rural lagoons, only eighteen correspond to the Ramsar site and are embedded and distributed in the urban matrix of the city (GEV 2018; Ramsar 2004b).

The system consists of lagoons from 0.52 up to 60.21 ha, corresponding to the coastal dune ecosystem of Veracruz, rare in the rest of the country and therefore of exceptional regional value (GEV 2016; Peralta-Peláez et al. 2019). These lagoons are a remnant of possibly more than 200 that existed in the region, many of which disappeared with urban and population growth (Cruz 2005). They provide nesting and resting habitats for species of shorebirds and waterfowl, local and migratory, and are on the route of the world's largest migratory corridor of raptors (Cruz 2005; Ramsar 2004b). The vegetation in the lagoons is composed of floating and hydrophyte species, tulars, typhals, and popals (GEV 2016; Ramsar 2004b).

These lagoons provide ecosystem services such as groundwater recharge, sediment retention, water quality maintenance, microclimate regulation, water storage, and flood protection for urban areas. Located in low to high-density habitational areas, they are used for recreational activities, environmental education, and meeting places for social interaction between neighbors and for the interaction of the people with nature. As places with a high aesthetic value, for some lagoons, the land in nearby neighborhoods has a higher capital gain (GEV 2016; Ramsar 2004b). In some lagoons, tilapia was introduced for subsistence fishing (Peralta-Peláez et al. 2019; Ramsar 2004b). However, after becoming an NPA, fishing is not allowed in the lagoons corresponding to the restoration zone (GEV 2018). Nevertheless, fishing occurs, causing irregular settlements and damage to wildlife (Ruiz 2023; Serrano 2022).

Because of pressures such as illegal human settlements, domestic water, and waste discharges, urban lagoons are also at constant risk of reduction and total desiccation (GEV 2016; Ramsar 2004b), are deteriorated, and all of them have some degree of pollution, eutrophication, clogging, and general degradation (GEV 2016; Peralta-Peláez et al. 2019). The leading causes of the deterioration of the Interdunary Lagoons System include the lack of knowledge about their functions and services, the lack of environmental education, and the lack of surveillance of these water bodies by decision-makers and the general population (GEV 2016; Ramsar 2004b).

However, part of the population, especially those living near the lagoons, recognizes their benefits and services, is interested in their conservation, and is even willing to pay for the conservation of the lagoons and their integrity as a system (Torres 2022). Furthermore, desilting, construction of walkways on the banks for recreational purposes, drainage construction to prevent sewage discharges, reforestation with native species, and maintenance and cleaning in general, have been carried out (GEV 2016; Ramsar 2004b). Such efforts must be continuous and include monitoring the evolution of each lagoon and the system as a whole (Torres 2022).

3.2 Non-Ramsar Sites

In Mexico, there are still many relevant urban wetlands that could meet some of the Ramsar criteria as they support vulnerable, endangered, or critically endangered species, or threatened ecological communities, support plant and/or animal species when they are in a critical stage of their biological cycle and provide them with a refuge when they are in adverse conditions (CW 2023d). These wetlands also provide benefits and services to the people and the cities they are located (Table 1).

3.2.1 'Laguna de las Ilusiones' in Villahermosa, Tabasco

The Laguna de las Ilusiones has been an NPA since February 1995, with the category of state Ecological Reserve (GET 1995). Extended over 259 ha, it is part of a more extensive system of urban lagoons. However, it is the city's most important and largest natural one (GET 2019; Rodríguez et al. 1997). Its constitution as an NPA is to protect, conserve, and restore the natural ecosystems of Tabasco, as well as to provide spaces for recreation, environmental education, use of natural resources, and ecological research (GET 2019).

The protected vegetation, including a 10 m buffer from the lagoon, is hydrophyte, riparian vegetation, secondary vegetation, grasslands, and palm groves (GET 2019), which provide refuge and food for aquatic, semiaquatic, and terrestrial species that thrive in the city. The two most emblematic species inhabiting the natural lagoon are the manatee (*Trichechus manatus*—at risk of extinction), with an isolated wild population, and the Morelet's crocodile (*Crocodylus moreletii*—special protection), with possible reproductive problems (Pérez-Garduza et al. 2023; Rueda et al. 2017). Other no less important wildlife groups are fishes, turtles, birds, and mammals such as bats and otters, including species with some conservation status (GET 2019; Hidalgo et al. 2019).

Recognized as an element of the city's identity (Ricárdez et al. 2016), it is linked to its beauty and scenic value, as well as to tourism, cultural, and recreational activities, for which parks, museums, docks, walkways, and other facilities are located on its edges, in the southern part. Among the recognized ecosystem services that the lagoon provides are carbon sequestration, regulation of the water flow, and contribution to regulating the city's microclimate (GET 2019).

Two water regulation vessels are part of Las Ilusiones. They are the first to receive the water entering the lagoon, along with sediment and pollutant inputs from runoff and wastewater from the hotel zone and adjacent residential areas, which is why the vessels and the lagoon are hypereutrophic, polluted, and in the continuous process of clogging. The lagoon has lost all its flow from tributaries, so it depends on rainfall and the volume of water provided by drains and sewers in the urban area (Flores et al. 2018; GET 2019; Hansen et al. 2007).

Actions of recovery of the lagoon have been carried out over the years aiming to maximize the resilience of the lagoon in coping with natural phenomena and anthropogenic activities (GET 2019, 2022; van Afferden et al. 2008). However, to prevent the complete deterioration of the lagoon, promote and achieve its restoration, and conservation of the flora and fauna that find a habitat in this system, it is necessary to count on the compromise and participation of the population and the corresponding authorities and stakeholders, and the effective implementation of the management plan and regulations that protect the lagoon. To achieve these goals, it is necessary to raise public awareness of the ecological, biological, environmental, economic, and social importance of Las Ilusiones (GET 2019).

3.2.2 'El Estero El Salado' in Puerto Vallarta, Jalisco

The estuary El Salado has been a state NPA since July 2000 in the category of Ecological Conservation Zone (GEJ 2000). When first established, its area was 168 ha; however, in 2020, 40 ha were added to 208 ha (GEJ 2020). The NPA protects and conserves a mangrove ecosystem, marsh vegetation, remnants of semi-deciduous forest, thorny forest, secondary vegetation, and aquatic and subaquatic vegetation, as well as wildlife groups such as birds, including migratory, mammals, reptiles including crocodiles (*Crocodylus acutus*), and amphibians; some of the species in these groups are in some level of vulnerability in official Mexican standards (GEJ 2020; Molina et al. 2012; Rubio 2018).

According to the objective of environmental education on the lagoon systems and coastal estuaries, tourism, and recreational activities are allowed in El Salado. The mangroves and the channel that permanently flows into the Pacific Ocean are used for interpretive trails and nature observation areas with guided visits (GEJ 2020; Ortiz et al. 2015). El Salado seeks to strengthen the integrity of the water body and protect species of fishing importance that complete some part of their life cycle in the estuary, aiming finally to benefit the fishers' quality of life. To achieve the objectives, activities of restoration and reforestation have been carried out since its decree as NPA (GEJ 2020).

The estuary, 80% of its basin perennial and intermittent runoff, 20% floodable areas, is an essential regulating vessel that allows runoff from upstream to be integrated, filtered, and diverted to the estuary, especially during the rainy season. Likewise, under extreme conditions, both the permeable soil system and the mangrove canopy serve as a natural barrier against the onslaught of tropical storms and hurricanes (GEJ 2020). In addition, the mangroves of the El Salado estuary are undeniable carbon sequestration reservoirs, a benefit that escalates to the municipal and national levels (Kumagai et al. 2020).

In 2019, a residential residual water collector collapsed, and the sewage was discharged into the estuary, affecting the estuary, the channel, the port dock, and the beaches adjacent to the estuary entrance, helping, however, to minimize the risk of flooding with sewage waters, especially for the most vulnerable population in Puerto Vallarta (CA 2019; GEJ 2020). According to the monitoring and due to the recuperation measures implemented, recovery of the estuary from the discharge would be in the short and medium term (CA 2019). However, according to El Salado's management program, education and information about all the benefits that this wetland provides to the population, as well as a 'great social conscience', are essential to achieve its conservation along with its contribution to the sustainability of Puerto Vallarta (GEJ 2020).

3.2.3 'Laguna La Escondida' in Reynosa, Tamaulipas

La Escondida, a natural lagoon and its surroundings, constitute a state NPA of 320 ha since 1997 in the category of Urban Park (GET 1997). La Escondida has been protected thanks to the concern of the population, ecologist groups, and the scientific community due to the high pressure that the accelerated urban expansion and the oil industry were exerting on this water body (Camacho et al. 2009; GET 1997).

The NPA represents a relic of the region's original vegetation harboring important biodiversity (GET 1997) since it is located in a coastal plain in the xerophytic region of Mexico, initially dominated by Tamaulipan thorn scrub. The potential distribution of the different vegetation types that are characteristic of this area in the north of the state and that may be present in La Escondida are halophytic vegetation and grassland, tular, mezquital, cultivated grassland, and riparian vegetation (GET 2015).

This park is home to amphibians, reptiles, mammals, fishes, and birds (breeding residents and winter migrants) (GET 2015). La Escondida is located on the route of migratory birds from Canada and Alaska, constituting a resting and nesting area for this taxonomic group (Camacho et al. 2009; Sánchez and Batres 2007). Seventeen of the species in the lagoon are under a conservation category according to official Mexican standards and international regulations (GET 2015).

La Escondida aims to promote the ecological recovery of the aquatic environment that allows its use for the improvement of the urban environment, recreation, and scientific research; to maintain ecological integrity and facilitate the development of environmental education and recreation, minimizing the impact on natural resources; and to allow the opening of a protected natural area for the exercise of an environmental interpretation and knowledge of natural processes and their equilibrium (GET 1997). It also aims to be transformed into a regulating basin to avoid flooding during the rainy season, thus protecting the people living in the peripheral neighborhoods (Arcos 2009). However, hydraulic infrastructure and the continuous maintenance of the drains and canals are still needed to avoid the risk of flooding during the rainy season (GET 2015; Jiménez 2020; Rivera 2011).

It is a place for recreational, sports, and tourist activities, linked to the development of environmental education and awareness (GET 2015). However, the NPA status of La Escondida has not been enough to stop the continuous deterioration of the lagoon. The different municipal administrations have attempted to restore the lagoon's ecosystem. However, they have not succeeded due to the lack of continuity and coordination between governmental areas (Camacho et al. 2009). The general situation of the lagoon represents challenges and priority lines of action for the scientific, environmental, and political community at the regional level, as well as the agreement between different governmental entities so that the recovery of La Escondida can be carried out (GET 2015).

3.2.4 The Artificial Wetland 'El Palote' Dam, in León, Guanajuato

El Palote dam is the main element of the 'Metropolitan Park from the Municipality of Leon, Gto.' a state NPA in the category of Ecological Park, decreed in September 2000 at the request of the Technical Committee of the León Metropolitan Park Trust (GEG 2000). The objectives of the Metropolitan Park are to preserve the water body of the dam (IEEG 2001), to provide recreational and leisure opportunities that contribute to the formation of an environmental culture that allows people to participate in the protection and conservation of the NPA, to stop the degradation of the natural resources of the area and to maintain the quality of the landscape (GEG 2000).

The dam was built to solve the water problems of León, both droughts and floods, with contributions from the people and the government (Córdova 2022; Navarro 2012). The main intention was to supply the stored water for the use of the population and industry (Córdova 2022; Navarro 2012). Today, the El Palote dam regulates rainwater runoff from the three original catchment micro-basins in the mountains to the north, northwest, and northeast, helping to prevent flooding in León (IEEG 2001; Navarro 2012). However, although the dam is part of the intermediate recharge zones (IEEG 2001) and is one of the largest dimensions in the municipality of León (GEG 2000; IEEG 2001), supplies water to only 3% of the urban population. The water from the dam is mainly used to maintain the Metropolitan Park (GEG 2000). However, in March 2023, the dam was at 34% of its capacity, resulting in a water shortage for 97 neighborhoods in León (SAPAL 2023). Faced with such a situation before, the management program of the NPA proposes the suspension of the extraction of water for public supply (IEEG 2001).

The Metropolitan Park, inaugurated in 1993, has an area of 337 ha, of which ~ 85% corresponds to the water body of the dam (Córdova 2022). Its flora is composed of native and introduced species (DRN 2016) and provides a habitat for species of wildlife groups: birds, fishes, amphibians, reptiles, and mammals. Twenty of these fauna species are endemic or in some status of protection in the Mexican official standards (SMAOT 2021). It is also located on the migratory route of the center of the country and provides a habitat for migratory birds (GEG 2000).

In this NPA, the economic activities allowed are mainly those related to recreation, education, culture, and tourism. Recreational and subsistence fishing are allowed, the latter only for a few family groups; fish re-stocking, if necessary, is allowed during periods of maximum dam flooding. Soil extraction (with the aim of de-silting the dam) is also allowed as long as it does not breach the impermeable layers, respects the total storage capacity of the dam, and does not disturb or increase the depth of the (shallow) feeding areas of the birds (GEG 2000; IEEG 2001).

To preserve the El Palote dam, it is necessary to implement a comprehensive management strategy that includes the micro-basins that feed it as they have various degrees of erosion problems, as well as the sanitation and cleaning of the streams that feed the dam, as problems with garbage and pollutants persist (IEEG 2001). Such a management strategy, together with the controlled activities of the visiting population, aims to reduce as much as possible the negative impacts on the NPA and to maintain its ecological integrity and functionality for the benefit of the people, the wildlife, and the city of León. In addition, there is a proposal to expand the NPA to maintain ecological quality, preserve connectivity between ecosystems (i.e., with a non-urban NPA), and limit productive activities, land use changes, and urban growth trends in the area (IMPLAN 2016).

4 Conclusions

Mexico's natural conditions have favored the presence of wetlands across a significant portion of its territory resulting in different types of urban wetlands. These spaces contribute to the sustainability of Mexican cities in different ways, depending on their origin (i.e., natural, artificial), location within the different regions (e.g., climatic, vegetation), protection category, conservation status, level of pollution and eutrophication, hydrology, and intermittency, among other factors. Mexican urban wetlands are home to populations of native, endemic, vulnerable, endangered, and/or critically endangered species of flora and fauna, constitute conservation areas that contribute to flood control, water storage, and filtration, and are important spaces for leisure and recreational activities that also promote tourism in the cities.

The urban wetlands described here are NPAs, mainly of state attribution, and represent the different types of Mexican ecosystems they protect. They are a sample of the categories and statuses of conservation and protection resulting from the application of Mexican laws on protecting natural areas. All of them are, to a greater or lesser extent, public spaces in which environmental education is considered, and in all of them, tourist and recreational activities are carried out. As a result, all of them have been modified to accommodate infrastructure and equipment, in most cases at the expense of the extent of the wetlands themselves.

Except for Xochimilco and San Gregorio, in which people are an integral part of the wetland, and the Interdunary Lagoons in Veracruz, scattered throughout the city and in direct contact with people since they are parts of residential areas; in the rest of the wetlands human settlements are not allowed, yet parts of their borders limit with residential areas or are subject to illegal settlements in their margins or zones of influence. Regarding tourism in these protected natural wetlands, it is an ambivalent factor since nature-based tourism, along with recreational and leisure activities, brings economic benefits and is an essential tool for environmental education and awareness; however, poor planning and execution of these activities can lead to the depletion of the same protected resources and areas designated as tourist sites (Mora-Olivo et al. 2009). This represents a critical situation and high vulnerability for these systems, as they are the first to receive the impacts of anthropic activities and the urbanization process. As a result, the systems are polluted, eutrophic, and in different stages of degradation, even though, they continue to provide services to the population, biodiversity, and cities.

The management programs of these protected wetlands include knowledge, monitoring, restoration, and research, yet such activities are carried out with different intensities or not at all, depending on the NPA. For some, there are no complete or updated data on the species that live and visit these wetlands, nor on their pollution and eutrophication dynamics. For successful management, conservation, and sustainable use for the benefit of the population, monitoring and periodic studies are needed. For some, restoration activities have been carried out depending on the interest of the authorities in turn and their available funds. However, they have been intermittent, sometimes insufficient, or carried out only in case of emergency. It is common in the news sites to find reports on the deterioration of the wetlands, death of fauna, or damage to flora.

Despite their protection status and importance to people for all the benefits they provide,

Mexico's urban wetlands are still vulnerable and under constant threat of disappearance. Non-protected wetlands are at even greater risk. The conservation status is not the last step as one might hope, but the continuation of a process to progressively achieve effective restoration and conservation of urban wetlands to benefit the population and the city environment.

Monitoring and restoration in and of protected wetlands can be expensive and not an easy task, yet necessary. To achieve adequate management and effective protection and conservation of these essential environments are necessary operational urban planning around the wetlands, compliance with laws and regulations at the three levels of government related to NPAs, and coordination between institutions and areas of government to carry out the actual implementation of management plans, reevaluate the social and economic pressures on their land and biodiversity, generate better policies of restoration and management, involvement and cooperation of educational institutions, and civil societies (Camacho et al. 2009; GEJ 2000). All actions are important, but above all, we need to understand that environmental care and protection of spaces such as urban wetlands are for the benefit of all. In times of global warming and water scarcity, the protection and conservation of urban wetlands is not an option but a necessity for the sustainability of cities and the well-being of their populations.

Acknowledgments Ina Falfán acknowledges the support of the "Postdoctoral Fellowship Program at the National Autonomous University of Mexico" ("Programa de Becas Posdoctorales en la Universidad Nacional Autónoma de México").

References

Alikhani, S., Nummi, P., & Ojala, A. (2021). Urban Wetlands: A Review on Ecological and Cultural Values. *Water*, *13*(22), Article 22. https://doi.org/10.3390/w13223301

Almazán-Núñez, R. C., & Hinterholzer-Rodríguez, A. (2010). Dinámica temporal de la avifauna en un parque urbano de la ciudad de Puebla, México. *Huitzil, Revista Mexicana de Ornitología*, *11*(1), 26–34.

Ampatzidis, P., & Kershaw, T. (2020). A review of the impact of blue space on the urban microclimate. *Science of The Total Environment*, *730*, 139068. https://doi.org/10.1016/j.scitotenv.2020.139068

Arcos, N. G. (2009). Preservar y restaurar la naturaleza: Objetivo de la UAT. *CienciaUAT*, *4*(1), 38–45.

Birch, S., & McCaskie, J. (1999). Shallow urban lakes: A challenge for lake management. In D. M. Harper, B. Brierley, A. J. D. Ferguson, & G. Phillips (Eds.), *The Ecological Bases for Lake and Reservoir Management* (pp. 365–377). Springer Netherlands. https://doi.org/10.1007/978-94-017-3282-6_31

Birch, W. S., Drescher, M., Pittman, J., & Rooney, R. C. (2022). Trends and predictors of wetland conversion in urbanizing environments. *Journal of Environmental Management*, *310*, 114723. https://doi.org/10.1016/j.jenvman.2022.114723

CA (Corporativo Ambiental). (2019). *Evaluación ambiental y caracterización de posible afectación del Área Natural Protegida, Zona de Conservación Ecológica Estero El Salado, en la localidad de Puerto Vallarta, por contingencia ambiental derivada de la ruptura del colector centro norte* (REQ-65575; p. 171). Corporativo Ambiental. https://www.seapal.gob.mx/Downloads/transparencia/XI/REPORTE%20FINAL%20EVALUACION%20ESTERO%20EL%20SALADO.pdf

Camacho, H., Ortíz, G., & Donath, E. (2009). *Coordinación de los ámbitos gubernamentales para la gestión integral de la laguna La Escondida, Reynosa Tamaulipas*. https://www.agua.org.mx/wp-content/uploads/2009/05/04_hector_camacho.pdf

Castelán, C. R., Contreras, S. A. I., & Tapia, S. F. O. (2015). Los últimos humedales en el Distrito Federal: Xochimilco y Tláhuac, servicios ambientales y la ruta hacia su preservación. En M. A. Sastre, C. I. A. Díaz, & H. J. Ramírez (Eds.), *Gestión de humedales españoles y mexicanos. Apuesta conjunta por su futuro* (pp. 43–69). Universidad de Alcalá.

CONANP (Comisión Nacional de Áreas Naturales Protegidas). (2023). *Áreas Naturales Protegidas Decretadas* [Governmental]. Gobierno de México. http://www.gob.mx/conanp/acciones-y-programas/areas-naturales-protegidas-decretadas

Contreras, V., Martínez-Meyer, E., Valiente, E., & Zambrano, L. (2009). Recent decline and potential distribution in the last remnant area of the microendemic Mexican axolotl (*Ambystoma mexicanum*). *Biological Conservation*, *142*(12), 2881–2885. https://doi.org/10.1016/j.biocon.2009.07.008

Córdova, M. (2022, septiembre 29). *Exhacienda El Palote, los vestigios de su historia bajo la presa* [Periódico]. El Sol de León. https://www.elsoldeleon.com.mx/local/exhacienda-el-palote-los-vestigios-de-su-historia-bajo-la-presa-fotos-8959018.html

Cruz, C. (2005). Sistema Lagunar de Veracruz: Quinto sitio Ramsar en el estado. *UniVerso: El Periódico de los Universitarios*, *5*(168), 68.

Cui, M., Ferreira, F., Fung, T. K., & Matos, J. S. (2021). Tale of Two Cities: How Nature-Based Solutions Help Create Adaptive and Resilient Urban Water Management Practices in Singapore and Lisbon.

Sustainability, *13*(18), Article 18. https://doi.org/10.3390/su131810427

CW (Convention on Wetlands). (2023a). *The List of Wetlands of International Importance*. Ramsar. https://www.ramsar.org/sites/default/files/documents/library/sitelist.pdf

CW (Convention on Wetlands). (2023b). *Ramsar Sites Information Service | North America: Mexico* [Institutional]. Ramsar. https://rsis.ramsar.org/ris-search/?language=en&f%5B0%5D=regionCountry_en_ss%3ANorth%20America&f%5B1%5D=regionCountry_en_ss%3AMexico

CW (Convention on Wetlands). (2023c). *Wetland City Accreditation* [Institutional]. Ramsar. https://www.ramsar.org/activity/wetland-city-accreditation

CW (Convention on Wetlands). (2023d). *The Ramsar Sites Criteria* [Institutional]. Ramsar. https://www.ramsar.org/document/the-ramsar-sites-criteria

Dallimer, M., Rouquette, J. R., Skinner, A. M. J., Armsworth, P. R., Maltby, L. M., Warren, P. H., & Gaston, K. J. (2012). Contrasting patterns in species richness of birds, butterflies and plants along riparian corridors in an urban landscape: Species richness in an urban landscape. *Diversity and Distributions*, *18*(8), 742–753. https://doi.org/10.1111/j.1472-4642.2012.00891.x

Darrah, S. E., Shennan-Farpón, Y., Loh, J., Davidson, N. C., Finlayson, C. M., Gardner, R. C., & Walpole, M. J. (2019). Improvements to the Wetland Extent Trends (WET) index as a tool for monitoring natural and human-made wetlands. *Ecological Indicators*, *99*, 294–298. https://doi.org/10.1016/j.ecolind.2018.12.032

Davidson, N. C. (2014). How much wetland has the world lost? Long-term and recent trends in global wetland area. *Marine and Freshwater Research*, *65*(10), 934. https://doi.org/10.1071/MF14173

DRN (Dirección de Recursos Naturales). (2016). *Listado florístico del Inventario de las Áreas Naturales Protegidas del Estado de Guanajuato*. Instituto de Ecología del Estado de Guanajuato. https://smaot.guanajuato.gob.mx/sitio/areas-naturales-protegidas/14/Parque-Metropolitano

Falfán, I., & Zambrano, L. (2023). Lacustrine urban blue spaces: Low availability and inequitable distribution in the most populated cities in Mexico. *Land*, *12*(1), 228. https://doi.org/10.3390/land12010228

FAO (Organización de las Naciones Unidas para la Alimentación y la Agricultura). (2017). *FAO reconoce oficialmente a las chinampas como Patrimonio Agrícola Mundial* [Institucional]. Organización de las Naciones Unidas para la Alimentación y la Agricultura | FAO en México. https://www.fao.org/mexico/noticias/detail-events/es/c/992183/

Figueroa, F., Puente-Uribe, M. B., Arteaga-Ledesma, D., Espinosa-García, A. C., Tapia-Palacios, M. A., Silva-Magaña, M. A., Mazari-Hiriart, M., Arroyo-Lambaer, D., Revollo-Fernández, D., Sumano, C., Rivas, M. I., Jiménez-Serna, A., Covarrubias, M., & Zambrano, L. (2022). Integrating Agroecological Food Production, Ecological Restoration, Peasants' Wellbeing, and Agri-Food Biocultural Heritage in Xochimilco, Mexico City. *Sustainability*, *14*(15), Article 15. https://doi.org/10.3390/su14159641

Flores, C. M., Del Ángel, E., Frías, D. M., & Gómez, A. L. (2018). Evaluación de parámetros fisicoquímicos y metales pesados en agua y sedimento superficial de la Laguna de las Ilusiones, Tabasco, México. *Tecnología y ciencias del agua*, *09*(2), 39–57. https://doi.org/10.24850/j-tyca-2018-02-02

Forman, R. T. T. (2014). *Urban Ecology: Science of Cities*. Cambridge University Press.

GDF (Gobierno del Distrito Federal). (2006). Acuerdo por el que se aprueba el Programa de Manejo del Área Natural Protegida con carácter de Zona de Conservación Ecológica "Ejidos de Xochimilco y San Gregorio Atlapulco". *Gaceta Oficial del Distrito Federal*, 2–41.

GEG (Gobierno del Estado de Guanajuato). (2000). Decreto Gubernativo No. 174, mediante el cual, se declara como Área Natural Protegida en la Categoría de Parque Ecológico, la zona conocida como Parque Ecológico Metropolitano del Municipio de León, Gto. *Periódico Oficial del Estado de Guanajuato*, 12925–12937.

GEJ (Gobierno del Estado de Jalisco). (2000). Decreto de Área Natural Protegida bajo la categoría de Zona de Conservación Ecológica al Estero "El Salado". *Periódico Oficial del Estado de Jalisco*, 1–20.

GEJ (Gobierno del Estado de Jalisco). (2020). Decreto del gobernador del Estado de Jalisco por el que se establece como área natural protegida de competencia estatal como Parque Estatal "Estero El Salado", con una superficie de 208-80-45 hectáreas, ubicada en el municipio de Puerto Vallarta, Jalisco. *Periódico Oficial del Estado de Jalisco*, 1–33.

GET (Gobierno del Estado de Tabasco). (1995). Acuerdo por el que se declara la Reserva Ecológica Laguna de las Ilusiones. *Periódico Oficial del Estado de Tabasco*. https://tabasco.gob.mx/sites/default/files/users/sbstabasco/DRE-LI-Reserva%20Ecol%C3%B3gica%20Laguna%20de%20las%20Ilusiones.pdf

GET (Gobierno del Estado de Tabasco). (2019). Programa de Manejo y Conservación. Reserva Ecológica Laguna de las Ilusiones. *Periódico Oficial del Estado de Tabasco*. https://tabasco.gob.mx/sites/default/files/users/sbstabasco/PM-RE-Lagunadelasilusiones-2019.pdf

GET (Gobierno del Estado de Tabasco). (2022). Declaratoria de Emergencia para Salvar la Laguna de las Ilusiones. *Periódico Oficial Del Estado de Tabasco*, 1–10.

GET (Gobierno del Estado de Tamaulipas). (1997). Acuerdo gubernamental mediante el cual se declara área natural protegida, clasificada como parque urbano al área denominada "Laguna La Escondida", ubicada en el municipio de Reynosa, Tamaulipas. *Periódico Oficial del Estado de Tamaulipas*, 1–13.

GET (Gobierno del Estado de Tamaulipas). (2015). Programa de Manejo del Área Natural Protegida, Parque Estatal "Laguna La Escondida", ubicada en el municipio de Reynosa, Tamaulipas. *Periódico Oficial del Estado de Tamaulipas*, 76–145.

GEV (Gobierno del Estado de Veracruz). (2016). Decreto por el que se declara como Área Natural Protegida en la categoría de Corredor Biológico Multifuncional la superficie denominada Archipiélago de Lagunas Interdunarias de la Zona Conurbada de los Municipios de Veracruz y La Antigua del Estado de Veracruz. *Gaceta Oficial del Estado de Veracruz*, 56–85.

GEV (Gobierno del Estado de Veracruz). (2018). Resumen del programa de Manejo del Área Natural Protegida Corredor Biológico Multifuncional Archipiélago de Lagunas Interdunarias de la Zona Conurbada de los Municipios de Veracruz y La antigua, Ver. *Gaceta Oficial del Estado de Veracruz*. http://repositorio.veracruz.gob.mx/medioambiente/wp-content/uploads/sites/9/2019/04/Resumen-Lagunas-interdunarias.pdf

Ghermandi, A., van den Bergh, J. C. J. M., Brander, L. M., de Groot, H. L. F., & Nunes, P. A. L. D. (2010). Values of natural and human-made wetlands: A meta-analysis. *Water Resources Research*, *46*(12), 1–12. https://doi.org/10.1029/2010WR009071

González, A. (2019). Condiciones de habitabilidad en el paisaje cultural chinampero de Xochimilco, Ciudad de México. *Revista I+A, Investigación más Acción*, *22*, 12–35.

Haase, D. (2015). Reflections about blue ecosystem services in cities. *Sustainability of Water Quality and Ecology*, *5*, 77–83. https://doi.org/10.1016/j.swaqe.2015.02.003

Haase, D. (2017). Urban wetlands and Riparian forests as a nature-based solution for climate change adaptation in cities and their surroundings. En *Nature-based solutions to climate change adaptation in urban areas* (pp. 111–121). Springer, Cham.

Hansen, A. M., van Afferden, M., & Torres-Bejarano, F. (2007). Saneamiento del vaso Cencali, Villahermosa, Tabasco. I. Contaminación y reúso de sedimentos. *Ingeniería Hidráulica en México*, 17.

Hidalgo, M. M. G., Bello, G. J., Arriaga, W. S. L., & López, L. M. A. (2019). Fauna silvestre de la ciudad de Villahermosa. En *La biodiversidad en Tabasco. Estudio de Estado* (Vol. 11, pp. 375–381). CONABIO.

Ibrahim, I., Aminudin, N., Young, M. A., & Yahya, S. A. I. (2012). Education for wetlands: Public perception in Malaysia. *Procedia—Social and Behavioral Sciences*, *42*, 159–165. https://doi.org/10.1016/j.sbspro.2012.04.177

IEEG (Instituto de Ecología del Estado de Guanajuato). (2001). Programa de Manejo del Área Natural Protegida "Parque Metropolitano" de la ciudad de León Guanajuato. *Periódico Oficial del Gobierno del Estado de Guanajuato*, 81–95.

IMPLAN (Instituto Municipal de Planeación). (2016). *Plan maestro del Parque Metropolitano de León, Guanajuato*. https://implan.gob.mx/pdf/estudios/sustentabilidad/plan-maestro-del-parque-metropolitano-de-leon.pdf

Jiménez, J. (2020). *Canales y lagunas se desbordan en Reynosa* [News]. El Sol de Tampico. https://www.elsoldetampico.com.mx/local/regional/canales-y-lagunas-se-desbordan-en-reynosa-5547934.html

Johansson, M., Pedersen, E., & Weisner, S. (2019). Assessing cultural ecosystem services as individuals' place-based appraisals. *Urban Forestry & Urban Greening*, *39*, 79–88. https://doi.org/10.1016/j.ufug.2019.02.011

Kumagai, J. A., Costa, M. T., Ezcurra, E., & Aburto-Oropeza, O. (2020). Prioritizing mangrove conservation across Mexico to facilitate 2020 NDC ambition. *Ambio*, *49*(12), 1992–2002. https://doi.org/10.1007/s13280-020-01334-8

Landgrave, R., & Moreno, C. P. (2012). Evaluación cuantitativa de la pérdida de humedales en México. *Investigación Ambiental. Ciencia y Política Pública*, *4*(1), 19–35.

Lee, T. S., Randall, L. A., Kahal, N. L., Kinas, H. L., Carney, V. A., Rudd, H., Baker, T. M., Sanderson, K., Creed, I. F., Moehrenschlager, A., & Duke, D. (2022). A framework to identify priority wetland habitats and movement corridors for urban amphibian conservation. *Ecological Solutions and Evidence*, *3*(2), e12139. https://doi.org/10.1002/2688-8319.12139

LRE-UNAM (Laboratorio de Restauración Ecológica de la UNAM). (2023). *Etiqueta Chinampera* [Institutional]. LRE UNAM. https://www.restauracionecologica.org/etiquetachinamperaxochimilcoo

Mancini, M., Crichigno, S., & Ortiz, M. (2012). Lagos urbano: Importancia, dinamismo y multiplicidad de usos. El caso del Lago Villa Dalcar (Córdoba, Argentina). *Biología Acuática*, *27*, 175–189.

MEA (Millennium Ecosystem Assessment). (2005). *Ecosystems and human wellbeing: Wetlands and water synthesis*. World Resources Institute.

Molina, D., Torres-Guerrero, J., & Avelarde-Gómez, M. de la L. (2012). Riqueza de aves del Área Natural Protegida Estero El Salado, Puerto Vallarta, Jalisco, México. *Huitzil*, *13*(1), 22–38.

Molina-Prieto, L. F., & Rubio, F. D. (2016). Elementos de urbanismo azul: Lagos naturales y artificiales. *Revista de Investigación*, *9*(2), 22–44. https://doi.org/10.29097/2011-639X.52

Mora-Olivo, A., Martínez-Ávalos, J. G., González-Rodríguez, L. H., & Garza-Torres, H. A. (2009). El turismo en áreas naturales protegidas en Tamaulipas. *CienciaUAT*, *4*(1), 30–35.

Moreno, C. P. (2008). Los humedales en México: Tendencias y oportunidades. *Cuadernos de Biodiversidad*, *28*, 10–18. https://doi.org/10.14198/cdbio.2008.28.02

Navarro, V. C. A. (2012). Antecedentes Históricos de la Presa "El Palote" (concluye). *Tiempos. Órgano de Divulgación Del Archivo Histórico Municipal de León, Gto.*, *131*, 3–16. 2020-07-23.

Ortiz, L. D. G., Maldonado, I. O. A., & Rosas, P. R. J. R. (2015). Actividades de ecoturismo en zonas de manglar, la experiencia del área natural protegida Estero "El salado" Puerto Vallarta, Jalisco. En A. E. Gámez, M. A. Palafox, & M. Gutiérrez (Eds.), *La construcción del espacio turístico: Procesos, actores e impactos* (pp. 276–285). Instituto Sud-Californiano de Cultura, Academia Mexicana de Investigación Turística, Universidad Autónoma de Baja California Sur. https://www.researchgate.net/publication/328103174_Actividades_de_ecoturismo_en_zonas_de_manglar_la_experiencia_del_area_natural_protegida_Estero_El_salado_Puerto_Vallarta_Jalisco

Ortiz-Arrona, C. I., Flores-Díaz, A. C., & Quijas, S. (2022). Ríos urbanos en México: Espacios de encuentro entre sociedad y naturaleza. *Boletín de la Sociedad Científica Mexicana de Ecología*, *2*(9), 86–94.

Palacios, E., & Castillón, M. (2013). El Salado: Impactos y potencial de desarrollo turístico sustentable. *Visión Docente con Ciencia*, *XII*(67), 28–41.

Peralta-Peláez, L. A., Moreno-Casasola, P., & Neri-Flores, I. (2019). Management and restoration of dune lakes in Veracruz, Gulf of Mexico. *Journal of Coastal Conservation*, *23*(2), 445–460. https://doi.org/10.1007/s11852-018-0674-1

Pérez-Garduza, F., Olivera-Gómez, L. D., Cueva, H. de la, Hidalgo-Mihart, M. G., & Jiménez-Domínguez, D. (2023). Density and abundance estimate of Antillean manatees (*Trichechus manatus*) within a landlocked lake in Southeastern México. *Therya*, *14*(2), Article 2. https://revistas-conacyt.unam.mx/therya/index.php/THERYA/article/view/253

Poó, R. A. (2012). *Origen es destino. La Ciudad de México. De la ciudad lacustre a la metrópoli de los desastres hídricos*. 36.

Ramírez-Jiménez, H. H., Olivera-Gómez, L. D., & de la Cueva, H. (2017). Habitat use by the Antillean manatee (*Trichechus manatus*) during an extreme dry season in an urban lake in Tabasco, Mexico. *Therya*, *8*(1), 19–26. 10.12933/therya-17-411

Ramsar. (2004a). *Ficha informativa de los Humedales Ramsar—MX1363RIS*. Ramsar. https://rsis.ramsar.org/RISapp/files/RISrep/MX1363RIS.pdf?language=en

Ramsar. (2004b). *Ficha informativa de los Humedales Ramsar—MX1450RIS*. Ramsar. https://rsis.ramsar.org/RISapp/files/RISrep/MX1450RIS.pdf

Ricárdez, de la C. G., López, O. G., Bautista, M. R. G., & Torres, B. C. A. (2016). Laguna de las Ilusiones y su entorno urbano: Aguas residuales, urbanas y sedimentos. *Kuxulkab'*, *22*(43), Article 43. 10.19136/kuxulkab.a22n43.1461

Rivera, J. (2011). *Cicatrices de Reynosa* [Press]. laprensa.mx. http://laprensa.mx/notas.asp?id=64667

Rodríguez, R. E., Ruíz, R. M., & Vertíz, P. L. A. (1997). *Procesos de eutroficación en siete lagunas urbanas de Villlahermosa, Tabasco, México*. 10. http://www.bvsde.paho.org/bvsaidis/ puertorico/xli.pdf

Rubio, M. J. E. (2018). Aspectos poblacionales de neonatos de *Crocodylus acutus* (reptilia: Crocodylidae) en el estero El Salado, Puerto Vallarta Jalisco, México. *Biocyt: Biología, Ciencia y Tecnología*, *11*(41–42), 757–769.

Rueda, C. B. A., López, L. M. A., & Olivera, G. L. D. (2017). Uso de hábitat del cocodrilo de pantano *Crocodylus moreletii* en una laguna urbanizada en México. *Quehacer Científico en Chiapas*, *12*(2), 35–40.

Ruiz, I. R. (2023). *Invadieron laguna Olmeca pero la limpian todos los días, aseguran habitantes*. Diario de Xalapa. https://www.diariodexalapa.com.mx/local/laguna-olmeca-de-veracruz-habitantes-viven-de-la-pesca-y-la-limpian-9841070.html

Sánchez, G. D., & Batres, G. J. J. (2007). Planeación y desarrollo del espacio turístico urbano de la Laguna del Carpintero, Municipio de Tampico, Tam. (México). En M. M. C. Montemayor, H. F. Bijarro, & B. P. C. Estrada (Eds.), *Políticas y gestión pública para el estudio municipal: Óptica académica* (pp. 32–47). Universidad Autónoma de Tamaulipas. https://www.eumed.net/libros-gratis/2008a/356/356.pdf#page=32

SAPAL (Sistema de Agua Potable y Alcantarillado de León). (2023). *SAPAL* [Institucional]. SAPAL. https://www.sapal.gob.mx/

Schueler, T., & Simpson, J. (2001). Introduction: Why urban lakes are different. *Watershed Protection Techniques*, *3*(4), 747–750.

Serrano, M. (2022). *Grulla queda atrapada en redes de pesca de fraccionamiento Laguna Real, en Veracruz* [News]. Imagen de Veracruz. https://imagendeveracruz.mx/veracruz/grulla-queda-atrapada-en-redes-de-pesca-en-laguna-real/50265582

SMAOT (Secretaría de Medio Ambiente y Ordenamiento Territorial). (2021). *Listado de fauna Área Natural Protegida Parque Metropolitano V.2.04.20.* https://smaot.guanajuato.gob.mx/sitio/areas-naturales-protegidas/14/Parque-Metropolitano

Stefanakis, A., Akratos, C. S., & Tsihrintzis, V. A. (2014). *Vertical Flow Constructed Wetlands*. Elsevier. https://doi.org/10.1016/B978-0-12-404612-2.00001-5

Thornhill, I., Hill, M. J., Castro-Castellon, A., Gurung, H., Hobbs, S., Pineda-Vazquez, M., Gómez-Osorio, M. T., Hernández-Avilés, J. S., Novo, P., Mesa-Jurado, A., & Calderon-Contreras, R. (2022). Blue-space availability, environmental quality and amenity use across contrasting socioeconomic contexts. *Applied Geography*, *144*, 102716. https://doi.org/10.1016/j.apgeog.2022.102716

Torres, J. R. M. C. (2022). *Valor social, económico y ambiental de las Lagunas Interdunarias de la Ciudad de Veracruz, México* [Master]. Colegio de Postgraduados.

UNESCO. (2023). *Historic Centre of Mexico City and Xochimilco* [Institutional]. UNESCO World Heritage Centre. https://whc.unesco.org/en/list/412/

van Afferden, M., Hansen, A. M., & Torres-Bejarano, F. (2008). Saneamiento del vaso Cencali, Villahermosa, Tabasco. II. Diseño del proceso. *Ingeniería Hidráulica en México*, 15.

Vázquez-Medina, J. A., Covarrubias, M., Jiménez-Serna, A., & Medina, F. X. (2023). Tourism, gastronomy and sustainability in world heritage sites. The enhancement of the gastronomy of agroecological products as a tourist alternative for environmental recovery in the chinampas of Xochimilco (Mexico City). *Journal of Tourism and Heritage Research*, *6*(1), Article 1.

White, M. P., Elliott, L. R., Gascon, M., Roberts, B., & Fleming, L. E. (2020). Blue space, health and well-being: A narrative overview and synthesis of potential benefits. *Environmental Research*, *191*, 110169. https://doi.org/10.1016/j.envres.2020.110169

Yang, X., Liu, W., Li, S., Ma, Z., Chen, C., Gu, W., Qu, M., Zhang, C., Tao, J., Ding, Z., Xu, Y., & Hu, H. (2022). Restoration of urban waterbird diversity: A case study of the construction of a waterbird ecological corridor in the Guangdong-Hong Kong-Macao Greater Bay Area, Southern China. *Global Ecology and Conservation*, *39*, e02277. https://doi.org/10.1016/j.gecco.2022.e02277

Zambrano, G. L., & Rojas, R. (2021). *Xochimilco en el siglo XXI*. Turner. https://books.google.com.mx/books?id=7LmdzgEACAAJ

Zambrano, L., Rivas, M. I., Uriel-Sumano, C., Rojas-Villaseñor, R., Rubio, M., Mena, H., Vázquez-Mendoza, D. L., & Tovar-Garza, A. (2020). Adapting Wetland Restoration Practices in Urban Areas: Perspectives from Xochimilco in Mexico City. *Ecological Restoration*, *38*(2), 114–123. https://doi.org/10.3368/er.38.2.114

Urban Wetlands Protection Law in Chile. A Successful Tool for Urban Planning

Carolina Rojas Quezada

Contents

Abstract

Urban wetlands in Chile have been recognized as spaces of biodiversity, recreation, flood mitigation, and other benefits. They have also been highly valued by the population and defended by the community, especially in the face of alterations to their hydrological cycle and loss of surface area, giving rise to a strong and organized movement in defense of urban wetlands. At the same time, one of the most significant laws in environmental terms for Chile, considered one of the most neoliberal countries in the world, was being developed. The law for the protection of Urban Wetlands 21.202, enacted in 2020, aims to halt indiscriminate real estate development on wetlands and integrate them into city planning.

From a scientific perspective, this chapter addresses the role of scientific evidence in achieving this law, presenting various case studies that allowed for a focus on sustainability and territorial planning in the law's development. Nowadays, more than 10,000 ha of wetlands have been declared, presenting a significant opportunity for the development of green infrastructure projects and nature-based solutions.

Keywords

Urban wetlands · Law · Protection · Urban planning

1 Urban Wetlands in Chile: Evidence of Their Destruction

We live in cities that have not respected nor valued nature in their urban development. Many ecosystems, such as rivers, wetlands, and lagoons,

C. Rojas Quezada (✉)
Instituto de Estudios Urbanos y Territoriales,
Centro de Desarrollo Urbano Sustentable CEDEUS,
Instituto Milenio SECOS, Santiago, Chile
e-mail: carolina.rojas@uc.cl

C. Rojas Quezada (ed.), *Urban Wetlands in Latin America*, Sustainable Development Goals Series,
https://doi.org/10.1007/978-3-031-69590-2_4

have been altered and intervened in due to weak urban planning that allows the intervention of vulnerable and flood-prone areas. One of the most mistreated ecosystems has been wetlands, which, despite being recognized as significant allies against climate change for their role as water reservoirs during droughts and mitigators of disasters like floods or ecosystems services, are disappearing. Even in developed countries, it is very difficult to implement policies for the protection, management, and restoration of wetlands, promoting the integration of ecosystem services into legislation and policies through reforms (Bell-James et al. 2023).

Urban wetlands, or wetlands within cities, are unique ecosystems that promote ecological balance, the carbon cycle, and the city's hydrological cycle, providing a diversity of ecosystem services (Zhang et al. 2023). However, in Chile and Latin America, they are vanishing due to filling, drying, and alteration (Rojas et al. 2019). Only a few Latin American cities have prioritized them as spaces for urban biodiversity. Currently, our continent leads the global loss of wetlands (Darrah et al. 2019) due to increased urbanization and coastal erosion, coupled with deficient and weak planning that often fails to consider the environmental values necessary to address the environmental crises affecting the region's cities.

Chile, being one of the most urbanized countries in Latin America, suffers serious consequences for urban biodiversity (Pauchard et al. 2006; Martínez et al. 2020) due to a planning system that does not consider ecosystem services and the protection of biodiversity areas. These wetlands and their ecosystem services have not been incorporated into city planning instruments because they are not properly zoned according to the value of the ecosystems and their benefits. Consequently, even in cities with current planning and legal figures for biodiversity protection, urban wetlands are exposed to various pressures and threats from unsustainable real estate projects and transportation infrastructures that fragment them (Rojas et al. 2019; Martínez et al. 2020). Despite being heavily intervened, these ecosystems still provide many environmental benefits to Chilean cities and their population, such as flood mitigation, protection against coastal erosion, habitat for various species, heat island buffering, and carbon sequestration, among others (Navarro et al. 2021; Rojas et al. 2022), and, therefore, they should be recognized by urban and environmental regulations.

Multiple reasons justified the enactment of Chile's first Law for the Protection of Urban Wetlands on Thursday, January 23, 2020. This pioneering and unprecedented environmental law aims to protect urban wetlands declared by the Ministry of the Environment, either at the ministry's initiative or upon request from the respective municipality (local government). The uniqueness of this law lies in the support it received from citizen participation and the strength it gained from scientific evidence. The law was celebrated with much enthusiasm and hope by various sectors of society, especially environmental organizations that advocate for wetland conservation. However, it has not been exempted from controversies, particularly regarding private property discussions and the balance between wetland protection and the common good.

2 The Scientific Evidence as Support for Legislation

Since 2015, formal interdisciplinary research with a socio-environmental focus called URBANCOST (urbanization costs on the coast) has been developed. It aims to investigate the rapid urbanization process affecting urban wetlands in coastal cities of Chile, mainly in the Metropolitan Concepción area. Today, URBANCOST has become a scientific platform that measures the multiple effects of urbanization on the urban wetland system and provides technical advice on conservation and restoration measures in city wetlands. Scientific evidence published in various articles between 2015 and 2019, and from 2015 to 2022, played a crucial role in the discussion and development of the law. These findings were presented and discussed in sessions of the Chamber of Deputies and the

Senate, fueling intense debates about whether urban wetlands should be protected.

What Have We Discovered?

Among the most relevant evidence are the intersections between disciplines such as urban planning, biodiversity, ecosystem services, and geography, which have led to the following scientific findings:

Delimiting wetlands is a complex process due to the dynamic temporal variation of vegetation and water in these ecosystems. The vegetation criterion has shown the highest accuracy and acts as an indicator that helps reduce temperatures (Munizaga et al. 2022).

Vegetation serves as an indicator of biodiversity, showing that the plant richness in urban wetlands is related to the type of wetland. For example, wetlands with reeds or bullrushes (*Schoenoplectus californicus*) predominate in palustrine or freshwater wetlands, while marshes have a low number of species due to high salinity, resulting in less diverse and more homogeneous vegetation. Native plant distribution is also lower in wetlands with greater vegetational heterogeneity (Rojas et al. 2015).

The naturalness of vegetation, another biodiversity indicator, shows a higher degree of intervention in wetlands surrounded by more urbanized areas, higher population density, and more roads, compared to wetlands surrounded by natural environments (Rojas et al. 2015).

Spatial relationships between urban and biodiversity variables have shown that housing density has a greater influence on plant richness than the distance to roads or the urbanizable surface area in wetlands due to the introduction of exotic plants resulting from human occupation. However, the percentage of urbanizable surface area defined in planning instruments positively influences the naturalness in less developed wetland environments, while distance to roads and housing density have a negative effect (Rojas et al. 2015).

Ecological Connectivity reveals that urban wetlands are one of the least connected ecosystems in the urban landscape, despite their potential as ecological corridors. They show connectivity with areas transitioning from stable to very regressive and regressive changes due to loss of connectivity (Rojas et al. 2017).

Urban wetlands have undergone significant interventions in the last 20 years, rapidly losing surface area due to the construction of housing and transportation infrastructure. Some wetlands, even while providing flood mitigation functions against tsunamis (Rojas et al. 2019) and river overflow (Rojas et al. 2022), have experienced concerning surface area loss, validated by planning instruments (Rojas et al. 2019).

The future outlook is not promising, even with territorial planning instruments. For instance, the Rocuant-Andalién marsh wetland is projected to lose 32% of its area, primarily for mixed residential, commercial, and transportation use (Rojas et al. 2019). However, recent delimitation efforts driven by the "GEF Coastal Wetlands" government project and the Law for Urban Wetlands could change this situation.

Regarding the urban environment surrounding wetlands, it is characterized by scattered, discontinuous, and low-density residential fabrics with buildings not integrated into wetland design. These fabric types promote motorized mobility at the expense of ecosystem integration. Only medium-density urbanizations with elements of garden cities prior to the 1980s included walkable and green areas to visit water bodies like lagoons. In contrast, recent urbanizations have favored discontinuous and scattered fabrics dependent on automobiles, associated with unsustainable urban models (Rojas et al. 2022).

Urban environments with medium densities, significantly surrounded by open vegetation, offer more opportunities for incorporating and designing green-blue infrastructure (Rojas and Jorquera 2021).

Concerning the role of urban wetlands in floods, it has been demonstrated that in a scenario of increased urbanization and decreased wetland surface area during an extreme return period (500 years), the volume of water stored by an urban marsh wetland would decrease by more than 38%, and the flooded area within the wetland would increase by 30%. This would result in increased flooding volume and vulnerability in the residential fabrics surrounding the wetlands, with various implications for neighborhoods and infrastructure (Rojas et al. 2022).

Regarding the values perceived by people living in the proximity of wetlands, their understanding of "what a wetland is" varies depending on the neighborhood's density (Villagra et al. 2023).

People perceive higher levels of biophilia when their homes are closer to ecosystems. Additionally, residents of compact, green, and walkable neighborhoods show a greater awareness of the value of wetlands.

For those living further away from accessible wetlands, the biophilic values, such as aesthetic appreciation, are less perceived. Spatial factors and participation in environmental organizations also influence the frequency of visits to wetlands (Villagra et al. 2023).

This evidence highlights the pressure of urbanization and infrastructure on ecosystems, despite the benefits provided by urban wetlands and the values perceived by people. This validates the need for at least regulating real estate pressure for the common good, meaning regulating surface loss, construction over wetlands, and recognizing them as green infrastructure in urban planning (Photos 1 and 2).

The Law on the Protection of Urban Wetlands

The enactment of the unprecedented Law on the Protection of Urban Wetlands, led by Senator Alfonso de Urresti, is the first regulation that emphasizes the protection and conservation of these ecosystems. Additionally, as its authors point out, it is a pioneering and citizen-oriented law in Chile and Latin America, as it defines an "urban wetland" and accomplishes a virtuous blend of politics, science, and society. The work on this law began in June 2017 and was finalized with its enactment in January 2020.

The law consists of 5 articles, summarized as follows:

Photo 1 Buildings on urban wetlands. (Source: Carolina Rojas, Urbancost 2021)

Photo 2 Presentation in the Senate of Chile. (Fuente: Center of Urban Development, June of 2018)

Article 1: The purpose of this law is to define ecosystems as "all extensions of marshes, swamps, and bogs, or surfaces covered by water, whether natural or artificial, permanent or temporary, stagnant or flowing, fresh, brackish, or salted, including extensions of marine water, whose depth at low tide does not exceed six meters and which are partially or totally located within the urban boundary." An important point for urban planning is that the urban boundary is defined by the Regular Plan or planning instrument. Surprisingly, and supported by scientific evidence, the planning instrument has often been a major obstacle to protecting such areas, as it lacked the necessary competencies and repeatedly zoned wetlands for urban expansion or industrial development, among others.

Article 2: This article states that a regulation issued by the Ministry of the Environment, also signed by the Minister of Public Works, will define the minimum criteria for the sustainability of urban wetlands to safeguard their ecological characteristics and functioning, as well as to maintain the hydrological regime, both on the surface and underground. The regulation that establishes the minimum criteria ensures the protection of the ecological characteristics, functioning, and maintenance of the hydrological regime, both on the surface and underground of urban wetlands, integrating social, economic, and environmental dimensions, i.e., sustainability for the city. Although the Ministry of the Environment declares a wetland as "urban," the request will be made by the Municipality, which will now have a tool to incorporate urban wetlands into their planning, and they will no longer have excuses to fill or develop housing and roads on these areas. Additionally, municipalities must establish, through a general ordinance, the criteria for the protection, conservation, and preservation of urban wetlands located within their commune's limits, using the guidelines established in the regulation. One of the subsequent criticisms is that municipalities lack the economic and technical resources to create these ordinances; so far, only 3 municipalities have published them.

Article 3: From the submission of the request for recognition of the status of an urban wetland until the Ministry of the Environment's pronouncement, the respective municipality may postpone the issuance of permits for subdivision, division, or land development and construction on the land where these wetlands are located. This postponement will be carried out using, as appropriate, the procedure established in Article 117 of the General Law on Urbanism and Construction. This article defines the procedure through which the municipality may request the recognition of the status of an urban wetland. It is essential because there have been complaints from private individuals regarding the declaration of an urban wetland on privately owned land.

Article 4: Amends Article 10 of Law No. 19,300, on General Environmental Bases, by incorporating urban wetlands. This is the first law that changes the focus of the Environmental Bases Law, placing the emphasis on the wetland ecosystem rather than the typology of the investment project, as previously done.

Article 5: Incorporates a modification to the General Law on Urbanism and Construction, stating:

> Every territorial planning instrument must include existing urban wetlands at each territorial scale as areas of protection of natural value, for the purpose of establishing the conditions under which permits for urbanization or construction to be developed in them shall be granted.

This is crucial for improving urban planning processes since plans have favored urbanization of wetlands by not recognizing them as areas of natural value. There are many examples of this; for instance, the construction of real estate projects in flood-prone and coastal wetland areas in Los Molles in the Valparaíso Region, where the intercommunal regulatory plan was modified. Another example is the Rocuant-Andalién wetland in the Metropolitan Area of Concepción, part of the GEF Coastal Wetlands project, an ecosystem that helped mitigate the tsunami of February 27, 2010. Unfortunately, under the current metropolitan plan, a portion of the wetland is designated for commercial areas and residential expansion, infrastructure for transportation linked to port use. Similar situations can be found from north to south in cities throughout the country, with emblematic cases of destruction in Algarrobo, Coquimbo, Concepción, Valdivia, and Puerto Montt, among others. These cases demonstrate that urban plans with authority over land allocation have significantly failed to consider the biodiversity of local ecosystems and their role in resilience and climate change. Article 5 will have a significant impact on how we plan cities by incorporating wetlands as areas of natural value in territorial planning instruments. This becomes even more relevant when there are still voices that consider wetlands as ecologically worthless and, therefore, suitable for construction. It is necessary to strengthen governance and the idea of economic valuation of nature, as proposed by the economics of diversity.

3 Discussion

Regulating the processes of urbanization in ecosystems in Latin America is not easy, given the region's inequality, the dilemma between the development of housing, the construction economy sector, and employment, among others, versus nature conservation, and even recent studies show negative relationships that the economy does not judge. In other words, as ecosystems are lost, poverty increases, and environmental quality decreases, with repercussions on people's well-being (Let and Pal 2023). But other legislative proposals suggest changes in land tenure that include compensation and financial assessments of ecosystem services (Bell-James et al. 2023); an aspect still pending in Chilean legislation.

Based on the scientific evidence that helped strengthen the arguments for the Chilean law, it is evident that urban development and urban planning have been one of the main threats to wetland protection in cities. This is not only common in Chile but also in Latin America. Changes in land use and wetland loss are evident and supported by studies, not only from this research group but also from other regions and cities

(Bonansea et al. 2021; Campos et al. 2020). It is common to observe an increase in mostly low and medium-density housing on and around urban wetlands in Chilean cities, regulated by urban plans, which made regulation necessary. In most cases, these urban developments have been legally approved without considering their environmental effects on biodiversity, urban structure homogenization, and urban biodiversity loss, as they were environmentally evaluated based on the dimension and volume of housing, rather than their impact on wetlands. Today, this situation changes as any project affecting an urban wetland must undergo an environmental impact assessment. These aspects make the law a successful process, given its urgency to protect wetlands. Another strong point was the evidence regarding floods; wetlands' contribution to saving resources by preventing infrastructure damage was crucial for a positive reception by legislators.

On the other hand, while the research did not aim to have an impact on public policy, it aligned well with the common goal of protection, with a direct impact on the articles of the regulation, especially those related to delimitation and territorial planning. This has generated many challenges in new planning processes. After the law, research efforts have focused on measuring the effects of accessibility and how proximity to urban wetlands directly influences positive valuations of ecosystem services, attachment, and restoration. There is still a lot of ignorance and lack of environmental education to strengthen protection arguments.

4 Conclusion

Just 4 years ago, we had no instruments to protect wetlands in our cities, and there was no Climate Strategy that incorporated them into its goals. Fortunately, this has been changing, thanks to this legislative instrument. Nowadays, there are 100 recognized wetlands in settlements from north to south, with over 10,000 ha declared for protection. However, there are still areas for improvement in delimiting wetlands and integrating their influence areas into urban and territorial planning. The law provides a starting point for the future implementation of Nature-Based Solutions, but of course, new modifications and advances must also be accompanied not only by normative changes but also by aspects of the economic model, such as the valuation of natural capital, which fosters sustainable development.

The process of developing this law demonstrates the importance of scientific evidence when successfully linked to public policy. Research helped strengthen the idea of legislating and justified the articles related to delimitation and planning criteria. In the country today, there is a dilemma between the scarcity of space for housing and infrastructure development and the suitability of the land for such purposes. In this regard, it is essential for scientific evidence to focus on ecosystem services in cities, flood-prone areas, settlement vulnerability to flooding, ecosystem restoration, green infrastructure, urban drainage, and ultimately Nature-Based Solutions, among others.

The law has shown that urban wetlands can be used rationally and that new criteria and paradigms for sustainable development can be adopted. Research should continue and take advantage of opportunities to influence the development of public policies for more sustainable cities. Additionally, the regulation includes criteria for sustainable management and governance, involving effective participation and governance in the protection of these spaces, adaptive management, and environmental education as spaces for comprehensive training and promoting research on these valuable areas. For those of us working in sustainability and urban planning, the tools provided by this law open enormous possibilities to achieve more sustainable cities through plans and projects, meeting the challenges of the 17 Sustainable Development Goals and climate change scenarios to create more inclusive, safe, resilient, and sustainable human settlements, a source of pride for all who participated in this great milestone.

Acknowledgment ANID FONDECYT N°1190251 "Efectos de urbanización y accesibilidad en los Humedales

Urbanos del área metropolitana de Concepción (URBANCOST II)".

"Center for Sustainable Urban Development" (CEDEUS). Fondo de Financiamiento de Centros de Excelencia en Investigación ANID FONDAP 1522A0002.

"Instituto Milenio en Socio-Ecología Costera" SECOS (2021-2025). Programa Milenio ANID Millennium Science Initiative Program–ICN2019_015.

References

Bell-James, J., Foster, R., & Lovelock, C. (2023). Identifying priorities for reform to integrate coastal wetland ecosystem services into law and policy. (2023). *Environmental Science & Policy, 142*, 164–172. https://doi.org/10.1016/j.envsci.2023.02.013

Bonansea, M., Bazán, R., Germán, A., Ferral, A., Beltramone, G., Cossavella, A., & Pinotti, L. (2021). Assessing land use and land cover change in Los Molinos reservoir watershed and the effect on the reservoir water quality. *Journal of South American Earth Sciences, 108*, 103243. https://doi.org/10.1016/j.jsames.2021.103243

Campos, J., Huachaca, J., & Morales, P. (2020). South sustainability. *South Sustainability, 2*(2), e037–e037. https://doi.org/10.21142/SS-0202-2021-ac001

Darrah, S., Shennan-Farpón, Y., Loh, J., Davidson, N., C., Finlayson, C. M., Gardner, R. C., & Walpole, M. J. (2019). Improvements to the Wetland Extent Trends (WET) index as a tool for monitoring natural and human-made wetlands. *Ecological Indicators, 99*, 294–298. https://doi.org/10.1016/j.ecolind.2018.12.032

Let, M., & Pal, S. (2023). Socio-ecological well-being perspectives of wetland loss scenario: A review. *Journal of Environmental Management, 326*, 116692. https://doi.org/10.1016/j.jenvman.2022.116692

Martínez, C., López, P., Rojas, C., Qüense, J., Hidalgo, R., & Arenas, F. (2020). A sustainability index for anthropized and urbanized coasts: The case of Concón Bay, central Chile. *Applied Geography, 116*, 102166. https://doi.org/10.1016/j.apgeog.2020.102166

Munizaga, J., García, M., Ureta, F., Novoa, V., Rojas, O., & Rojas, C. (2022). Mapping coastal wetlands using satellite imagery and machine learning in a highly urbanized landscape. *Sustainability, 14*(9). https://doi.org/10.3390/su14095700

Navarro, N., Abad, M., Bonnail, E., & Izquierdo, T. (2021). The arid coastal wetlands of Northern Chile: Towards an integrated management of highly threatened systems. *Journal of Marine Science and Engineering, 9*(9). https://doi.org/10.3390/jmse9090948

Pauchard, A., Aguayo, M., Pena, E., & Urrutia, R. (2006). Multiple effects of urbanization on the biodiversity of developing countries: The case of a fast-growing metropolitan area (Concepción, Chile). *Biological Conservation, 127*(3), 272–281. https://doi.org/10.1016/j.biocon.2005.05.015

Rojas, C., Barrera, F. de la, Aranguíz, T., Munizaga, J., & Pino, J. (2017). Efectos de la urbanización sobre la conectividad ecológica de paisajes metropolitanos. *Revista Universitaria de Geografía, 26*(2), 155–182.

Rojas, C., & Jorquera, F. (2021). Urban fabrics to eco-friendly blue–green for urban wetland development. *Sustainability, 13*(24). https://doi.org/10.3390/su132413745

Rojas, C., Munizaga, J., Rojas, O., Martínez, C., & Pino, J. (2019). Urban development versus wetland loss in a coastal Latin American city: Lessons for sustainable land use planning. *Land Use Policy, 80*, 47–56. https://doi.org/10.1016/j.landusepol.2018.09.036

Rojas, C., Sepúlveda-Zúñiga, E., Barbosa, O., Rojas, O., & Martínez, C. (2015). Patrones de urbanización en la biodiversidad de humedales urbanos en Concepción metropolitano. *Revista de Geografía Norte Grande, 61*, 181–204. https://doi.org/10.4067/S0718-34022015000200010

Rojas, O., Soto, E., Rojas, C., & López, J. J. (2022). Assessment of the flood mitigation ecosystem service in a coastal wetland and potential impact of future urban development in Chile. *Habitat International, 123*, 102554. https://doi.org/10.1016/j.habitatint.2022.102554

Villagra, P., Rojas, C., Alves, S., Rojas, O. (2023). Spatial interactions between perceived biophilic values and neighborhood typologies in wetlands. City and Built Environment. https://doi.org/10.21203/rs.3.rs-1919228/v1 (*Under Review*)

Zhang, Z., Jiang, W., Peng, K., Wu, Z., Ling, Z., & Li, Z. (2023). Assessment of the impact of wetland changes on carbon storage in coastal urban agglomerations from 1990 to 2035 in support of SDG15.1. *Science of The Total Environment, 877*, 162824. https://doi.org/10.1016/j.scitotenv.2023.162824

Urban Wetlands in Latin America as Support to 17 SDG: A Guideline to Sustainable Cities to SDG 11

Carolina Rojas Quezada,
Juanita Aldana-Dominguez, Paola Moschella,
Caroline Stamm, and Evelyn Soto

Contents

Abstract

Wetlands support all 17 Sustainable Development Goals, but they are particularly relevant to SDG 11 on "Sustainable Cities and Communities" because their preservation reinforces the ecosystem services that are essential for cities, such as flood protection, recreation, and freshwater provision, among other benefits. This chapter proposes the integration of wetlands into SDG 11 through the development of a participatory action guide with key stakeholders. The objectives of this guide are (i) to define sustainability criteria for wetlands, and (ii) to review all the targets of SDG 11 by 2030, to

C. Rojas Quezada (✉)
Instituto de Estudios Urbanos y Territoriales, Centro de Desarrollo Urbano Sustentable CEDEUS, Instituto Milenio SECOS, Santiago, Chile
e-mail: carolina.rojas@uc.cl

J. Aldana-Dominguez
Departamento de Química y Biología, Universidad del Norte, Barranquilla, Colombia
e-mail: ajuanita@uninorte.edu.co

P. Moschella
Departamento de Humanidades, Geografía y Medio Ambiente, Pontificia Universidad Católica del Perú, Lima, Peru
e-mail: pmoschella@pucp.edu.pe

C. Stamm
Instituto de Estudios Urbanos y Territoriales, Pontificia Universidad Católica de Chile, Centro de Desarrollo Urbano Sustentable CEDEUS, Instituto Milenio SECOS, Santiago, Chile
e-mail: castamm@uc.cl

E. Soto
Facultad de Ciencias Ambientales, Centro Eula, UDEC, Concepción, Chile

C. Rojas Quezada (ed.), *Urban Wetlands in Latin America*, Sustainable Development Goals Series,
https://doi.org/10.1007/978-3-031-69590-2_5

propose guidelines, and to introduce new indicators that consider urban wetlands to achieve sustainable cities in Latin America. Results show that the urban wetlands could contribute to goals 11.1 and 11.2 by playing a crucial role in neighborhood improvements, especially in providing green areas, the results indicated that the most relevant goals for the inclusion of wetlands are 11.3, 11.4, 11.5, 11.6, and 11.7 targets of SGD 11.

Keywords

Urban wetlands · 17 SDG · SDG 11 · Sustainable cities

1 The Contribution of Urban Wetlands to Achieving SDG 11: Sustainable Cities and Communities

Urban wetlands are natural transitional areas, rich in water and vegetation, located within urban zones. Despite being considered marginal and neglected areas in Latin America and leading in their global loss (Darrah et al. 2019), these valuable ecosystems can provide a wide range of ecosystem services to our cities, such as flood control, water purification, recreation, tourism, and cultural activities. Therefore, it is important to establish and direct measures toward greater restoration, conservation, protection, and integration in Latin American cities, significantly contributing to the sustainability of the communities and cities they are in, as stated in SDG 11 on sustainable cities and communities.

The United Nations' 17 Sustainable Development Goals (Agenda 2030) were defined in 2015 with specific targets to eradicate poverty, protect the planet, and ensure prosperity for all by 2030. SDG 11 is of special relevance to wetlands, as its goal is to make cities and human settlements inclusive, safe, resilient, and sustainable. In this context, the project "Urban Wetlands in Latin America: A Solution for Sustainable Cities SDG 11" has been carried out with the objective of "Strengthening research and interdisciplinary research cooperation in urban wetlands as key and fundamental ecosystems to enhance more sustainable and resilient cities by 2030, as mandated by SDG 11 in Latin American countries, and thus strengthen their connection with public policy. As part of the project, an analysis of the SDG targets was conducted to observe where urban wetlands can be integrated.

Below are some of the contributions that urban wetlands can provide to achieve SDG 11:

Improvement of the quality of life of urban communities

Urban wetlands can offer a wide range of ecosystem services that enhance the quality of life of urban communities. For example, they can help reduce air and water pollution, improve water quality, and provide green spaces for recreation and leisure activities. These ecosystem services are particularly valuable due to the location of urban wetlands, as they can have a direct impact on the physical and mental health of people living in urban areas.

Protection against natural disasters

Urban wetlands can act as natural buffers against extreme natural phenomena such as floods, tsunamis, and heatwaves. By acting as natural sponges, wetlands can retain and filter large amounts of water during heavy rainfall, thus reducing the amount of water entering the sewer systems and decreasing the risk of floods. Additionally, wetlands can help reduce soil erosion and protect coastal areas from erosion.

Contribution to climate change mitigation

Urban wetlands can significantly contribute to climate change mitigation by acting as carbon sinks. Wetlands can store large amounts of carbon in their soil and vegetation, making them important carbon sinks. Furthermore, in the balance of emissions from different greenhouse gases, wetlands are estimated to have a negative radiative forcing in the long term, despite methane emissions (Neubauer 2014; Whitting and Chanton 2001).

Promotion of urban biodiversity

Urban wetlands can promote biodiversity in urban areas by providing habitat for fauna and flora. Wetlands can support a wide variety of plant, animal, and microbial species that are not found in other urban habitats. Additionally, wetlands can act as ecological corridors.

It is important to recognize that in Latin America, cities are unequal, and in many cases, prioritizing the economy and employment has come at the expense of nature conservation, particularly wetlands. In fact, 86 million people still live in conditions of extreme poverty, which worsened with the social and health crisis of the COVID-19 pandemic (CEPAL 2022), and nature has not been considered as an indicator of well-being or quality of life. This is despite recent research showing that psychological well-being, human development, environmental quality, social progress, and economic situation are negatively associated with wetland loss (Let & Pal 2023). Additionally, the socio-economic effects of the COVID-19 pandemic in the region have been devastating, which predicts a probable setback in SDGs and conservation efforts and will also depend on the ecosystem services provided by nature (Hernández-Blanco et al. 2022), as they provide resources for human well-being and economic development. This is further exacerbated as new urban growth occurs in vulnerable areas and floodplains in the region (Rojas et al. 2019). On the other hand, Latin America has a negative environmental performance, with increasing CO_2 emissions, air pollution, and water contamination (Koengkan et al. 2023), areas in which wetlands and their restoration can significantly contribute to improvement. In this sense, it is commendable that communities and governments in Latin America are increasingly investigating how to effectively mitigate impacts and protect and restore these unique ecosystems through protected areas, water resource planning, environmental assessment, and rehabilitation. Therefore, through participatory work, we propose to develop sustainability criteria for wetlands for the SDG 11 targets, to introduce new indicators to achieve sustainable cities in Latin America that consider wetland ecosystems.

2 A Participatory Method to Strengthen the Role of Urban Wetlands in SDG 11

2.1 Proposal of Sustainability Criteria

In the development of sustainability criteria or standards to promote the integration of wetlands in urban planning, various scientific, management, and regulatory sources were reviewed concerning the contribution of these aquatic ecosystems in different cities around the world. Based on this review, we defined four categories of criteria:

(A) Physicochemical and/or Biological Criteria: Actions, tasks, organizations, constructions, and/or activities that influence the physical, chemical, or biological characteristics of urban wetlands.

1. Sustainable management of biodiversity: Identification of flora, fauna, and microorganisms in wetlands to enhance knowledge for their conservation and recovery.
2. Ecological management of wetlands based on the life cycle: Safeguarding the physical integrity of wetlands to maintain the balance of water, nutrients, and other elements within specific areas, preventing habitat fragmentation or destruction that the wetland provides.
3. Maintenance of the hydrological regime: Proper maintenance of water quality, quantity, and hydraulic characteristics of the wetland.
4. Control and monitoring of invasive species: Wetlands and other peri-urban ecosystems have been destroyed, fragmented, or invaded by non-native and invasive species over the years. There is an urgent need to conserve urban ecosystems.
5. Mitigation of harmful processes to the ecosystem health of wetlands: Control of

pollution, overexploitation, water extraction; biological pollutants are eliminated by plant life and deposited in sediment, significantly improving the quality of runoff.

(B) Green Infrastructure Criteria: Actions, tasks, organizations, constructions, and/or activities that promote the use of green infrastructure over gray infrastructure.
  1. Cost-benefit evaluation of mangrove and flooded forest reforestation: Monetization of the benefits provided by wetlands becomes a comparative point for decision-making at different levels.
  2. Climate change mitigation strategies through Nature-Based Solutions: Based on the ecosystem services provided by wetlands (ESPs), such as carbon absorption and fixation, flood and overflow mitigation, and drought buffering, among others.
  3. Promotion of healthy and happy lifestyles: Wetlands provide residents with green spaces where they can recreate and enjoy diverse plant and animal life.
  4. Hydrological storage for water control and management in urban areas: Urban wetlands efficiently and affordably purify urban water. They also facilitate sustainable urban drainage to store and purify urban rainwater runoff.

(C) Regulatory/Public Policy Criteria: Actions, tasks, organizations, constructions, and/or activities that depend on the establishment of institutional instruments for their applicability and implementation.
  1. Utilization of wetland ecosystems for natural disaster reduction: Integrating wetlands as flood buffering zones leads to a reduction in costs associated with potential disasters. Implementing preventive measures and using green infrastructure reduces the costs of constructing gray infrastructure.
  2. Development of urban planning instruments that maintain ecological parameters of urban wetlands: Elaborating strategic plans and public projects with investments and constructions involving green infrastructure.
  3. Recognition of conservation areas: Identifying priority areas for biodiversity conservation and legal protection.
  4. Evaluation of cities' climate vulnerability: Systematizing the understanding of threats, exposure, and vulnerability through surveys, characterization, and scenario-building to facilitate better decision-making. Evaluating the use of ecosystems to mitigate threats, exposure, and reduce vulnerability.
  5. Promoting ecotourism or sustainable tourism, birdwatching, and contemplation areas: Tourism can be a mechanism for promoting awareness, care, and protection of urban wetlands, benefiting the city and fostering the economy.

(D) Citizen Participation Criteria: Actions, tasks, organizational, constructions, and/or activities that involve the participation of different sectors of the public, whether public, private, or civil society, for their development.
  1. Inclusion of science and knowledge: Promoting scientific and technical cooperation by supporting joint research on the management and rational use of wetlands.
  2. Community education and awareness for ecosystem maintenance: Designing and implementing education strategies to address the main threats and forces of change in the wetland. As well as joint strategies for the management and rational use of wetlands.
  3. Facilitating multi-level governance processes and citizen participation: Facilitating collaboration and dialogue among multiple actors and organizations for participatory management of urban wetlands (governance and participation).
  4. Creating alliances for cooperativeness and capacity building: Promoting action and success on the ground in the conservation and management of wetlands internationally through public-state-international alliances.

5. Integrating a gender perspective in wetland preservation: Guiding research, education, and wetland management strategies considering gender equity relations and promoting active participation of women.
6. Use and consideration of citizen science in wetland conservation and protection: Positive results can be achieved when local community awareness, conservation measures, cultural values/traditions, tourism, and silviculture are proactively integrated to achieve the rational use of wetlands.

2.2 Workshop on Sustainability Criteria

Based on criteria that consider quantitative and qualitative aspects of urban wetlands in the sustainability of cities, an online workshop was conducted on October 29, 2021, involving 15 key actors or experts in wetland conservation, urban planning, and public policies in Latin America. The workshop aimed to review and propose guidelines in a participatory manner to promote more sustainable Latin American cities through the recognition and protection of aquatic ecosystems within the goals of SDG11.

The workshop involved three exercises:

EXERCISE 1: Inspiration

Participants were invited to think about the following question: "How do you imagine the future of urban wetlands in Latin America?" Eight images of different urban wetlands from around the world were presented, and the result indicated that the most voted image was the wetland in the center of London (Fig. 1).

EXERCISE 2: Evaluation of Goals

The ten goals of SDG 11 were reviewed, and the alignment and relevance of each of the urban wetland sustainability criteria were assessed for each of the SDG 11 goals (Fig. 2).

EXERCISE 3: Development of New Indicators

Subsequently, the indicators for measuring the SDG 11 goals were assessed, and in the case of existing associated sustainability criteria, new indicators were proposed for some goals regarding the integration of urban wetlands into SDG 11 (Fig. 3).

Fig. 1 Pictures of urban wetlands in cities. (Source: Workshop SDG in 2021)

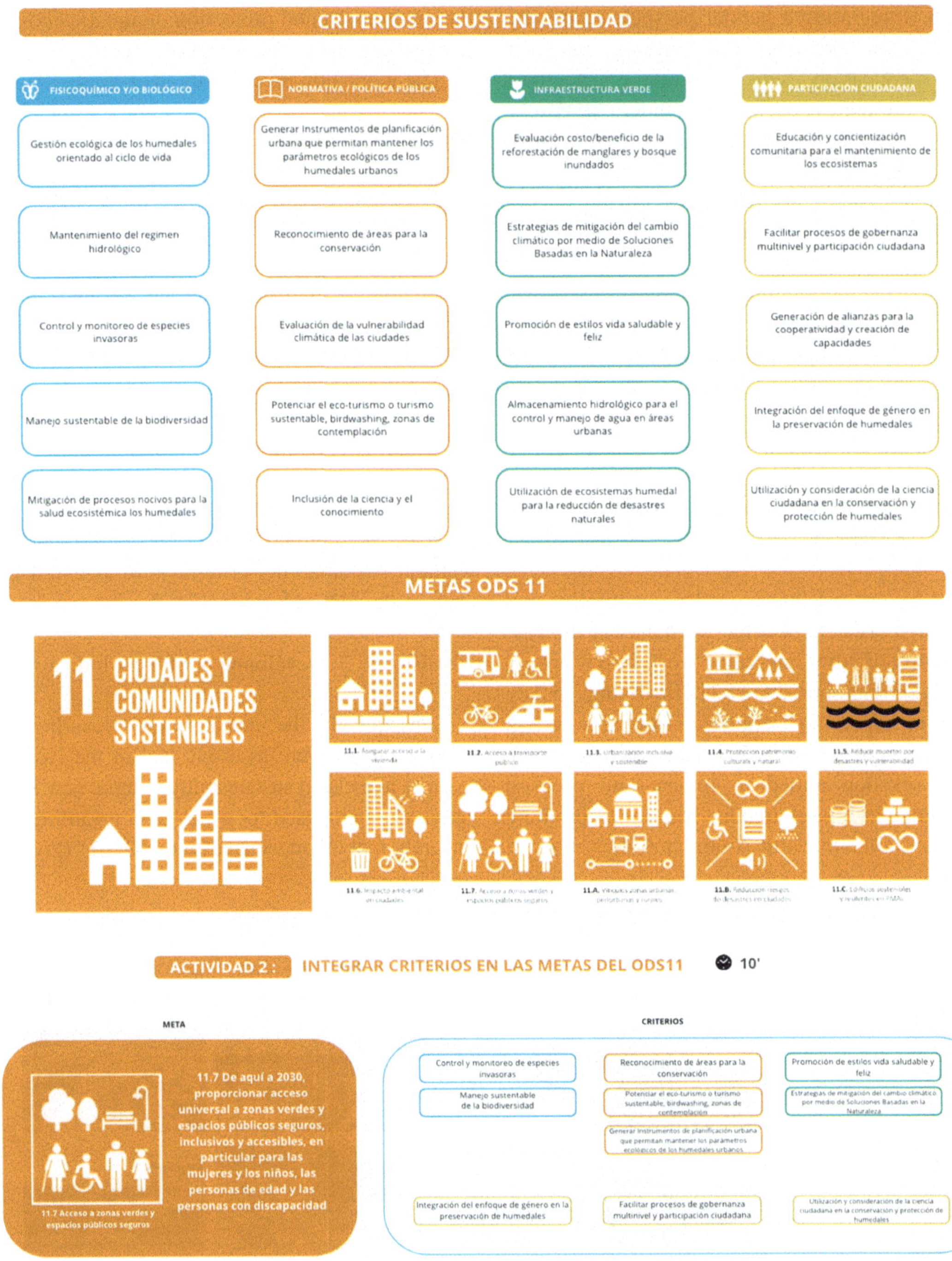

Fig. 2 Sustainability criteria, SDG 11 goals, and example of relevance evaluation for each criterion in one of the goals. (Source: Workshop SDG in 2021)

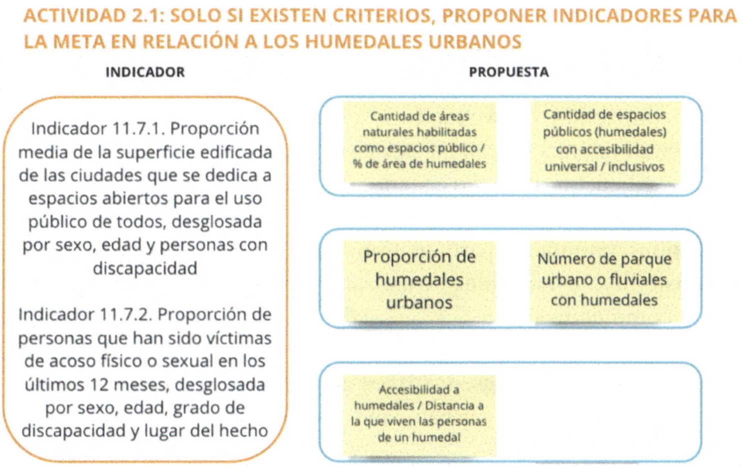

Fig. 3 Proposal of indicators. (Source: Workshop SDG in 2021)

3 Results of SDG 11: Sustainable Cities and Communities That Incorporate Urban Wetlands

Subsequently, the indicators for measuring the SDG 11 goals were assessed, and in the case of existing associated sustainability criteria, new indicators were proposed for some goals regarding the integration of urban wetlands into SDG 11.

Meta 11.3: By 2030, ensure access for all to adequate, safe, and affordable housing and basic services and upgrade slums.

The 20 proposed sustainability criteria were linked to this goal. The inclusion of urban wetlands in planning promotes sustainable urbanization and fosters inclusion of communities through local governance of these ecosystems. The sense of identity and belonging promoted by wetlands becomes a tool for multilevel environmental education. The consideration of wetlands in urban planning boosts city protection against harmful impacts of climate change, such as mitigating heatwaves, floods, and droughts, among others.

Indicator 11.3.3: Percentage of cities with inventories of urban wetlands.

Therefore, a new indicator, 11.3.3, is proposed at the national level. It is essential to quantify the number and loss of urban wetlands to understand, measure, and value the ecosystem services and benefits they provide to cities. Estimating the contribution of wetlands in cities facilitates decision-making on investments in mitigation or water storage infrastructure, pollution control, and green spaces, among others.

Expected Impact: Increased generation of information and data to quantify the value of urban wetlands. Increased area of protected wetlands in urban and peri-urban regions.

Meta 11.4: Strengthen efforts to protect and safeguard the world's cultural and natural heritage.

In Meta 11.3, all 20 developed criteria were linked. Wetlands play a fundamental role in the natural and cultural heritage of cities, forming part of their heritage and environmental culture. Wetlands, as natural spaces, contribute to the physical and mental well-being of the population and host concentrations of resident and migratory biodiversity, making their conservation globally important in some cases.

Indicator 11.4.2: Percentage of surface area of wetlands protected in cities.

The protection of wetlands is essential for their conservation, care, and restoration, enabling increased benefits and ecosystem services they provide. The involvement of citizens in wetland protection promotes participation and civic responsibility, establishing trust and security within communities.

Expected Impact: Increased surface area of wetlands protected in urban and peri-urban regions.

Although wetland conservation and protection should be a cross-cutting aspect among actors involved in urban development, evidence indicates that wetlands located in urban or peri-urban areas are the most vulnerable to disappear in the short to medium term due to the demand for land for urbanization (housing and infrastructure). Thus, protecting these ecosystems is urgent.

Meta 11.5: By 2030, significantly reduce the number of deaths and the number of people affected and substantially decrease the direct economic losses caused by disasters, including water-related disasters, with a focus on protecting the poor and people in vulnerable situations.

This goal was linked to 10 criteria, which are related to the conservation and use of wetlands for disaster protection and mitigation. Additionally, the maintenance of the hydrological regulation functions of these ecosystems enables the economic valuation of the ecosystem services provided, based on disaster modeling and quantifying losses avoided during catastrophes.

Indicator 11.5.3: Per capita investment in Nature-Based Solutions with Urban Wetlands for flood mitigation.

Implementing Nature-Based Solutions has substantially lower economic costs than conventional gray infrastructure solutions. Furthermore, nature-based infrastructure requires less maintenance and is more cost-effective. Additionally, it can provide other services such as water supply during droughts or wildfires.

Expected Impact: Increased infrastructure for flood mitigation that incorporates Nature-Based Solutions (NBS) using urban wetlands.

The increase in this type of infrastructure will save costs associated with river overflows, tsunamis, tidal waves, and heatwaves, among other effects of climate change.

Meta 11.6: By 2030, reduce the adverse environmental impact per capita of cities, including by paying special attention to air quality and municipal and other waste management.

This goal was linked to only 3 criteria, 2 of which consider citizen participation to achieve the goal. Additionally, Nature-Based Solutions are considered again for goal achievement. Wetlands, as community articulators, promote participation and engagement in the management and conservation of these ecosystems.

Indicator 11.6.3: Proportion of Descontaminación Plans (improvement in water, air, and/or soil quality) that include measures with urban wetlands.

Wetlands can be included in atmospheric decontamination plans due to their role in carbon sequestration. In water decontamination plans, they help filter and capture pollutants and heavy metals. Additionally, they contribute to the biogeochemical cycles of elemental elements for life, such as Carbon, Nitrogen, and Oxygen.

Expected Impact: Increase in decontamination plans that include urban wetlands as means of prevention, mitigation, and reduction of contaminants in various environments.

Meta 11.7: By 2030, provide universal access to safe, inclusive, accessible, and green public spaces, particularly for women and children, older persons, and persons with disabilities.

Considering urban wetlands as part of city green spaces provides landscape and social benefits. Promoting inclusive access to these areas offers enjoyment and learning zones for the entire population.

Indicator 11.7.3: Percentage of urban wetlands with public access.

Formalizing access to wetlands promotes their protection and rational and conscious use in cities, fostering ecotourism development. This allows biodiversity and citizens to coexist. However, this usage must be accompanied by environmental education and governance to increase awareness and care for wetlands as fragile ecosystems.

Expected Impact: Facilitation of access to urban wetlands.

4 Conclusions

Urban wetlands are essential for achieving SDG 11 and significantly contribute to the sustainability of cities and communities in Latin America, as supported by compelling scientific evidence used in the development of the criteria presented in the sheets for each of the identified SDG goals. Their conservation and proper management require the integration of sustainability criteria, promotion of green-blue infrastructure, and citizen participation in collaboration with different stakeholders. While urban wetlands could contribute to goals 11.1 and 11.2 by playing a crucial role in neighborhood improvements, especially in providing green areas, the results indicated that the most relevant goals for the inclusion of wetlands are 11.3, 11.4, 11.5, 11.6, and 11.7.

Regarding the SDG 11 goals, we can conclude:

Goal 11.3: Urban wetlands can be an effective tool to promote inclusive and sustainable urbanization and participatory planning and management of human settlements, involving communities in the conservation and restoration of these ecosystems and fostering citizen participation in decision-making related to urban development.

Goal 11.4: Urban wetlands are part of the natural and cultural heritage of cities. Their conservation and restoration contribute to protecting and safeguarding biodiversity and cultural diversity while providing opportunities for education and sustainable tourism around these valuable ecosystems.

Goal 11.5: Urban wetlands can act as a measure to reduce natural disaster risks by providing buffering areas and protection against floods and mitigating the negative impacts of water-related disasters in cities.

Goal 11.6: Urban wetlands can play a significant role in reducing the negative environmental impact of cities. These ecosystems act as natural filters by purifying water and air, reducing pollution, and regulating nutrient cycles. Incorporating and preserving urban wetlands in urban areas can improve air quality and reduce waste generation.

Goal 11.7: Urban wetlands can play a crucial role in providing green and public spaces that are safe, inclusive, and accessible to all people. These ecosystems offer opportunities for recreation, leisure, environmental education, and connection with nature in the urban environment. By promoting the conservation and restoration of urban wetlands, universal access to safe and inclusive green spaces can be ensured, considering the specific needs of vulnerable groups such as women, children, the elderly, and people with disabilities.

In summary, urban wetlands play an essential role in achieving the goals related to sustainable urban development. Integrating them into urban planning can improve quality of life, strengthen citizen participation, safeguard natural and cultural heritage, and reduce disaster-related risks. Furthermore, recent research points to a link with sustainable mobility by promoting recreation through walking (Rojas et al. 2022). Urban wetlands offer green and public spaces that are safe and accessible to all, contributing to ensuring universal access to green areas and promoting the inclusion of vulnerable groups in enjoying these spaces.

Anexxe N°1. Targets of SDG 11

Target SDG 11	11.1 From here to 2030, ensure access for all people to adequate, safe, and affordable housing and basic services and upgrade slums
Criteria	Generate urban planning instruments that allow maintaining the ecological parameters of wetlands Hydrological storage for water control and management in urban areas Utilize wetland ecosystems for the reduction of natural disasters
Indicator	No indicator proposed
Expected impact	Not applicable
Scientific evidence	Not applicable
Target SDG 11	"11.2 By 2030, provide access to safe, affordable, accessible, and sustainable transportation systems for all, and improve road safety, particularly by expanding public transport, with special attention to the needs of vulnerable populations, women, children, persons with disabilities, and the elderly"
Criteria	Climate vulnerability assessment of cities Develop urban planning instruments that allow maintaining the ecological parameters of urban wetlands Inclusion of science and knowledge Utilization of wetland ecosystems for the reduction of natural disasters Utilization and consideration of citizen science in the conservation and protection of wetlands
Indicator	No indicator proposed
Expected impact	Not applicable
Scientific evidence	Not applicable
Target SDG 11	"11.3 By 2030, enhance inclusive and sustainable urbanization and capacity for participatory, integrated, and sustainable planning and management of human settlements in all countries"
Criteria	Ecological management of wetlands focused on the life cycle Maintenance of the hydrological regime Control and monitoring of invasive species Sustainable management of biodiversity Mitigation of processes harmful to wetland ecosystem health Climate vulnerability assessment of cities Development of urban planning instruments that allow maintaining the ecological parameters of urban wetlands Enhancement of ecotourism or sustainable tourism, birdwatching, and contemplation areas Recognition of areas for conservation Inclusion of science and knowledge Cost-benefit evaluation of mangrove and flooded forest reforestation Hydrological storage for water control and management in urban areas Strategies for climate change mitigation through Nature-Based Solutions Utilization of wetland ecosystems for the reduction of natural disasters Promotion of healthy and happy lifestyles Formation of alliances for cooperation and capacity-building Community education and awareness for ecosystem maintenance Facilitating multi-level governance processes and citizen participation Utilization and consideration of citizen science in wetland conservation and protection Integration of a gender approach in wetland preservation
Indicator	Indicator 11.3.3. Percentage of cities with inventories of urban wetlands
Expected impact	Increase in the number of protected wetlands in urban and peri-urban areas

Target SDG 11	"11.3 By 2030, enhance inclusive and sustainable urbanization and capacity for participatory, integrated, and sustainable planning and management of human settlements in all countries"
Scientific evidence	Buldrini, F., Simoncelli, A., Accordi, S., Pezzi, G., & Dallai, D. (2015). Ten years of citizen science data collection of wetland plants in an urban protected area. Acta Botanica Gallica, 162(4), 365–373 Cherkaoui, S., Selmi, S., Amhaouch, Z., & Hanane, S. (2018). Assessment of the effectiveness of wetland protection in improving waterbird diversity in a Moroccan wetland system. Environmental Monitoring and Assessment, 190(12), 1–11 McInnes, R., Davidson, N., Rostron, C., Simpson, M., & Finlayson, C. (2020). A Citizen Science State of the World's Wetlands Survey. Wetlands (Wilmington, N.C.), 40(5), 1577–1593 Polaz, C., Ferreira, F., & Petrere Júnior, M. (2017). The protected areas system in Brazil as a baseline condition for wetlands management and fish conservancy: The example of the Pantanal National Park. Neotropical Ichthyology, 15(3), Neotropical ichthyology, 2017-09-28, Vol.15 (3)
Target SDG 11	11.4 Double efforts to protect and safeguard the world's cultural and natural heritage
Criteria	Ecological management of wetlands oriented toward the life cycle Maintenance of the hydrological regime Control and monitoring of invasive species Sustainable management of biodiversity Mitigation of harmful processes to wetland ecosystem health Climate vulnerability assessment of cities Development of urban planning instruments that allow maintaining the ecological parameters of urban wetlands Strengthening ecotourism or sustainable tourism, birdwatching, and contemplation areas Recognition of areas for conservation Inclusion of science and knowledge Cost-benefit evaluation of mangrove and flooded forest reforestation Hydrological storage for water control and management in urban areas Climate change mitigation strategies through Nature-Based Solutions Utilization of wetland ecosystems for the reduction of natural disasters Promotion of healthy and happy lifestyles Creation of alliances for cooperation and capacity building Community education and awareness for ecosystem maintenance Facilitation of multi-level governance processes and citizen participation Utilization and consideration of citizen science in wetland conservation and protection Integration of a gender approach in wetland preservation
Indicator	Indicator 11.4.2. Percentage of protected wetland area in cities
Expected impact	Increase in the area of protected wetlands in urban and peri-urban areas
Scientific evidence	Leitheiser, S. (2018). The Politics of "Better Than Nothing" in National Wetland Protection Law. Journal of International Wildlife Law and Policy, 21(1), 51–78 Kingsford, R., Basset, A., & Jackson, L. (2016). Wetlands: Conservation's poor cousins. Aquatic Conservation, 26(5), 892–916 Kleijn, D., Cherkaoui, I., Goedhart, P., Hout, van der, J, & Lammertsma, D. (2014). Waterbirds increase more rapidly in Ramsar-designated wetlands than in unprotected wetlands. The Journal of Applied Ecology, 51(2), 289–298 Thorslund, J., Jarsjo, J., Jaramillo, F., Jawitz, J., Manzoni, S., Basu, N., … Destouni, G. (2017). Wetlands as large-scale nature-based solutions: Status and challenges for research, engineering and management. Ecological Engineering, 108(Part B), 489–497

Target SDG 11	"11.5 By 2030, significantly reduce the number of deaths and the number of people affected by disasters, including those related to water, and substantially decrease the direct economic losses caused by disasters in relation to the global gross domestic product (GDP). This should be achieved while giving special attention to the protection of the poor and vulnerable individuals" If there's anything else you'd like to translate or if you have any other questions, feel free to ask!
Criteria	Maintenance of the hydrological regime Mitigation of harmful processes to wetland ecosystem health Climate vulnerability assessment of cities Development of urban planning instruments that allow maintaining the ecological parameters of urban wetlands Hydrological storage for water control and management in urban areas Climate change mitigation strategies through Nature-Based Solutions Utilization of wetland ecosystems for the reduction of natural disasters Community education and awareness for ecosystem maintenance Facilitation of multi-level governance processes and citizen participation Utilization and consideration of citizen science in wetland conservation and protection
Indicator	Indicator 11.5.3. Per capita investment in Nature-Based Solutions with Urban Wetlands for flood mitigation
Expected impact	Increase in infrastructure for flood mitigation that considers Nature-Based Solutions
Scientific evidence	Kelvin S.-H. Peh, Balmford, A., Field, R. H., Lamb, A., Birch, J. C., Bradbury, R. B., … Hughes, F. M. R. (2014). Benefits and costs of ecological restoration: Rapid assessment of changing ecosystem service values at a U.K. wetland. Ecology and Evolution, 4(20), 3875–3886. doi: https://doi.org/10.1002/ece3.1248 Patagua, Fundación Legado Chile & Pontifica Universidad Católica de Chile. (2021). Ciudades sensibles al agua. Guía de drenaje urbano sostenible para la Macro Zona Sur de Chile Leitheiser, S. (2018). The Politics of "Better Than Nothing" in National Wetland Protection Law. Journal of International Wildlife Law and Policy, 21(1), 51–78 Moomaw, W., Chmura, G., Davies, G., Finlayson, C., Middleton, B., Natali, S., … Sutton-Grier, A. (2018). Wetlands In a Changing Climate: Science, Policy and Management. Wetlands (Wilmington, N.C.), 38(2), 183–205 Thorslund, J., Jarsjo, J., Jaramillo, F., Jawitz, J., Manzoni, S., Basu, N., … Destouni, G. (2017). Wetlands as large-scale nature-based solutions: Status and challenges for research, engineering and management. Ecological Engineering, 108(Part B), 489–497 Zhou, T., & Penning-Rowsell, E. (2021). China's 'Sponge Cities': The role of constructed wetlands in alleviating urban pluvial flooding. Water and Environment Journal : WEJ, 35(3), 1133–1146
Target SDG 11	1.6 By 2030, reduce the negative environmental impact per capita of cities, including special attention to air quality and the management of municipal and other types of waste
Criteria	Inclusion of science and knowledge Climate change mitigation strategies through Nature-Based Solutions Utilization and consideration of citizen science in wetland conservation and protection
Indicator	Indicator 11.6.3. Proportion of Decontamination Plans (improvement of water, air, and/or soil quality) that incorporate measures with wetlands
Expected impact	Increase in decontamination plans that consider urban wetlands as means for prevention, mitigation, and reduction of pollutants in different environments
Scientific evidence	Dayathilake, D., Lokupitiya, E., & Wijeratne, V. (2021). Estimation of Soil Carbon Stocks of Urban Freshwater Wetlands in the Colombo Ramsar Wetland City and their Potential Role in Climate Change Mitigation. Wetlands (Wilmington, N.C.), 41(2), Wetlands (Wilmington, N.C.), 2021-02-24, Vol.41 (2). Moomaw, W., Chmura, G., Davies, G., Finlayson, C., Middleton, B., Natali, S., … Sutton-Grier, A. (2018). Wetlands In a Changing Climate: Science, Policy and Management. Wetlands (Wilmington, N.C.), 38(2), 183–205 Musser, W., & Bomar, M. (2017). Reclaiming Water to Rehydrate Wetlands: Panama City Beach's Conservation Park. Journal - American Water Works Association, 109(4), 55-63. Pozo, K., Sarkar, S., Estellano, V., Mitra, S., Audi, O., Kukucka, P., … Corsolini, S. (2017). Passive air sampling of persistent organic pollutants (POPs) and emerging compounds in Kolkata megacity and rural mangrove wetland Sundarban in India: An approach to regional monitoring. Chemosphere (Oxford), 168, 1430–1438

Target SDG 11	11.7 By 2030, provide universal access to green and public spaces that are safe, inclusive, and accessible, particularly for women and children, the elderly, and persons with disabilities
Criteria	Control and monitoring of invasive species Sustainable management of biodiversity Development of urban planning instruments that allow maintaining the ecological parameters of urban wetlands Promotion of ecotourism or sustainable tourism, birdwatching, and contemplation areas Recognition of areas for conservation Climate change mitigation strategies through Nature-Based Solutions Promotion of healthy and happy lifestyles Facilitation of multi-level governance processes and citizen participation Utilization and consideration of citizen science in wetland conservation and protection Integration of a gender approach in wetland preservation
Indicator	Indicator 11.7.3. Percentage of urban wetlands with public access
Expected impact	Facilitating access to urban wetlands
Scientific evidence	Chen, V., Lin, J., & Tzeng, G. (2019). Assessment and improvement of wetlands environmental protection plans for achieving sustainable development. Environmental Research, 169, 280–296 Gonçalves, L., Tsuge, M., Borghi, V., Miranda, F., Sales, A., Lucchetti, A., & Lucchetti, G. (2018). Spirituality, Religiosity, Quality of Life and Mental Health Among Pantaneiros: A Study Involving a Vulnerable Population in Pantanal Wetlands, Brazil. Journal of Religion and Health, 57(6), 2431–2443 Polaz, C., Ferreira, F., & Petrere Júnior, M. (2017). The protected areas system in Brazil as a baseline condition for wetlands management and fish conservancy: The example of the Pantanal National Park. Neotropical Ichthyology, 15(3), Neotropical ichthyology, 2017-09-28, Vol.15 (3) Pedersen, E., Weisner, S., & Johansson, M. (2019). Wetland areas' direct contributions to residents' well-being entitle them to high cultural ecosystem values. The Science of the Total Environment, 646, 1315–1326
Target SDG 11	11.a Support positive economic, social, and environmental linkages between urban, peri-urban, and rural areas by strengthening national and regional development planning
Criteria	Maintenance of the hydrological regime Control and monitoring of invasive species Development of urban planning instruments that allow maintaining the ecological parameters of urban wetlands Promotion of eco-tourism or sustainable tourism, birdwatching, and contemplation areas Recognition of areas for conservation Climate change mitigation strategies through Nature-Based Solutions Utilization of wetland ecosystems for the reduction of natural disasters Formation of alliances for cooperation and capacity building Community education and awareness for ecosystem maintenance Facilitation of multi-level governance processes and citizen participation
Indicator	No indicator proposed
Expected impact	Not applicable
Scientific evidence	Not applicable
Target SDG 11	11.b By 2030, significantly increase the number of cities and human settlements adopting and implementing integrated policies and plans to promote inclusion, resource efficiency, climate change mitigation and adaptation, and disaster resilience, and develop and implement, in line with the Sendai Framework for Disaster Risk Reduction 2015-2030, holistic disaster risk management at all levels
Criteria	Climate vulnerability assessment of cities Development of urban planning instruments that allow maintaining the ecological parameters of urban wetlands Climate change mitigation strategies through Nature-Based Solutions Utilization of wetland ecosystems for the reduction of natural disasters Formation of alliances for cooperation and capacity building Facilitation of multi-level governance processes and citizen participation

Target SDG 11	11.b By 2030, significantly increase the number of cities and human settlements adopting and implementing integrated policies and plans to promote inclusion, resource efficiency, climate change mitigation and adaptation, and disaster resilience, and develop and implement, in line with the Sendai Framework for Disaster Risk Reduction 2015-2030, holistic disaster risk management at all levels
Indicator	No indicator proposed
Expected impact	Not applicable
Scientific evidence	Not applicable
Target SDG 11	11.c Provide support to the least developed countries, including financial and technical assistance, to enable them to build sustainable and resilient buildings using local materials
Criteria	Inclusion of science and knowledge Cost-benefit evaluation of mangrove and flooded forest reforestation Hydrological storage for water control and management in urban areas Promotion of healthy and happy lifestyles Formation of alliances for cooperation and capacity building
Indicator	No indicator proposed
Expected impact	Not applicable
Scientific evidence	Not applicable

References

CEPAL (2022). Comisión Económica para América Latina y el Caribe (CEPAL), Panorama Social de América Latina, 2021(LC/PUB.2021/17-P), Santiago, 2022

Darrah, S., Shennan-Farpón, Y., Loh, J., Davidson, N., C., Finlayson, C. M., Gardner, R., C., & Walpole, M., J. (2019). Improvements to the Wetland Extent Trends (WET) index as a tool for monitoring natural and human-made wetlands. Ecological Indicators, 99, 294–298. https://doi.org/10.1016/j.ecolind.2018.12.032

Desarrollo urbano: Panorama general. (2021). World Bank. https://www.bancomundial.org/es/topic/urbandevelopment/overview#1

Hernández-Blanco, M., Contanza, R., Chen, H., deGroot, D., Jarvis, D., Kubiszewski, I., Montoya, J., Sangha, K., Stoeckl, N., Turner, K., van't Hoff, V. (2022). Ecosystem health, ecosystem services, and the well-being of humans and the rest of nature. Global Change Biology, 28, 5027–5040. https://onlinelibrary.wiley.com/doi/full/10.1111/gcb.16281

Koengkan, M., Fuinhas, J. A., Tavares, A. I. P., & Gonçalves Silva, N. M. B. (2023). Environmental degradation in the Latin American and Caribbean region. In Obesity Epidemic and the Environment (pp. 147–172). Elsevier. https://doi.org/10.1016/b978-0-323-99339-5.00007-8

Let, M., & Pal, S. (2023). Socio-ecological well-being perspectives of wetland loss scenario: A review. Journal of Environmental Management, 326, 116692. https://doi.org/10.1016/j.jenvman.2022.116692

Neubauer, S. C. (2014). On the challenges of modeling the net radiative forcing of wetlands: reconsidering Mitsch et al. 2013. Landscape Ecology, 29, 571–577. https://doi.org/10.1007/s10980-014-9986-1

Rojas, C., Munizaga, J., Rojas, O., Martínez, C., & Pino, J. (2019). Urban development versus wetland loss in a coastal Latin American city: Lessons for sustainable land use planning. Land Use Policy, 80, 47–56. https://doi.org/10.1016/j.landusepol.2018.09.036

Rojas, C., Jorquera, F., Steiniger, S. (2022). Acceder caminando a los humedales urbanos: Una oportunidad de recreación y bienestar. Revista Urbano N°46, 56–67. https://revistas.ubiobio.cl/index.php/RU/article/view/5118/4502.

Whiting, G. J., & Chanton, J. P. (2001). Greenhouse carbon balance of wetlands: methane emission versus carbon sequestration. Tellus B, 53(5), 521–528. https://doi.org/10.1034/j.1600-0889.2001.530501.x

Part II

Innovation and Restoration

Exploring the Behavior of Llanquihue's Urban Wetlands: Lessons Learnt and Perspectives for Water-Sensitive Urban Design

Camila Teutsch, Teresita Scheuch, Antonio Fernández, Felipe Briceño, Catalina Vattuone, Juan Carlos Domínguez, Andrés Riveros, Enzo Pichunleo, Osvaldo Moreno, and Jorge Gironás

Contents

Abstract

Wetlands play a significant role in preventing floods and providing access to nature in urban contexts, among many other benefits. However, in southern Chile, they have been severely threatened by an urban growth paradigm that has traditionally overlooked the valuable ecosystem services they provide. This has led to frequent flooding, water quality issues, and biodiversity loss in several urban areas within the region. Llanquihue, a small city in Los Lagos district with a rich wetland system, is no exception to these problems. The approach proposed by Patagua et al. (Ciudades Sensibles al Agua: Guía de Drenaje Urbano Sostenible

The methodology and results presented here are part of the guide "Ciudades Sensibles al Agua", Volumen III [Water-Sensitive Cities, Volume III] (Patagua et al. 2024) available at www.ciudades-sensibles-al-agua.cl.

C. Teutsch (✉) · A. Fernández · C. Vattuone
Patagua, Puerto Varas, Chile
e-mail: camila@patagua.cl; antonio@patagua.cl

T. Scheuch · J. Gironás
Department of Hydraulic and Environmental Engineering, Pontificia Universidad Católica de Chile, Santiago, Chile
e-mail: jgironas@uc.cl

F. Briceño
NIVA Chile Research Foundation, Santiago, Chile
e-mail: felipe.briceno@nivachile.cl

J. C. Domínguez
Mejores Prácticas, Santiago, Chile
e-mail: jcdominguez@mejores-practicas.com

A. Riveros · E. Pichunleo
Legado Chile Foundation, Santiago, Chile
e-mail: ariveros@legadochile.cl

C. Rojas Quezada (ed.), *Urban Wetlands in Latin America*, Sustainable Development Goals Series,
https://doi.org/10.1007/978-3-031-69590-2_6

para la Macrozona Sur de Chile [Water sensitive cities: a guide to sustainable urban drainage for the Southern Macrozone of Chile]. https://www.ciudades-sensibles-al-agua.cl/, 2021) is applied to Llanquihue as a case study to understand the wetlands' current behavior within the urban drainage system and to explore possibilities in terms of water-sensitive urban design. To simulate the flooding patterns and the spatial dynamics of surface/subsurface flux exchanges, a coupled model was adopted. Results suggest that continuous and integrated surface-subsurface simulations might be a key tool for urban stormwater planning in cities characterized by its strong interaction with a shallow water table. Based on these model results, one Llanquihue's critical flooding areas is further analyzed and green infrastructure solutions are explored, showing great potential to mitigate flooding. Therefore, from a water-sensitive urban design perspective, it is suggested that stormwater management in Llanquihue should focus on preserving and restoring natural wetlands, as well as on developing additional green-blue infrastructure. This approach is meant to enhance current hydrological functions through a sustainable drainage system, increasing urban resilience within a global change context.

Keywords

Water sensitive urban design · Sustainable urban drainage · Hydrological modelling

1 Introduction

It is widely recognized that wetlands play an important role in the hydrological cycle (Acreman and Holden 2013), as they fulfill functions such as water storage, storm protection and flood mitigation, erosion control, aquifer recharge, and water purification, among others (Ramsar Convention Secretariat 2013). In addition, wetlands provide other ecosystem services (ES), such as habitat for flora and fauna, biodiversity conservation, climate regulation and adaptation to climate change, scenic beauty, and spaces for scientific research and education (Ramsar Convention Secretariat 2013).

When it comes to urban wetlands (UWs) this role becomes even more relevant, as they constitute an essential part of the infrastructure for runoff management in cities. As green-blue infrastructure (GBI), wetlands have the potential to become multifunctional natural systems that contribute to the social and cultural well-being of communities in their surroundings, in addition to providing the aforementioned ES (Fischer et al. 2009; Alikhani et al. 2021). Unfortunately, these ecosystems are highly vulnerable to the effects of urban development resulting from land use change and the implementation of gray infrastructure for rainwater management, among others. These effects refer mainly to the alteration of the natural hydrological cycle as peak flows and total runoff increase, to a decrease in water quality due to runoff pollution, and to the loss of the hydrological services they naturally provide (Hammer and Bastian 1989; Ampatzidis and Kershaw 2020).

Given the need to move toward a more sustainable approach of urban drainage, the Water-Sensitive Urban Design (WSUD) offers an approach that integrates water management into urban planning in order to reduce the impacts of urbanization on the water cycle and to improve urban resilience to the impacts of climate change (Whelans et al. 1994; Wong and Brown 2009; Patagua et al. 2021). In this context, this chapter takes the city of Llanquihue (Los Lagos district, southern Chile) as a case study to study the role of wetlands in urban drainage and to explore opportunities for WSUD as a way to solve existing flooding problems and move toward climate adaptation and resilience.

O. Moreno
Escuela de Arquitectura, Pontificia Universidad Católica de Chile, Santiago, Chile
e-mail: omorenof@uc.cl

Llanquihue is located by the lake of the same name—the second largest lake in Chile—and is crossed by Maullín River. It is home to a population of 14,222 people (Instituto Nacional de Estadística, [National Statistics Institute] 2017), distributed over an area of 370.2 ha which has experienced an increase in housing of over 60% between 1992 and 2017 (Ilustre Municipalidad de Llanquihue [Municipality of Llanquihue] 2019). Like many small and intermediate cities in the southern macrozone of Chile, this growth has been poorly planned and has resulted in a loss of almost 40% of wetland area between 1962 and 2018 (Legado Chile Foundation 2020). Even so, the northern part of the city has still an extensive network of urban wetlands, which currently represents 14% of its total area. As shown in Fig. 1, five main wetlands stand out in this network: El Loto (LW) (3.39 ha), Baquedano (BW) (0.35 ha), Teodosio Sarao (TSW) (4.88 ha), Los Helechos (HW) (3 ha), and Las Ranas (RW) (5.99 ha) (Legado Chile Foundation 2020).

In a place where annual rainfall reaches 1500 mm on average (Scheuch 2021) and there is no rainwater infrastructure except for a few specific solutions (Ilustre Municipalidad de Llanquihue [Municipality of Llanquihue] 2015), wetlands' loss and increasing land impermeabilization have led to recurrent flooding. These floods occur even with frequent rainfall events and seriously affect people's quality of life. It is therefore of interest to understand the role of the remaining natural wetland network in urban drainage and the opportunities that Llanquihue presents for WSUD.

In order to adopt a stormwater management strategy following the WSUD principles in Llanquihue, hydrological modeling is key to

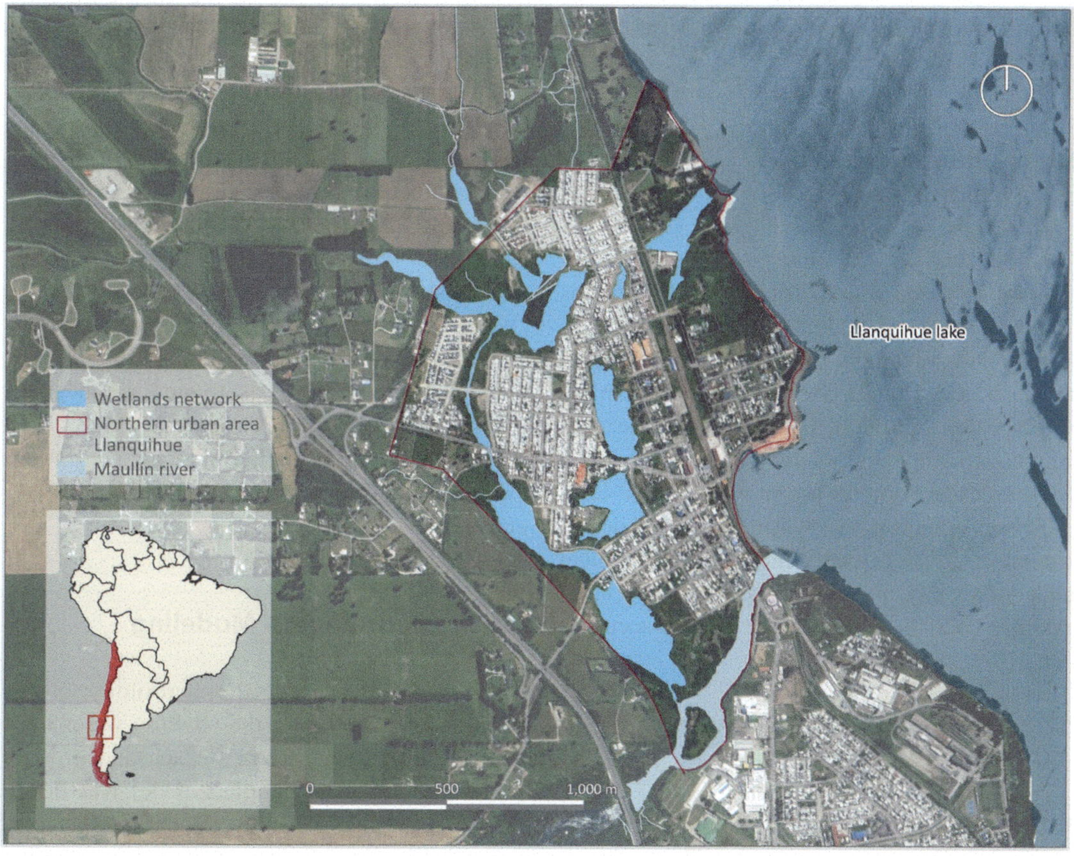

Fig. 1 Territorial context. Urban boundary of the northern section of Llanquihue and its wetland network

understanding the UWs regime and assessing critical flooding points in the city. An additional challenge presented in this case is addressing surface/subsurface interactions due to the presence of a shallow water table. In this context, the implementation of a coupled surface/subsurface model is proposed to simulate in a continuous and spatially distributed way the flow exchange. The results of the model will guide design and planning decisions, and will allow the evaluation of different stormwater management strategies.

2 Methodology

This work is based on the methodological approach proposed by Patagua et al. (2021), considering four main steps: (i) characterization of the study area; (ii) hydrological modeling; (iii) identification and characterization of flood zones; and (iv) pilot solutions conceptualization and response modeling. Each of these is described below.

2.1 Characterization of the Study Area

The first step was to characterize the city of Llanquihue based on environmental, hydrological, meteorological, and urban information, both primary and secondary. All the geographic information was systematized in a geographic information system through the QGIS software. The main information components, with their respective sources and outputs are shown in Table 1.

This information served as input for the development of a water quality index, a hydrological model of the city, and the subsequent conceptualization of WSUD solutions in a specific sector.

2.2 Water Quality Monitoring and CCME-WQI

Water quality monitoring of Llanquihue's UWs was evaluated on an annual cycle, which considered four seasonal sampling campaigns (one per season). For this study, the Canadian Council of the Ministers of the Environment (CCME) Water Quality Index (WQI) was applied (CCME 2017). This index was calculated for each wetland, considering the following variables: temperature, pH, dissolved oxygen, chemical oxygen demand (COD), total dissolved solids, turbidity, chlorophyll a, alkalinity, total phosphorus, nitrate, fecal coliforms, total copper, and total zinc. Other compounds such as total nitrogen, hydrocarbons, oils and greases, heavy metals (e.g., lead and chromium), and metalloids (e.g., arsenic), among other parameters included in Chilean Water Quality Standard NCh 1333, were analyzed but not considered in the index calculation. Target concentrations were based on NCh 1333 (specifically, Table 2, which refers to water quality requirements for the protection of aquatic life), as well as on the Australian and New Zealand guide for the protection of aquatic life (ANZECC & ARMCANZ 2000) and the Canadian guide of limits for the protection of aquatic life (CCME 2017).

To further characterize the water quality of the UWs network, the trophic status of each wetland was evaluated based on previous studies conducted in other urban wetlands of southern Chile (Almanza-Marroquín et al. 2016). Reference values came from the Organization for Economic Cooperation and Development (Vollenweider and Kerekes 1982), which indicates that a water body is identified as eutrophic when total phosphorus is found in concentrations between 35 and 100 μg/L and chlorophyll-a between 8 and 25 μg/L. When concentrations exceed these values, it is recognized as a hyper-eutrophic body.

2.3 Hydrological Modeling

To correctly represent the hydrology of the study area considering the shallow water table and, therefore, the strong surface and groundwater interaction, a coupled model was developed that allows the bidirectional exchange of flows. SWMM (Storm Water Management Model, developed by the United States

Table 1 Main information used, with their respective sources and outputs

Component	Input	Source	Output
Topography	DEM (12.5 m × 12.5 m)	ALOS PALSAR (2016)	Topography map
Meteorology	(a) Precipitation series (2017–2019; daily data) (b) Temperature series (1979–2016; daily data) (c) IDF (Intensity-duration-frequency) Curve	(a) National Water Directorate (multiple meteorological stations) (b) CR2MET data set (5 × 5 km gridded data) (c) Patagua et al. (2021)	2 years hourly precipitation data & graphs Monthly evapotranspiration index & graphs
Waterways	(a) Waterways basemap (b) Bathymetric curves of wetlands and cross sections of the Sarao stream	(a) Legado Chile Foundation (2020) (b) Primary information	Waterways map Topo-bathymetric planimetry
Urban features	(a) Road networks (b) Key infrastructure (c) Public space use	(a) PRC (Ilustre Municipalidad de Llanquihue 2009) (b) Primary information	Roads and key infrastructure map
Hydrology	(a) Lake Llanquihue water level series (2017–2019) (b) Maullín River pressure series (2017–2019) (c) Citizen information	(a) National Water Directorate (b) Previous studies (c) Primary information	Boundary conditions for the hydrological-hydraulic model Critical flooding areas
Hydrogeology	(a) Previous characterizations (b) Perfil estratigráfico asociados a derechos de aprovechamiento de aguas subterráneas	(a) Pizarro (2019) (b) National Water Directorate	Hydrogeological characterization of the study area
Urban wetland water quality	(a) Physical, chemical, and biological parameters contrasted with national and international water quality standards (b) Four seasonal sampling (spring 2020–winter 2021)	Primary information	Water quality index (Canadian Council of the Ministers of the Environment Water Quality Index [CCME-WQI])
Land use	(a) Aerial photographs (b) Permitted land uses (c) Previous land registries	(a) Primary information (b) Communal regulatory plan (Ilustre Municipalidad de Llanquihue 2009) (c) Legado Chile Foundation (2020)	Land use classification Green infrastructure potential

Environmental Protection Agency) and MODFLOW (developed by the United States Geological Survey), two widely used stormwater and groundwater models, respectively, were used. The Python libraries PySWMM (McDonnell et al. 2020) and FloPy (Bakker et al. 2016) were used for the integration of both models. Details of the coupling scheme constructed and the conceptualization of the model for Llanquihue are described in Scheuch (2021). A brief summary is described below.

2.3.1 SWMM Model Conceptualization

The conceptualization of the SWMM surface model considered 6 main steps following methodological recommendations of Gironás et al. (2009):

- Delimitation and characterization of subcatchments using digital elevation model (Centro de Información de Recursos Naturales [Natural Resources Information Center] 2016) and characteristics such as land use (Fig. 2a)

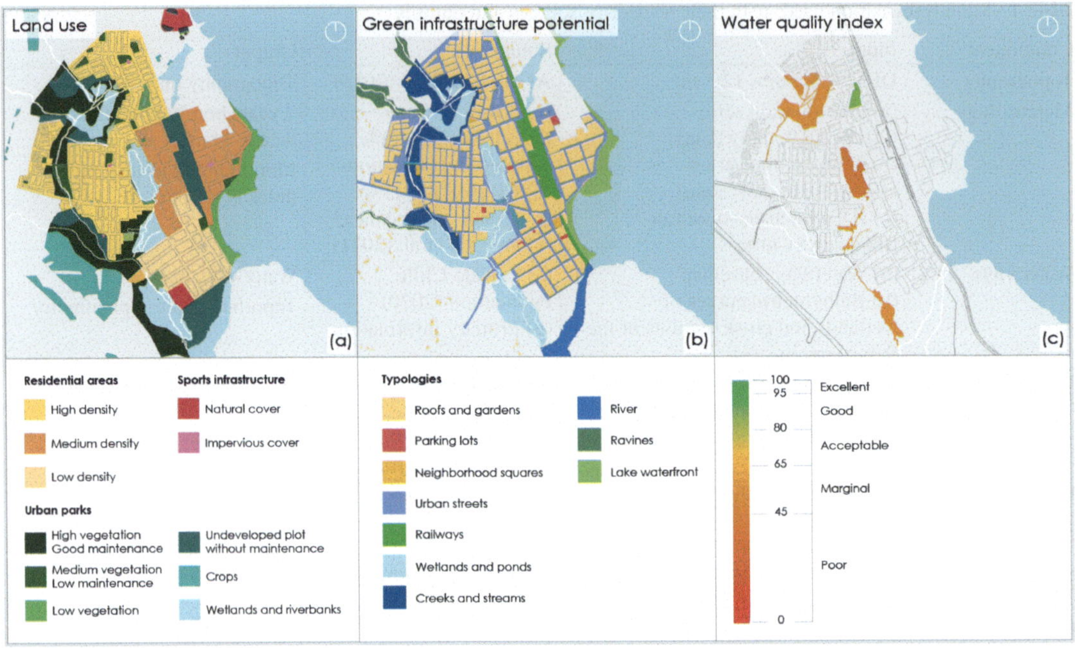

Fig. 2 Characterization of study area. (**a**) Land use; (**b**) Green infrastructure potential; (**c**) Water quality index

and road network. Six rural (338 ha) and 36 urban (162 ha) sub-basins were defined.

- Representation of water bodies based on bathymetries and topographies surveyed in the field. Five subcatchments and five storage units (SU) that operate jointly were used for their conceptualization.
- Representation of surface and subsurface conveyance elements of the road network and natural channels using mainly open conduits.
- Incorporation of the hourly precipitation series of a 2-year extension (from 01/03/2017 to 01/03/2019) for continuous modeling, obtained based on a stochastic disaggregation process (Socolofsky et al. 2001).
- Incorporation of the evapotranspiration series estimated using the WUCOLS III method (Costello et al. 2000). For this purpose, the Hargreaves method using the temperature series was previously used.
- Based on preliminary results and citizen information, six critical flooding points, explicitly represented as SU were defined.

2.3.2 MODFLOW Model Conceptualization

Due to the lack of quality data regarding groundwater, the MODFLOW subsurface model was constructed in a simplified way. Homogeneity assumptions were made in the definition of parameters, which were then studied through a sensitivity analysis. The grid was defined in a single layer of 10,000 cells with 50 × 50 m dimensions. The elevation of the cells responds to the ALOS PALSAR DEM. The constant water height of the cells associated with Llanquihue Lake and the Maullín River were defined based on both height time series available for the lake and pressure time series for the river.

2.3.3 SWMM-MODFLOW Coupled Model

A spatial association of SWMM elements and MODFLOW cells was carried out in order to transfer water fluxes in a spatially distributed manner. Infiltration from urban watersheds and wetlands was incorporated as recharge in the associated MODFLOW cells and; in the case of MODFLOW cells draining water to the surface, this was incorporated as exfiltration in the respective SWMM nodes or storage units. For the simulation, a temporal exchange scheme was adopted using a time step of one day to transfer flows between the surface and subsurface model. A continuous approach was used to model a 2-year period in order to characterize the seasonality of the wetland network, as opposed to the commonly used event-driven approach.

2.3.4 Sensitivity Analysis

Given the absence of a time series of flows or water heights to calibrate and validate the model, a one-at-a-time (OAT) sensitivity analysis was implemented for the main parameters that influence flow exchange in the models. The analysis of the results was done by selecting a single set of parameters whose hydrological response was consistent with field observations and with information reported by citizens.

2.4 Identification and Characterization of Flood Zones

After recognizing the behavior of rainwater within the city and understanding the socio-environmental dynamics of the study area (previous methodological steps), the main flood zones in Llanquihue and their contributing subcatchments were identified and characterized. The identification and characterization of flood zones was based on: (i) information gathering with neighbors; (ii) previous records from the Municipality of Llanquihue; and (iii) numerical modeling results. Then, the sub-basins contributing to these zones and their runoff paths were studied in greater detail, the main inputs being the city's topography—from the digital elevation model (CIREN 2016)—, and a high-resolution (5 cm) drone topographic survey for the preliminary contributing areas of each flood.

The model was used to understand the behavior, magnitude, and frequency of floods. In addition, the Capture Volume of each flood-contributing area was calculated, following the methodology recommended by the Ministry of Public Works (MOP 2013). Then, to facilitate the visualization of water movements and to distinguish the areas affected by major storms, a 2D version of the HEC-RAS model was built and simulated based on topographic inputs.

2.5 Pilot Solutions Conceptualization and Response Modeling

In order to understand the human dimension of the affected areas and to ensure the design of pertinent solutions, field visits and participatory activities were carried out. The information gathered in these instances, together with the previous results, nurtured the conceptual design of solutions.

The conceptualization of WSUD solutions was carried out for one of the floods as a pilot case through an iterative process of ideation and analysis of different rainwater management possibilities. This process aimed to enhance natural drainage infrastructure, prioritizing the implementation of nature-based solutions (NbS) and verifying their effectiveness through both continuous simulation and event modeling of different management scenarios.

Design criteria were established based on risk reduction recommendations by the national authority, particularly related to flood heights and widths and runoff velocity in urban environments (MOP 2013). These were complemented with general ecosystemic criteria, which seek to ensure the correct implementation of NbS solutions, improve water quality before it is conveyed to natural receiving bodies, and generate spaces in which nature and people coexist.

3 Results

3.1 Characterization of the Study Area

According to the land use analysis (Fig. 2a), a large part of the urban area shows high-density residential use (85% impervious surfaces). On the other hand, significant green areas were identified around urban water bodies, which show potential for the implementation of NbS that can contribute to flood management and runoff purification.

The Green-Blue Infrastructure (GBI) potential analysis (Fig. 2b) shows the great opportunity represented by the road network; given its key role in runoff conduction, GBI can be implemented to optimize its drainage function, seeking to replicate as best as possible the natural hydrological cycle. Along with this result, the presence of public squares near flood zones was noted, which can be intervened to become floodable green areas.

3.2 Water Quality Monitoring and CCME WQI

The calculation of the Canadian water quality index (CCME-WQI) (Fig. 2c) shows that LW, TSW, HW, and RW wetlands present 'marginal' or 'poor' water quality conditions, with LW showing the lowest value (44.4%). BW is highlighted as an exception, showing 'good' water quality conditions (81.1%). In terms of trophic status, the following classifications were determined: eutrophic-hypereutrophic (LW, TSW, and RW), eutrophic (HW), and oligotrophic-mesotrophic (BW). The annual calculation of the CCME-WQI was based on 100% of the campaigns in LW, TSW, and RW wetlands, while its estimation was partial in BW and HW wetlands due to the absence of superficial water during summer (HW and BW) and autumn (HW).

Regarding the disaggregated variables used in the CCME-WQI, it can be indicated that dissolved oxygen remained within the expected range in wetlands LW, RW, and TSW, reporting an anoxic condition in 33% of the measurements in BW and 100% in HW. COD results indicate deviations in all the measurements made for each of the wetlands, with the highest value reported in winter in LW (70 mg/L). As for fecal coliforms, the highest values were reported in HW (9200 NMP/100 mL) and LW (4450 NMP/100 mL), both during the fall season (2021) and exceeding NCh 1333. Nitrate presented deviations of 75% in both TSW and RW, with higher levels in the winter season with values of 9.68 and 8.22 mg/L, respectively. As for chlorophyll a, the highest values were reported in LW (fall 2021: 206.61 μg/L) and TSW (summer 2021: 312.2 μg/L). Total phosphorus was characterized by being out of expectation in 25% of the measurements in LW, as well as in 50% in RW, 75% in TSW, and 100% in HW. The highest value was recorded in TSW during winter with a value of 0.107 mg/L. The highest zinc concentration was reported in RW during spring 2019 (45.95 μg/L). For copper, deviations between 25% of measurements (TSW) and 50% were reported for HW and RW wetlands. The highest value of this metal was recorded in RW during spring 2020 (3.21 μg/L).

3.3 Hydrological Modeling

The SWMM-MODFLOW coupled model brings out useful results for decision-making and design of a rainwater management system, including the following: (i) water table height; (ii) duration curves for wetland heights; and (iii) flood frequency curves. These results are reviewed below.

Thanks to the modeling scheme adopted, it is possible to know the *water table height* in each of the MODFLOW cells and at any time point of the continuous simulation. This makes it possible to construct maps of the study area with this information and to analyze the changes that occur in the water table between the driest (summer) and the rainiest (winter) seasons.

Figure 3 shows the results for two months of simulation, representative of the rainy season and the dry season. These results reveal the importance of understanding groundwater behavior when designing rainwater management systems in urbanizations with shallow water tables. It is observed that most of the urban area of Llanquihue has water table depths below 1 m (shades of red) consistently during the rainy season. Therefore, the city's drainage system should take this into account and prioritize storage and conveyance as control techniques.

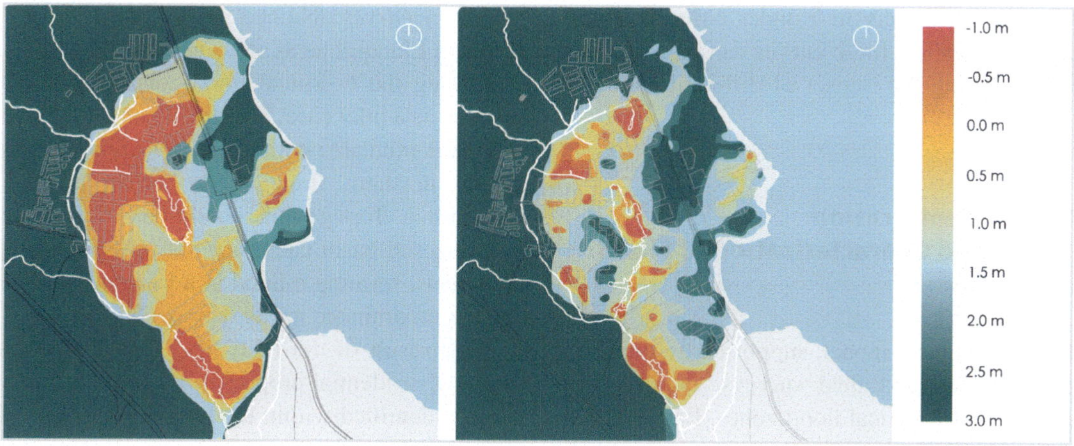

Fig. 3 Depth of the water table in August 2017 (left) and February 2018 (right). Orange, yellow, and green areas have positive depths, meaning that water is below ground level. In contrast, the red-shaded areas have negative depths, which can be interpreted as above-groundwater heights

Fig. 4 Duration curve for El Loto wetland for the period of the continuous simulation

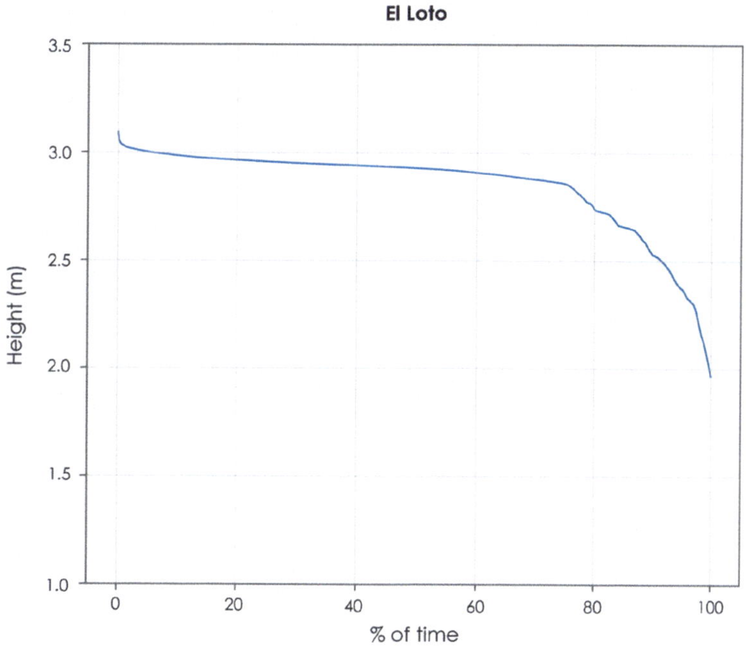

Secondly, water height *duration curves* were constructed for the wetlands. These make it possible to identify areas of flooding and the frequency with which they occur. Due to the continuous simulation adopted, it is possible to identify the percentage of time that the water height of a wetland is exceeded, which allows the appropriate design of infrastructure and landscaping. Figure 4 represents the water height duration curves of LW wetland for the simulation period.

A second application of these curves is that they allow characterizing the hydrological regime of the water bodies—which sustains the development of the flora and fauna typical of these ecosystems- in dry and wet conditions. In case of seeking to implement WSUD solutions that conserve this hydrological regime, or to quantify the variations that occur in it under different urbanization scenarios, these curves can be a key tool for evaluating the impact of different strategies adopted.

3.4 Identification and Characterization of Flood Zones

The hydrological model, supported by the background study and field surveys, indicates the existence of 6 critical flood zones. Each of them accumulates rainwater from one or more contributing subcatchments, as shown in Fig. 5. These subcatchments are the key areas where various rainwater control strategies should be implemented to avoid current flooding issues.

Frequency curves for the 6 critical points were obtained from the model. These curves indicate the number of times that the values of some parameter of interest are exceeded within a study time frame. Figure 6 presents the *flood volume frequency curves* obtained for 3 of the 6 critical points in Llanquihue as an example.

Among the 6 critical flood zones, SU4 was selected as a pilot case. This flood zone is located in a residential area with a high density of houses in the northern limit of the urban area of Llanquihue. Topographic conditions of the contributing sub-basin cause a rapid runoff through the streets, forming a flood at a low point where there is no drainage infrastructure (Fig. 7a).

As a result of the HEC-RAS model simulations, 18 residential properties affected by flooding are identified within the area of influence; at the critical point, the probability of rainwater exceeding the sill plate (15 cm) during a rainfall event is 70%. Moreover, at this critical point, there is a probability of about 17% that flooding will exceed the height of the sidewalk (40 cm) (Fig. 7b). This situation demonstrates how frequent the inconveniences generated by precipitation events are to the surrounding community.

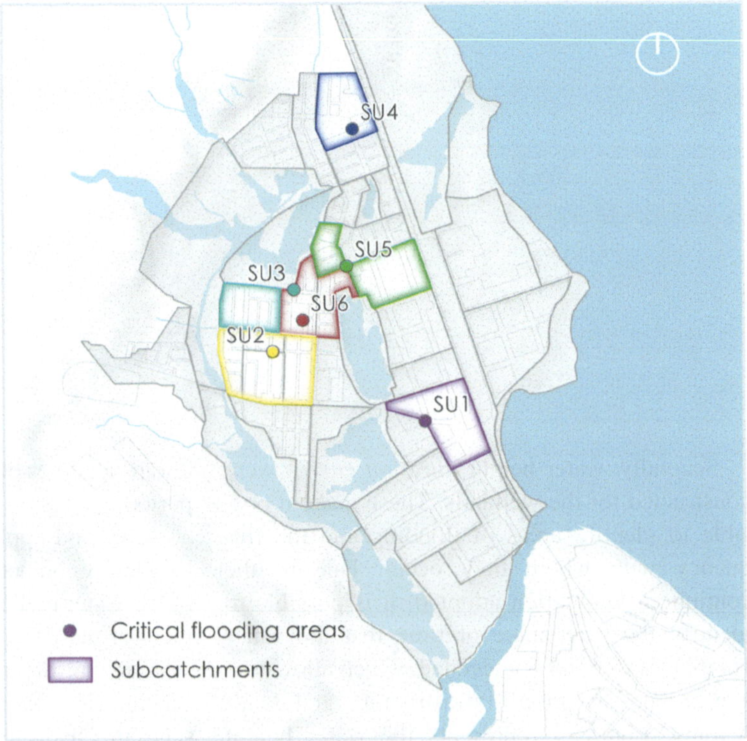

Fig. 5 Critical flooding areas in Llanquihue and their respective subcatchments

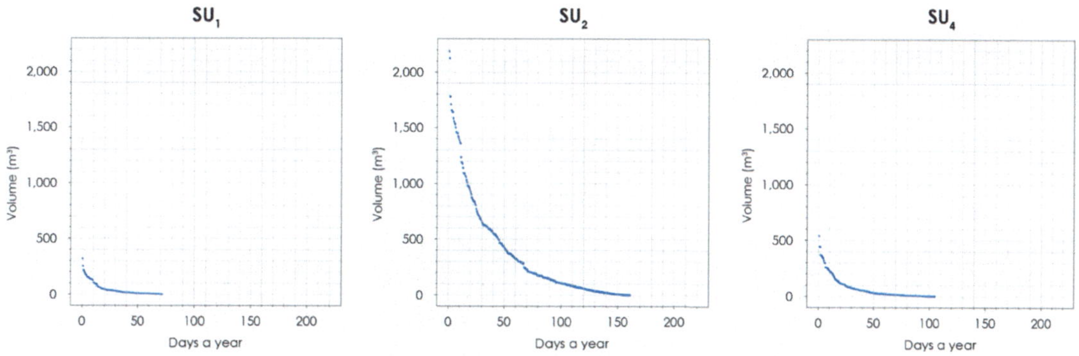

Fig. 6 Flood volume frequency curves (m^3) for 3 of the 6 critical points in Llanquihue

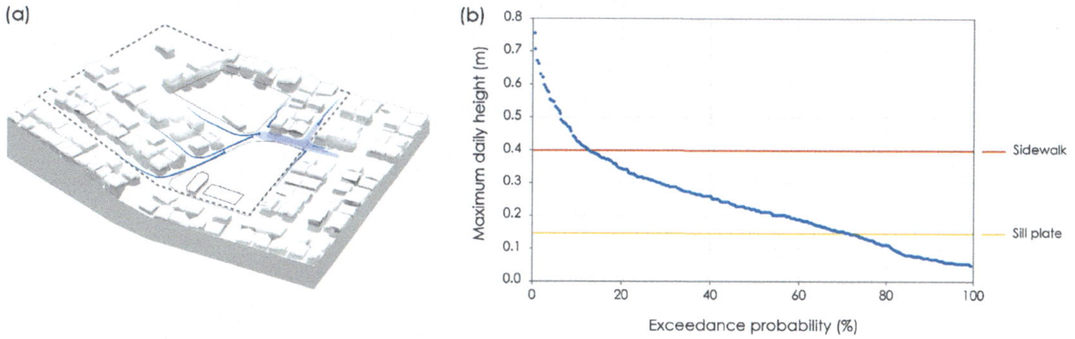

Fig. 7 (**a**) Water path and flood zones in the sub-basin contributing to SU4. (**b**) Probability curve of maximum daily height exceedance at critical point SU4

3.5 Pilot Solutions Conceptualization and Response Modeling

The ideation and design process was carried out iteratively until the design criteria were met, using SWMM-based hydrologic modeling as a tool. As a result, WSUD solutions that enhance local infiltration and storage were conceptualized as follows:

- A detention basin, as the main solution to be implemented in a square located right next to the flood zone. It is designed to receive the runoff accumulated at the critical point and then conduct it through a subsuperficial pipe to Teodosio Sarao wetland.
- A set of complementary local control solutions, including raingardens, vegetated channels, filter strips, and permeable pavements.

Figure 8 shows these solutions in plan and cross-sections. It is important to note that, depending on the context and design objectives, the detention basin could be replaced by a constructed wetland. In this case, given the proximity to natural wetlands and the need for green areas for public use in the study area, the detention basin was prioritized as a solution that makes the permanence of people compatible with local stormwater management.

It should also be noted that, as previously shown in Fig. 3, the water table in this sector is deeper than 1 m, so it is possible to implement some infiltration elements. This is why only the surface model was used at this stage for the design process of local solutions. To evaluate the impact of the implementation of these elements on the water table and wetlands, it would be necessary to perform the coupled simulation again.

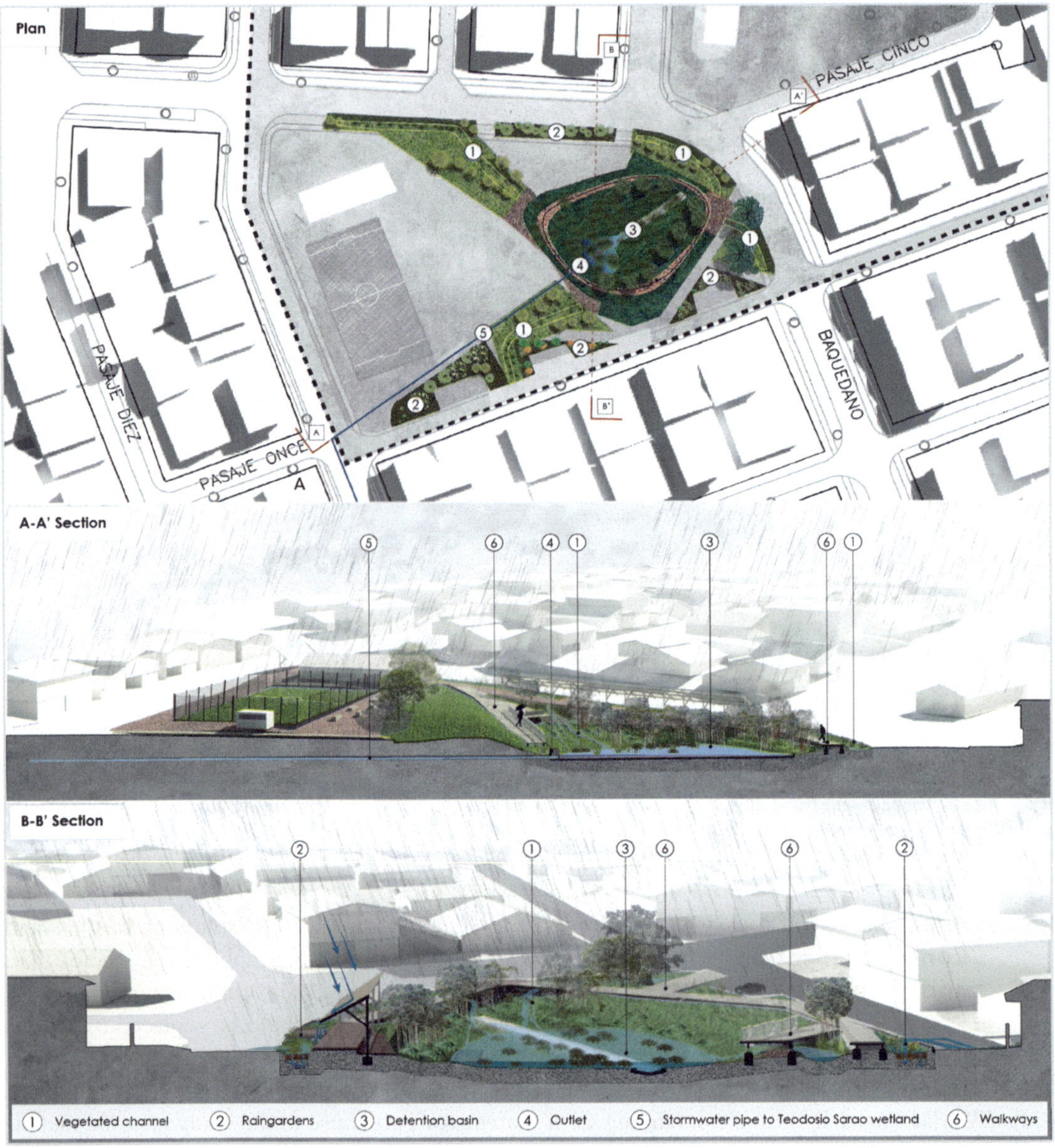

Fig. 8 Proposed solution SU4. Plan view, section A-A′ and section B-B′

For the design process, event-based modeling was adopted, using design storms associated with specific durations and return periods. It was found that with the implementation of this set of solutions, the flooding point is completely solved, since rainwater is collected before it reaches the identified critical zone. In addition, it is observed that the maximum flows delivered to Teodosio Sarao wetland do not exceed the maximum flows it would receive in a natural scenario (simulated), proving the flood attenuation functionality of the proposed solution. As an example, the response in the three conditions (natural, urbanized/current, and with WSUD strategies) for the 10-year return period storm and 24-h duration is presented (Fig. 9). Additionally, it is verified that the set of incorporated elements can take over the Capture Volume, releasing it in a period of

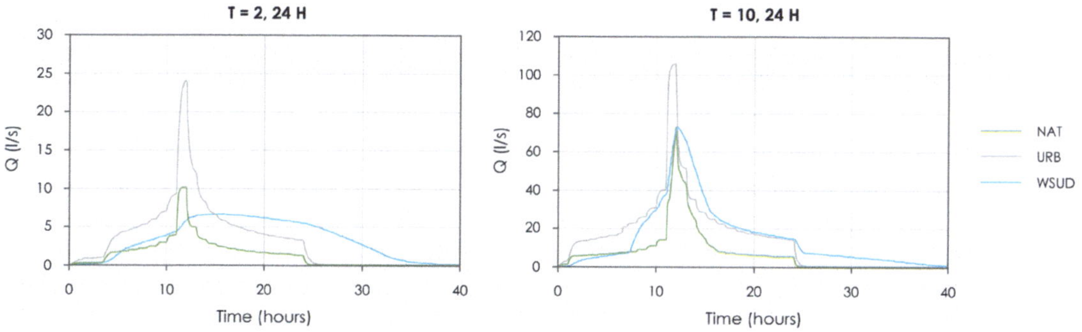

Fig. 9 Response of the subcatchment to point SU4 in three conditions: Natural (NAT), urbanized (URB), and with the implementation of WSUD strategies (WSUD)

Fig. 10 Water height duration curve in the detention basin

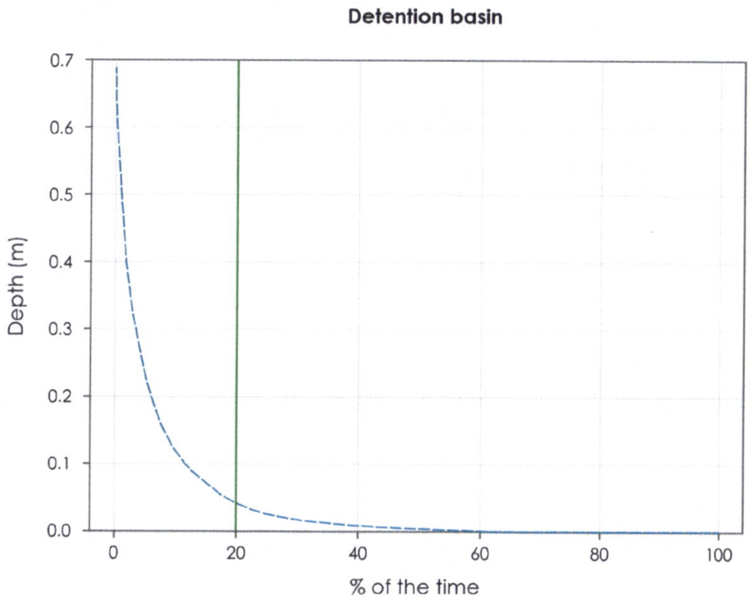

between 24 and 48 h, which is consistent with the recommendations of the Urban Drainage Manual (MOP 2013). Finally, based on the results of the continuous modeling, the duration curve of the water height in the detention basin was constructed (Fig. 10). This result shows that water levels remain under 10 cm over 80% of the time, allowing for its use as a public space.

4 Discussion and Conclusions

The analysis conducted in the city of Llanquihue reinforces the functional importance of wetlands as central elements of green-blue infrastructure in the urban drainage network. From the perspective of water-sensitive urban design, there is an opportunity to strengthen this network through the design of complementary GBI solutions, which can enhance hydrological and ecosystemic functions in other elements of the urban landscape (Patagua et al. 2021).

From the hydrological modeling work undertaken for the conceptualization of WSUD strategies in Llanquihue, the following lessons were learned that may be useful to address similar challenges in other territories:

- Continuous modeling—as opposed to the event-based approach commonly used for the design of urban drainage systems in Chile—is proposed as a pertinent approach to characterize the seasonality of wetlands and frequent rainfall-runoff events. These events, considered as "minor", are those that impact the quality of life of people on a daily basis, so their management through WSUD strategies is particularly significant (Meng 2022).

- Through the continuous modeling approach, useful results are obtained such as flood frequency curves and head duration curves in water bodies. Frequency curves might be used to compare different stormwater management alternatives against the uncontrolled situation, with a focus on reducing the frequency of flooding in a specific time frame and location. Duration curves allow to characterize the base hydrological regime in wetland. They might be a key tool to evaluate the impact of different urban developments and drainage systems on wetlands.
- Characterization, and optimally modeling, of groundwater is especially important in urban areas where the water table is shallow. These results enable the quantification of groundwater inputs to wetlands, as well as to other flood zones. In addition, they will guide decision-making for the design of WSUD strategies; in particular, they will provide insight into the feasibility of implementing infiltration solutions in specific areas of the city.
- The lack of information at the local scale is usually a limitation for modeling. In the case of Llanquihue, the information obtained from citizens made it possible to fill this gap to a large extent, providing relevant data for the conceptualization of the model. Relying on citizen science programs for data collection could be a useful and low-cost strategy for conceptualizing and calibrating models.
- The territorial unit for the ideation and conceptualization of solutions is the contributing catchment. The greatest opportunities for mitigating flooding at a critical point are located upstream, within the urban areas that drain to said point, and not in the zone where the problem is observed.

Although the conceptual design proposal developed for Llanquihue responds to the central objective of mitigating flooding to promote climate adaptation and improve people's quality of life, the initial analysis takes a broader view, integrating hydrologic, environmental, and urban variables. This integration will guide the design at the detailed level, a stage at which it will be important to incorporate complementary objectives, including the protection of water quality in natural receiving waters. This challenge is illustrated by the results of the water quality index, which indicates contamination risks in 4 of the 5 wetlands analyzed. The results show the high vulnerability of the Llanquihue wetland network to stormwater runoff, especially in terms of the high presence of fecal coliforms, which exceeds Chilean regulations. In this sense, WSUD strategies have the potential to contribute to the purification of urban runoff in the natural drainage network (Patagua et al. 2021).

In general, the work carried out supports the relevance and feasibility of applying the WSUD approach for sustainable stormwater management in Llanquihue. Given the importance of the natural network of wetlands still preserved in the city, it is suggested that efforts be focused on their conservation and restoration, as well as on the development of additional green-blue infrastructure such as that designed for the pilot case.

For the application of this approach in similar contexts, it is important not to lose sight of the fact that the proposed solutions should not only be effective from a hydraulic point of view, but also pertinent to each territory. In this sense, it will be relevant to understand the uses of the urban space and the characteristics of the ecoregion in which the city is inserted. Taking into account these variables, it will be possible to design multifunctional green spaces that provide various benefits (e.g., buffering of climatic extremes, support of flora-fauna interactions, environmental education, direct experience of natural ecosystems, opportunities for healthy social interaction) (Ahern 2007) and are fundamental to progress toward more sustainable cities that provide a better quality of life to their inhabitants (Ramaswami et al. 2016).

References

Acreman, M. & Holden, J. (2013). How Wetlands Affect Floods. *Wetlands* 33, 773–786. https://doi.org/10.1007/s13157-013-0473-2

Ahern, J. (2007). Green infrastructure for cities. The spatial dimension. In: Novotny, V. (Ed.), Cities of The Future. *Towards Integrated Sustainable Water and Landscape Management.* IWA Publ., London. 267–283.

Alikhani S, Nummi P, Ojala A. (2021). Urban Wetlands: A Review on Ecological and Cultural Values. *Water.* 13(22). https://doi.org/10.3390/w13223301

Almanza-Marroquín, V., Figueroa, R., Parra, O., Fernández, X., Baeza, C., Yañez, J., Urrutia. (2016). Bases limnológicas para la gestión de los lagos urbanos de Concepción, Chile [Limnological basis for the management of urban lakes in Concepción, Chile]. *Latin American Journal of Aquatic Research.* 44. 313–326.

Ampatzidis, P.; Kershaw, T. (2020). A review of the impact of blue space on the urban microclimate. *Sci. Total Environ.* 730, 139068. https://doi.org/10.1016/j.scitotenv.2020.139068

ANZECC & ARMCANZ. (2000). *Australian and New Zealand guidelines for fresh and marine water quality.*

Bakker, M., Post, V., Langevin, C. D., Hughes, J. D., White, J. T., Starn, J. J., & Fienen, M. N. (2016). Scripting MODFLOW model development using Python and FloPy. *Groundwater,* 54(5), 733–739. https://doi.org/10.1111/gwat.12413

Canadian Council of the Ministers of the Environment [CCME]. (2017). *CCME Water Quality Index User's Manual: 2017 Update.* https://www.ccme.ca/files/Resources/water/water_quality/WQI%20Manual%20EN.pdf

Centro de Información de Recursos Naturales [CIREN]. (2016). *DEM Alos Palsar Región de Los Lagos.* [Modelo de Elevación Digital]. https://www.ide.cl/index.php/imagenes-y-mapas-base/item/1567-dem-alos-palsar-region-de-los-lagos

Costello, L.R., Matheny, N.P., & Clark, J.R. (2000). WUCOLS III: *A guide to estimating irrigation water needs of landscape plantings in California: the landscape coefficient method.* San Mateo and San Francisco Counties: University of California Cooperative Extension, California Department of Water Resources, 160.

Fischer B, Turner R K, Morling P. (2009). *Defining and classifying ecosystem services for decision making.* Ecological Economics 68:643–653.

Gironás, J., Roesner, L. A., Davis, J., Rossman, L. A., & Supply, W. (2009). *Storm water management model applications manual.* Cincinnati, OH: National Risk Management Research Laboratory, Office of Research and Development, US Environmental Protection Agency.

Hammer, D.A. and Bastian, R.K. (1989). *Wetland Ecosystem Natural Water Purifiers.* [Conference: Constructed Wetlands for Wastewater Purifiers].

Ilustre Municipalidad de Llanquihue. (2015). *Plan de desarrollo comunal, comuna de Llanquihue, 2015–2018* [Communal Development Plan, Municipality of Llanquihue, 2015–2018].

Ilustre Municipalidad de Llanquihue. (2009). *Plan Regulador Comunal* [Communal Regulatory Plan].

Ilustre Municipalidad de Llanquihue. (2019). *Plan Regulador Comunal, informe ambiental complementario* [Communal Regulatory Plan, complementary environmental report].

Instituto Nacional de Estadísticas. (2017). *Resultados censo 2017* [2017 census results]. Instituto Nacional de Estadísticas. http://resultados.censo2017.cl/Home/Download

Legado Chile Foundation. (2020). *Plan de conservación de los humedales urbanos de Llanquihue* [Conservation plan for Llanquihue's urban wetlands].

McDonnell, B. E., Ratliff, K., Tryby, M. E., Wu, J. J. X., & Mullapudi, A. (2020). PySWMM: the python interface to stormwater management model (SWMM). *Journal of Open Source Software.* 5(52), 2292, https://doi.org/10.21105/joss.02292

Meng, X. (2022). Understanding the effects of site-scale water-sensitive urban design (WSUD) in the urban water cycle: a review. *Blue-Green Systems.* 4(1). 45–57. https://doi.org/10.2166/bgs.2022.026

Ministerio de Obras Públicas [MOP] (2013). *Manual de Drenaje Urbano: Guía para el diseño, construcción, operación y conservación de obras de drenaje urbano* [Urban Drainage Manual: Guide for the design, construction, operation and maintenance of urban drainage works]. Ministerio de Obras Públicas. http://www.doh.cl/manualdrenajeurbano/Paginas/default.aspx

Patagua, Fundación Legado Chile, Pontificia Universidad Católica de Chile. (2021). *Ciudades Sensibles al Agua: Guía de Drenaje Urbano Sostenible para la Macrozona Sur de Chile, Volumen II* [Water Sensitive Cities: A Guide to Sustainable Urban Drainage for the Southern Macrozone of Chile, Volume II]. https://www.ciudades-sensibles-al-agua.cl/

Patagua, Fundación Legado Chile, Pontificia Universidad Católica de Chile. (2024). Ciudades Sensibles al Agua: Guía de Drenaje Urbano Sostenible para la Macrozona Sur de Chile, Volumen III [Water Sensitive Cities: A Guide to Sustainable Urban Drainage for the Southern Macrozone of Chile, Volume III]. https://www.ciudades-sensibles-al-agua.cl/

Pizarro Pavez, S. A. (2019). *Caracterización hidrogeológica e hidrogeoquímica de la subcuenca del río Maullín (41° 05′-41° 46′ SY 72° 46′-73° 53′ W), Región de los Lagos, Chile* [Hydrogeological and hydrogeochemical characterization of the Maullín river sub-basin (41° 05′-41° 46′ SY 72° 46′-73° 53′ W), Los Lagos Region, Chile].

Ramaswami, A., Russell, A.G., Culligan, P.J., Sharma, K.R., Kumar, E. (2016). Meta-principles for developing smart, sustainable, and healthy cities. *Science.* 352 (6288) 940–943.

Ramsar Convention Secretariat. (2013). *The Ramsar Convention Manual: a guide to the Convention on*

Wetlands (Ramsar, Iran, 1971), 6th ed. Ramsar Convention Secretariat, Gland, Switzerland.

Scheuch, T. (2021). *Making SWMM and MODLOW dance together: Developing and applying an open-source model integration scheme in a city characterized by its wetland network* (Thesis for the Degree of Master of Science in Engineering). Pontificia Universidad Católica, Santiago de Chile. Advisor: Jorge A. Gironás.

Socolofsky, S., Adams, E. E., & Entekhabi, D. (2001). Disaggregation of Daily Rainfall for Continuous Watershed Modeling. *Journal of Hydrologic Engineering*. 6(4), 300–309. https://doi.org/10.1061/(ASCE)1084-0699(2001)6:4(300)

Vollenweider, R.A. & J. Kerekes. (1982). *Eutrophication of waters. Monitoring, assessment and control. OECD Cooperative program on monitoring of inland waters (Eutrophication control).* Environment Directorate. OECD. 154 pp.

Whelans and Halpern Glick Maunsell (1994). *Planning and Management Guidelines for Water Sensitive Urban (Residential) Design.* Report prepared for the Department of Planning and Urban Development of Western Australia, ISBN 0 64615 468 0.

Wong TH, Brown RR. (2009). The water sensitive city: principles for practice. *Water Sci Technol.* 2009. 60(3):673–82. https://doi.org/10.2166/wst.2009.436.

The Use of Urban Wetlands as an Integrating Element for Wastewater and Stormwater Solutions in an Ecosystem-Based Disaster Risk Reduction Approach: A Case Study in Rio de Janeiro, Brazil

Antonio Krishnamurti Beleño de Oliveira, Rodrigo Hoerner Morais Marques, Aline Pires Veról, Osvaldo Moura Rezende, and Marcelo Gomes Miguez

Contents

Abstract

This study addresses significant flaws in the separate sanitary sewer system of major Brazilian cities, where sanitary and storm sewers often face operational difficulties due to undersized systems or irregular connections. Such flaws result in the overloading of networks and environmental degradation of water bodies, particularly during dry periods due to untreated stormwater contamination. Additionally, climate change poses further risks with potential precipitation increases

A. K. B. de Oliveira (✉)
Department of Civil and Environmental Engineering, Pontifícia Universidade Católica do Rio de Janeiro, Rio de Janeiro, Brazil
e-mail: krishnamurti@puc-rio.br

R. H. M. Marques
Programa de Engenharia Naval e Oceânica, Universidade Federal do Rio de Janeiro, Rio de Janeiro, Brazil
e-mail: hoerner@poli.ufrj.br

A. P. Veról
Faculdade de Arquitetura e Urbanismo, Universidade Federal do Rio de Janeiro, Rio de Janeiro, Brazil
e-mail: alineverol@fau.ufrj.br

O. M. Rezende · M. G. Miguez
Escola Politécnica, Universidade Federal do Rio de Janeiro, Rio de Janeiro, Brazil
e-mail: omrezende@poli.ufrj.br; marcelomiguez@poli.ufrj.br

C. Rojas Quezada (ed.), *Urban Wetlands in Latin America*, Sustainable Development Goals Series,
https://doi.org/10.1007/978-3-031-69590-2_7

that exceed these systems' hydraulic capacity. To mitigate these challenges, the study proposes the Ecosystem-Based Disaster Risk Reduction Approach, which leverages ecosystems to support urban resilience. A key solution highlighted is constructing wetlands, which mimic natural wetlands, to restore water storage and improve environmental quality. The study applies a Multilayer Hydrodynamic Unified Simulation (MODCEL-MHUS) integrating urban stormwater, sanitary sewer, and surface flow systems. This method, coupled with an environmental quality model, was applied in a pilot basin to evaluate its effectiveness. The findings suggest that constructed urban wetlands play a crucial role in flood mitigation and enhancing wastewater treatment, contributing to biodiversity preservation, and offering ecosystem services within urban areas.

Keywords

Sanitary sewer system · Stormwater system · MODCEL-MHUS

1 Introduction

When considering major Brazilian cities, significant flaws in the installed separate sanitary sewer system (two separate networks for the sanitary sewers and the stormwater) are commonly observed. These flaws can be attributed either to the drainage system being undersized or to the presence of clandestine or irregular sewage connections to the drainage network (IRM 2017). There are also cases in which "parasitic" contributions from building installations add rainwater to the sanitary sewage system, overloading networks that were not designed for these flows. In addition to these operational difficulties, there is the risk of climate change, which may lead to possible increases in precipitation volumes, exceeding the hydraulic capacity, and exacerbating overflows of both storm and sanitary sewage networks.

It is worth highlighting that there is no provision for the treatment of stormwater in Brazil. Consequently, the irregular contribution of domestic sewage to the drainage networks directly affects the receiving water bodies, degrading their environmental quality.

In Brazil, there is thus a scenario of water body degradation during dry periods. Additionally, there is an additional scenario of flooding caused by the failure of the drainage system during rainy periods (Oliveira 2018). In this complex context of the interaction of two unforeseen networks, the possibility arises to utilize the concept of the Ecosystem-Based Disaster Risk Reduction Approach as a solution. The aim of this approach is to engage in decision-making activities that take into consideration both present and future urban requirements, as well as the biophysical needs of ecosystems. It explores how ecosystems can effectively support cities in addressing disaster situations (Gonsalves and Mohan 2011).

Therefore, the utilization of constructed wetlands, which simulate natural wetlands, can be viewed as a viable solution. These spaces for water storage and dispersion, suppressed by the urbanization process, can be restored to the drainage system. Additionally, the environmental quality of water bodies can be improved through the treatment processes that this type of structure can provide.

Confer and Niering (1992) had already proposed the use of wetlands for floodwater storage, with the potential to integrate this function with sediment accumulation and wildlife habitat functions. Hey and Philippi (1995) also demonstrated that employing wetlands can be a sustainable approach for addressing flood consequences, as an alternative to the traditional method of managing drainage systems.

More recently, Pattison-Williams et al. (2018) proposed that the use of wetlands represents an economically viable solution for mitigating the financial, social, and environmental damages caused by floods.

A separate study conducted by Yamanaka et al. (2020) proposes that employing wetlands as green infrastructure can help alleviate the impacts of climate change on flooding while preserving local natural characteristics, such as species diversity and abundance, which are on par with or exceed those found in most other water bodies. The research also highlights that wetlands used as flood control basins can serve as habitats for various species, particularly those adapted to hydrologically variable environments. Hence, utilizing these drainage structures as green infrastructure emerges as a viable choice for preserving regional biodiversity.

It is important to highlight that in some cases, due to the degree of watershed urbanization, it may be necessary to carry out pre-treatment of stormwater before its contribution to constructed wetlands, aiming to remove heavy sediment loads and other pollutants (Livingston 2020).

Additionally, it has been observed in recent decades that the use of constructed wetlands for wastewater treatment processes has yielded favorable results in improving water quality (Farooqi et al. 2008).

Constructed wetlands replicate natural wetland processes and have the potential to remove various pollutants from effluents. Within the category of constructed wetlands, the following types are observed: Horizontal; Vertical; Hybrid; and Free Surface Flow.

According to the study by Bakhshoodeh et al. (2020), the average removal of BOD5 for these types of wetlands can range from 60.1% to 80.6%, with the latter value being associated with constructed wetlands featuring free surface flow. Estimating the BOD5 value is crucial in determining the amount of dissolved oxygen required to biologically decompose biodegradable organic matter in the water.

In a context of multifaceted use of constructed wetlands for treatment and flood mitigation, it is important to consider that the selection of vegetation for these environments must account for hydrology and geomorphology (Jacobson et al. 2011) to mitigate the adverse effects of seasonal flooding on urban wetland flora (Kitanović et al. 2023; Thompson et al. 2009). It is worth highlighting that the selection of vegetation is crucial to tailor the use of constructed wetlands in the treatment of wastewater, which can be rich in both mineral nutrients and organic matter (Končalová et al. 1993).

Additionally, post-implementation or restoration monitoring becomes essential to prevent potential impacts on the local ecosystem (Charles et al. 2023).

Determining the quantity and quality of urban waters in a context where urban infrastructure networks intermingle despite being designed to operate independently is a significant challenge, given the complexity of the phenomena involved. However, mapping these connections is essential for proposing measures such as the one presented earlier.

Mathematical models have been widely used internationally to meet this challenge. However, there are clear differences in how various countries approach this issue. Von Sperling (2014) points out that while developed countries are concerned with identifying and tackling punctual events of disconformity, in developing countries like Brazil, the main usage to this tool is to recuperate highly degraded water bodies. However, the usual scenario for large cities is that the aimed standard is still far from reality, what validates any effort to improve the quality parameters of urban water bodies, even if it means the implementation of transitory or temporary solutions.

Consequently, a comprehensive understanding of the interdependent operations of networks is crucial for accurately interpreting the challenges they pose, both from a hydraulic and environmental perspective. Therefore, there is a growing demand for tools that enable the proper diagnosis and prognosis of these and other essential urban infrastructures in an integrated manner.

Numerous studies have investigated the integrated treatment of urban waters over the past four decades. However, unlike the approach taken in Brazil, the rest of the world has been treating sewage—both stormwater and sanitary—in a unified system with modifications to safeguard the natural and built environment (EPA 1983). Consequently, most international publications on the subject have focused on technical issues related to this strategy, offering little guidance to Brazil and other countries that continue to rely on the absolute separator solution. As a result, the separator technique has not benefited much from significant advancements.

Thereafter, integrated study of urban water does not necessarily mean the integration and simulation of two urban water networks. Several studies deal with the simulation of unitary systems, which are designed to operate with the inflow of both stormwater and sanitary wastewater in a single network since their conception (Gupta et al. 1996; Sztruhár et al. 2002). The concept of Combined Sewer Overflows (CSOs), present in several publications (Todeschini et al. 2012; Montserrat et al. 2014; Jean et al. 2021), is entirely associated with the operation and simulation of only one network. Furthermore, its overflow is not simulated by a surface layer of runoff, due to limitations in the hydraulic representation of the chosen models.

Numerous studies have primarily focused on analyzing the impact of failures in unitary systems on the environment and cities, which can have a negative effect (Gupta et al. 1996; Riechel et al. 2016; Sztruhár et al. 2002). While some studies have explored solutions such as Green Infrastructure and Sustainable Urban Drainage, they have predominantly focused on combined sewer overflow in unitary systems and lack a specific representation of a surface layer that simulates urban flows (Joshi et al. 2021; Jean et al. 2021). Nationally, sewage networks' interactions have been frequently studied (Dias 2003; Bahia 2010). However, the hydrodynamic representations of the phenomenon have not shown significant advances.

The work of Radesca (2018) stands out as an initial attempt to simulate the systems. The simulation is carried out by applying the Flow Cell Model—MODCEL (Miguez 2001, 2017). The work is a forerunner in the use of *Quasi*-2D models as support for decision-making regarding drainage and sewer systems. The usual output results from this model, such as flooding frequency and critical spots, work well for each individual system, but it typically does not perform the integrated simulation of these systems. The model results are only for the drainage system, and the impact on the sanitary sewer system is done in post-processing.

The MODCEL is characterized as a *Quasi*-2D flow cell model, in which each cell communicates with neighboring cells through one-dimensional equations. The flow between cells can be calculated through well-known hydraulic laws, such as the complete or simplified dynamic equation of *Saint-Venant*, the equation for flow over free or drowned weirs, the equation for flow through orifices, and the equation for flow through culverts, among several others, considering the effects of inertia on the flow that occurs in the main watercourses. Flow can also occur simultaneously in two layers, a surface layer, and a subsurface layer, in pipes or galleries, with surface cells and drainage gallery cells being linked together. In pipes or drainage galleries, flow is initially considered to be free surface flow, but may become drowned, becoming considered under pressure. A series of international papers consolidate the model for simulating drainage networks (Oliveira et al. 2019, 2022, 2023; Rezende et al. 2019; Guimarães et al. 2021; Fraga et al. 2021; Sousa et al. 2022).

Considering this context, the primary aim of this study is to introduce a methodology for the integrated simulation of urban stormwater drainage and sewage networks with the urban surface runoff layer. This methodology is designed to enable the proposal of surface-level solutions that effectively mitigate the interaction challenges observed within the subsurface networks.

To accomplish the main objective, this study developed a Multilayer Hydrodynamic Simulation method (MODCEL-MHS), which integrates urban drainage, wastewater, and surface runoff systems (including runoff on streets and open spaces) through the Quasi-2D Flow Cell Model—MODCEL. This model was coupled with an environmental quality model to incorporate hydrodynamic results, predicting long-term BOD contamination levels across the flow cells using the Streeter-Phelps model, and considering complete mixing between MODCEL connections. The methodology's potential will be evaluated through a pilot basin (Case Study) to explore the applicability of the proposed tool, which will include a set of measures, including a series of urban constructed wetlands, to mitigate flooding and contamination of the receiving watercourse.

2 Method

In the initial stage of the methodology, the watershed and the sewage basin of the system are defined to gain a comprehensive understanding of the problem and estimate the potential systemic and reciprocal impacts of the systems in operation.

Stage 2 involves the development of a Multilayer Hydrodynamic Simulation method (MODCEL-MHS) (Oliveira 2022; Oliveira et al. 2024) that integrates urban drainage, sanitary sewers, and surface runoff systems, including runoff on streets and open spaces, utilizing the Flow Cell Model (MODCEL) as described by Miguez (2001, 2017).

MODCEL is a *Quasi*-2D hydrodynamic model with a focus on simulating flows within drainage networks and considering potential failures and flows or storage outside the watercourse channels. This is essential for the analysis of urban drainage. Furthermore, this work suggests the application of the model to simulate urban water networks, including sanitary and storm sewers, in order to observe potential interaction issues.

As a first step of this stage, the territory is divided into flow cells, homogeneous compartments that integrate through unidimensional hydraulic laws. These cells represent the surface layer, discretized at the street and lot scale for compatibility with subsurface layers represented at the manhole scale. This layer employs flow cells to represent streets, lots, and watercourses and includes the representation of built lot roofs to predict connections to drainage or sewage systems.

The next step in Stage 2 involves preparing the subsurface layer, which will serve as the hydrodynamic foundation for representing urban stormwater networks. Storm sewers networks will be discretized into flow cells that represent drainage manholes. Characteristic connections between the subsurface drainage layer and the surface layer will be integrated into the model.

In the next step of this stage, with the surface and subsurface drainage layers working together, model calibration and validation are performed, considering the available information for this purpose. In the modeling process, calibration and validation are two essential steps that must precede simulation for prospects (Tejaswini and Sathian 2018).

With the hydrodynamic model of stormwater flow calibrated (and validated, if information or records are available for such), the next step in Stage 2 of the methodology involves simulating elements of the Sanitary Sewer System. The integrated representation of these elements with the layers developed previously constitutes the primary contribution of the work presented here.

To represent the entry of stormwater into the sewer system or the contribution of sewage to the surface, the modeled manholes will be connected to their respective surface cells, which represent urban areas. In the event of flooding, water can accumulate and enter the network, while in cases of hydraulic overload, the sewage networks will contribute from bottom to top. This connection can occur at various levels and through different points in the sanitary sewer system. The primary contributions will come from sewage manhole covers located on streets or sidewalks, depending on the layout of the local sewer system. Additionally, if flooding covers a larger area or the sanitary sewer network becomes overloaded, communication can take place at a higher level, corresponding to sewage points within buildings. This communication can occur in both directions through inspection boxes and house drains.

Stage 2 aims to quantify the impacts related to flow rates between systems, presenting responses in terms of levels, velocities, and flow rates between cells. The accuracy of the results will depend on the data available for calibrating the hydrodynamic model.

The next stage involves the development of Environmental Quality Modeling, running parallel to the hydrodynamic calculations. This modeling assesses the environmental quality of drainage systems and receiving bodies in a post-processing routine. It takes into account contributions from the sanitary sewer system to the storm sewer system, vice versa, and their interaction with the surface layer of the flow.

However, due to the logic of interpreting the territory in flow cells, the possibility of flows in multiple directions, and the potential for vertical interaction between layers. Figure 1 schematically illustrates the considerations made for the development of this methodology.

Initially, the model developed for the water quality study will focus on representing BOD parameters. BOD allows for the indirect estimation of the amount of organic matter in various accumulation surfaces, including watercourses, drainage and sewage networks, and even streets and lots. Thus, the determination of BOD also provides an indication of the consumption of dissolved oxygen (DO). Furthermore, areas with a higher accumulation of BOD are more likely to harbor pathogens responsible for waterborne diseases. Therefore, both BOD and COD are essential parameters for assessing the pollution level of water bodies (Von Sperling 2014). Reference values for BOD should be established for the specific case study in question, taking into account the characteristics of stormwater runoff and domestic sewage in the region under analysis.

In this final stage, we will explore the potential of the integrated MODCEL-SHIM simulation to identify urban solutions. Priority will be given to actions within the territory that impact the networks. Miguez (2001) highlights that the primary innovation of the flow cell model lies in using spaces within the territory to accommodate urban water volumes and seeking solutions outside the traditional networks to mitigate flood volumes, thereby reducing the load on drainage systems. Given the recognized need to work on multiple fronts, the alternatives proposed in this study should have a multifunctional character for both drainage and sewage systems.

The operation of the composite system, integrating storm and sanitary sewer simulations, can encompass three operational dynamics.

The first dynamic occurs during dry periods (rain-free), where the system receives input exclusively from irregular connections between the sewage and drainage systems. This contaminated water, without rainwater inflow, undergoes conventional treatment.

In addition, during rainy periods, these structures ensure the overflow of flows resulting from the mixture of dry weather flows and surface runoff into a treatment system featuring an urban constructed wetland. This configuration prevents the hydraulic entry of diluted sewage into either the conventional sanitary sewer system or a treatment plant.

The second operational dynamic corresponds to typical rainfall events or recurrent weather, which strain the dry weather captures mentioned earlier, resulting in overflows due to rainwater contributions. However, these events do not provide adequate sewage treatment capacity, maintaining the system flows in unfavorable environmental conditions. Consequently, the urban constructed wetland will receive the excess contaminated water. Although constructed wetlands in Brazil have been extensively studied for in-situ effluent treatment (Lana et al. 2013; Souza

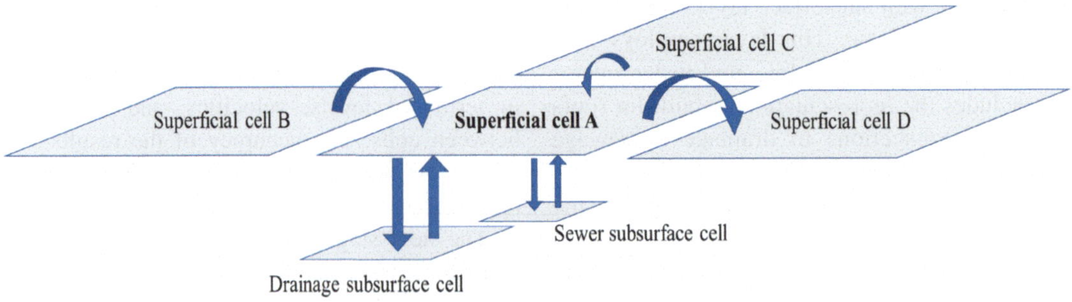

Fig. 1 Flow cells diagram for the consideration of complete mixing reactors in networks

2021; Shinohara 2021; Rodrigues et al. 2021), this study suggests their use for a configuration of combined waters, including both rainwater and sewage wastewater.

The design will follow the guidelines outlined in "Guidelines for Free Water Surface Wetland Design" (Bendoricchio et al. 2000), which have been updated by more recent studies cited in the introductory chapter.

The third dynamic pertains to the system's behavior during intense rainfall events. In this scenario, it's expected that the high volume of rainwater can purify contaminants to environmentally acceptable quality levels, even originating from irregular contributions. However, the volume of surface runoff may exceed the treatment system's capacity. Therefore, the constructed wetland will be sized to handle these extreme conditions, considering challenges related to discharging water into the receiving body, particularly due to upstream water level elevation. This configuration aims to evaluate a suitable solution for urban systems located in large river plains or coastal areas.

The simulation of the constructed wetlands is conducted in the surface flow layer, where contributions from the drainage or sewage systems vary depending on the chosen solution. The treatment efficiency of these solutions can be estimated by applying the contaminant reduction percentages obtained during the literature review stage.

3 Case Study

The Piraquê-Cabuçu River Basin is situated in the West Zone of Rio de Janeiro municipality. Its primary drainage course begins at Lameirão Mountain in the Campo Grande neighborhood, initially known as Gatos River. It flows toward Cabuçu Hill, taking the same name as the hill from that point on. In its final stretch, after merging with the Porcos River, it becomes the Piraquê River and continues for another 3.7 km before emptying into Sepetiba Bay.

The basin is notable for having 95% of households supplied with piped water and 99% with garbage collection. However, only 71% of households have proper sanitary sewage disposal through a general sewer or rainwater network (IBGE 2010), with indications that a significant portion of this sewage is discharged into the drainage system. This is evident when assessing the environmental quality of the main course of the basin, as the Piraquê-Cabuçu River has been classified as having poor or very poor water quality (ANA 2021). Marginal settlements along the main course of the basin discharge untreated sewage directly into the river.

Additionally, IBGE (2010) reports that open-air sewage can be observed around households in about 8% of cases throughout the basin. Moreover, only 62.3% of households have subsurface drainage structures. Concerning the macrodrainage system, there are undersized river stretches and urban development along riverbanks, which exposes the population residing there to river overflow during floods.

The selected sub-basin for applying the methodology developed in this study is located in the upper section of the Piraquê-Cabuçu basin. This region, situated within the Campo Grande neighborhood, features consolidated urbanization along with open spaces where nature-based drainage and sanitation solutions can be implemented without necessitating resettlement.

Figure 2 displays the location of the area of interest within the Piraquê-Cabuçu River watershed, covering approximately 0.50 km^2, entirely within the Campo Grande neighborhood.

Regarding the urban drainage and sanitation sectors, there is substantial evidence indicating interactions between both systems. Contributions of urban water are observed in the Prata do Cabuçu River, even during dry periods, likely originating from the sanitation system. Additionally, there are buildings with sewage pipes directly discharging into the Prata do Cabuçu River.

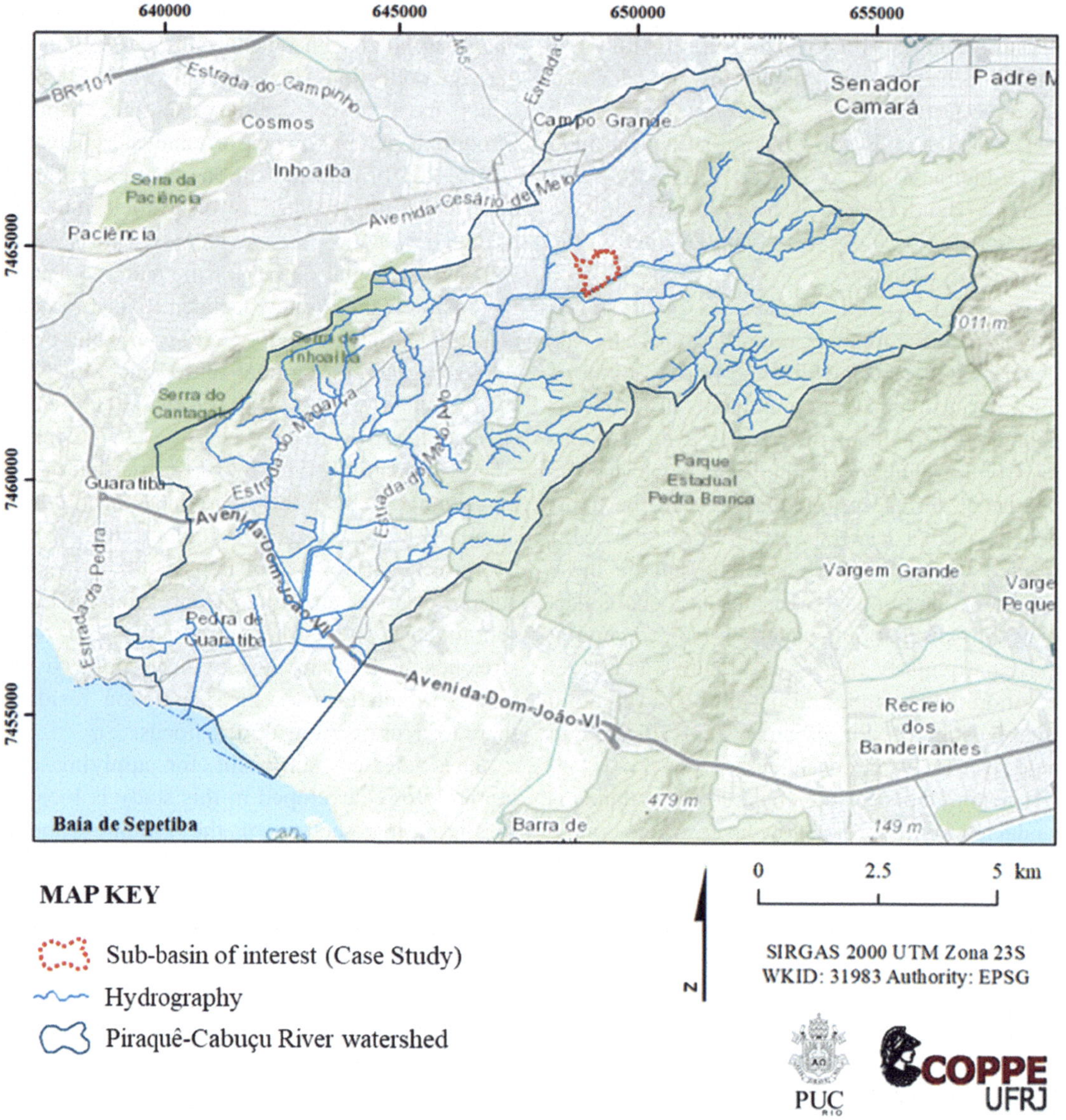

Fig. 2 Location of the sub-basin of interest, which will be treated as a specific study area for the development of the proposed methodology

4 Results and Discussion

Utilizing topographical information of the region, it was divided into flow cells. These interconnected cells, governed by hydraulic equations, simulate the flow of urban waters in the sub-basin. Among the 702 prismatic cells represented, 148 correspond to streets, 254 to buildings, 46 to non-urbanized areas, and 254 to rooftops.

The river cells of the Prata do Cabuçu River can receive flow boundary conditions determined by the integral simulation of the basin. This consideration is crucial for simulating potential backwater effects on the urban networks of the sub-basin during flood periods.

To incorporate the subsurface drainage layer into the complementary MODCEL-MHS simulation, it was necessary to estimate the subsurface drainage design for the area of interest. Although there are indications that the region already has a subsurface drainage system implemented, there is insufficient data for accurate representation.

Similar to the subsurface drainage system simulation, the sanitary sewer system begins with estimating the networks using classic sewer network sizing methods. Contributions from water accumulation on streets and lots were considered for the sanitary sewer system. At the first level, water accumulating in the street cell that reaches sidewalk level can enter the sewer system through manhole covers. At the second level, if flooding reaches the lots, another orifice connection was planned through the building sewer system structures, with a focus on inspection chambers, which are often implemented without proper sealing.

The results of this simulation are presented in Fig. 3, illustrating flood areas during an intense precipitation event with a 25-year recurrence. It is evident that there is significant overflow from the Prata do Cabuçu River into the lower region of the study basin. Additionally, overflows from manholes and contributions from flooding can be observed in the sanitary sewage system. Given the extent of the observed floods, there is interaction among the three simulated layers.

In line with the water quality modeling described in the methodology chapter, this section presents simulation results for BOD, considering the interaction between the flow cells corresponding to each of the simulated layers.

A value of 350 mg/L for domestic sewage BOD was adopted, following the recommendation of Von Sperling (1996). For rainwater, we referred to the publication by Weibel et al. (1964), which establishes varying BOD values based on rainfall duration, with higher values for the initial runoff waters. Table 1 displays the values used.

The model provides results for all represented flow cells, allowing the observation of BOD concentrations at 284 points. However, we will focus on presenting concentration results exclusively for the Prata do Cabuçu River to assess its contamination due to interactions with urban water systems.

Figure 4 illustrates the BOD variation in the Prata do Cabuçu River, particularly at the section closest to the sewage system outlet. It is evident that around minute 120, there is a decrease in BOD in the river, consistent with the specified parameters. However, following the response of surface runoff and the mixing of rainwater with domestic sewage, along with its discharge into the river, there is a subsequent increase in BOD levels in the watercourse.

After identifying the primary issues within each urban water network and recognizing potential interactions between these systems, proposing alternatives becomes imperative. As previously mentioned, our approach to proposing alternatives will align with a multifunctional strategy, adhering to the principles of the Ecosystem-Based Disaster Risk Reduction Approach. These solutions should be capable of mitigating floods and enhancing wastewater treatment capacity.

To achieve this, we conducted an assessment of the primary available open spaces within the study area. It is crucial to identify open areas with attributes conducive to flood attenuation and integration into the sanitary sewage treatment system. Therefore, the ideal locations should be in proximity to water flow paths and naturally flood-prone zones, facilitating the implementation of constructed wetlands.

Furthermore, we have pinpointed overflow points along the primary watercourse, necessitating the inclusion of traditional drainage infrastructure in our proposal. The system's enhancement stems from the need to prevent overflows from the primary Parta do Cabuçu River. To accomplish this, the construction of embankments along the riverbanks, serving as flood containment barriers, is essential. The elevated region near the intersection of Severino Ferreira da Costa Street and the watercourse is particularly prone to overflows.

As a complementary measure to prevent water from flowing back into the urban area via the subsurface drainage system from downstream to upstream, we recommend incorporat-

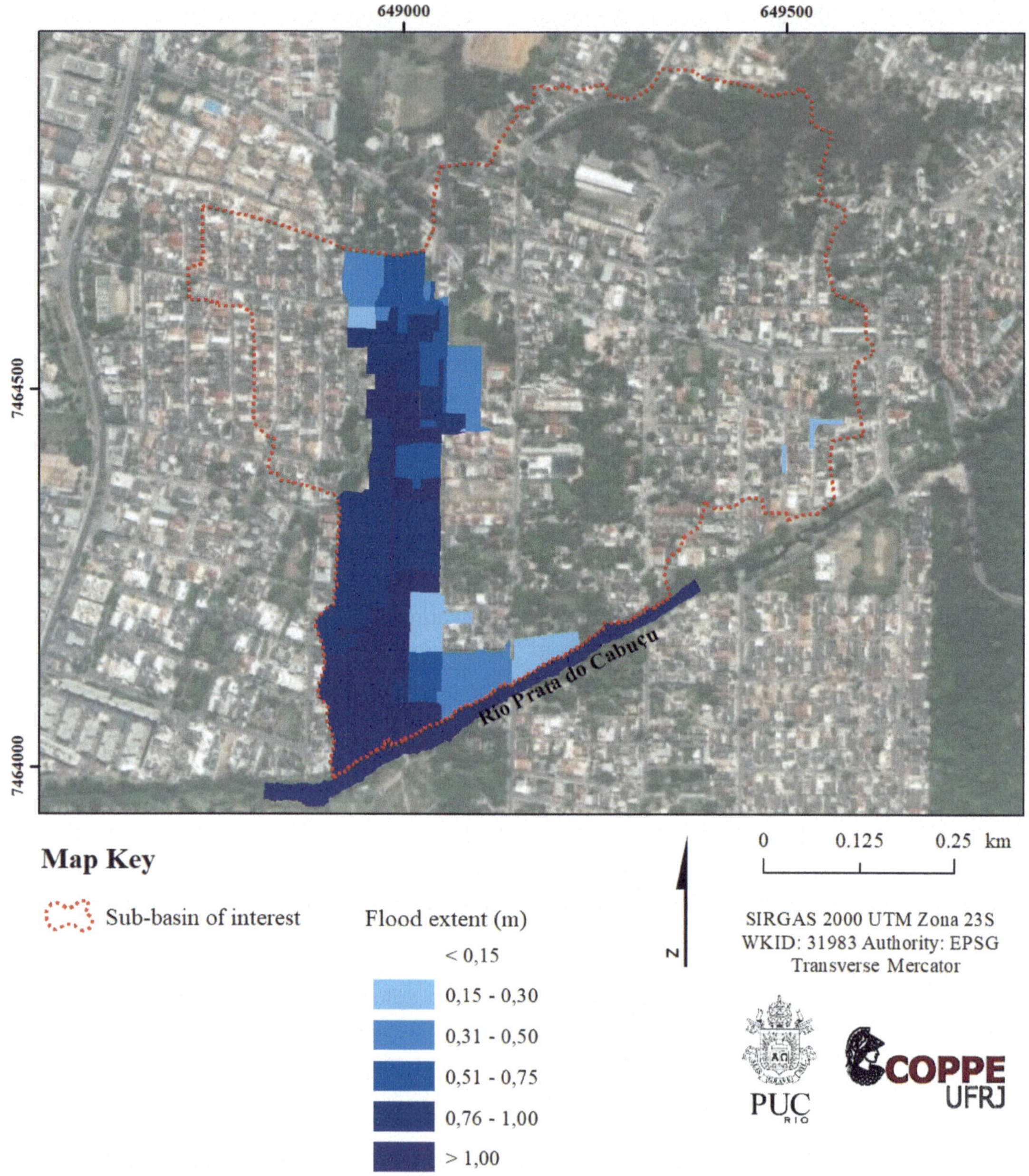

Fig. 3 Simulation results from MODCEL-MHS considering rainfall with a return period of 25 years

Table 1 Concentration values of BOD parameters over time of the rainfall event

Parameter	0–15 min	15–30 min	30–60 min	60–120 min	>120 min
BOD	28 mg/L	26 mg/L	23 mg/L	20 mg/L	12 mg/L

Source: Weibel et al. (1964)

ing unidirectional gates at the termination of each network. These gates not only prevent the river from contributing to the drainage system but also impede the emptying of the subsurface system during high water levels in the receiving watercourse.

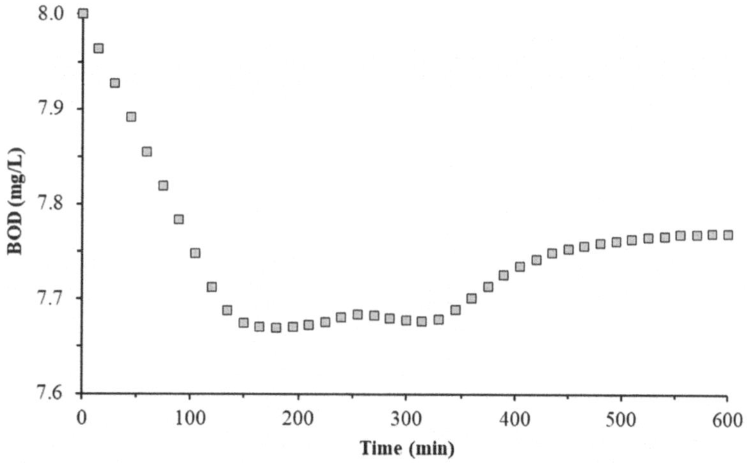

Fig. 4 Variation of BOD in the Prata do Cabuçu River at the section closest to the sewage system outlet

Consequently, the constructed wetland will function as a reservoir for the temporary accumulation of water from the drainage system, retaining runoff from the sub-basin until the main river's water level drops below the arrival point of the subsurface drainage system. Fig. 5b provides a schematic illustration of the final subsurface drainage manhole, adapted for this purpose.

During periods of low water levels in the main river, the unidirectional gate will only be engaged for return periods exceeding the capacity of the constructed wetland. For rainfall events in the sub-basin surpassing a 50-year return period and high main river levels, some flooding may occur at higher levels of the constructed wetland, necessitating flood park infrastructure. For more extended return periods and when the river is at full capacity, flooding near the manholes may occur but at lower intensities than presently observed.

To ensure the system operates during dry weather, another manhole should be implemented just upstream of the manhole contributing to the constructed wetland. It should be equipped with an overflow mechanism to direct sanitary sewage for treatment during rain-free periods. Consequently, in dry weather and periods of low precipitation, water is directed to the treatment system. For heavier rainfall, an overflow weir inside the manhole redirects flows to the manhole contributing to the wetland. Figure 5a schematically illustrates the manhole with an overflow weir for separating contributions during dry weather and low precipitation.

Figure 6 provides an overview of the locations of the three essential constructed urban wetlands for treating effluents from the pilot sub-basin. The sizing of these wetlands was determined based on the results of hydrodynamic simulations and data collected from the theoretical framework, as detailed in Table 2.

Regarding the results of this intervention scenario, it is possible to observe that the flows were managed effectively, both during dry weather conditions and in the case of extreme events. Figure 7 depicts the flood extent for the system correction scenario. It is noticeable that there are no unexpected inundations in the surface flow layer; the only areas with water accumulation are in the constructed wetlands. With the implementation of the system, the receiving watercourse no longer receives contributions from the sanitary sewage system. Additionally, the proposal attenuates the flow contributions to the main river, thereby reducing the total downstream runoff and mitigating other observed floods in the watershed.

In relation to the water quality of the Prata do Cabuçu River, the introduction of constructed urban wetlands results in the maintenance of BOD decay rates in the watercourse.

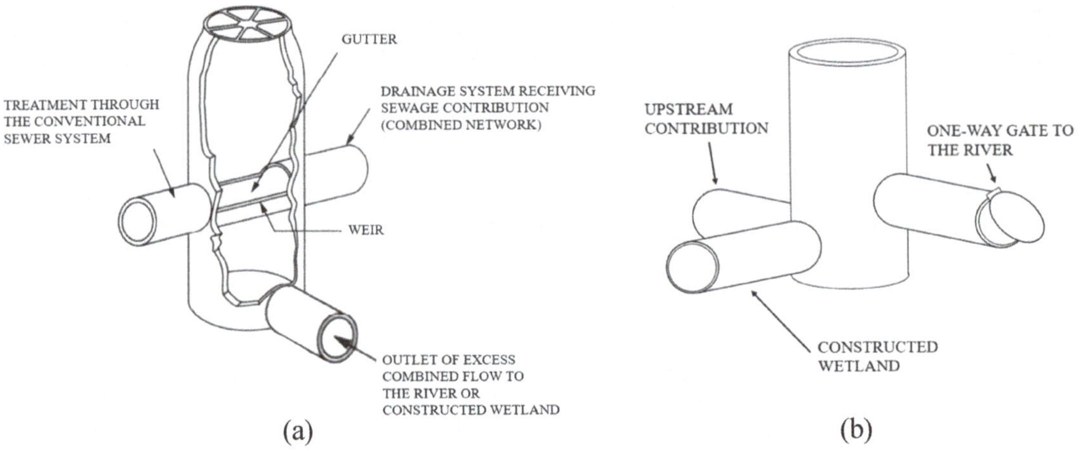

Fig. 5 (**a**) Manhole for separation of dry weather sewage contribution and for sewage contribution with rainwater in rainfall events. (**b**) Manhole for routing waters to the constructed wetland and one-way gate to the river to be utilized in extreme rainfall events. (Source: adapted from Tchobanoglous et al. 1991; Madeira 2012)

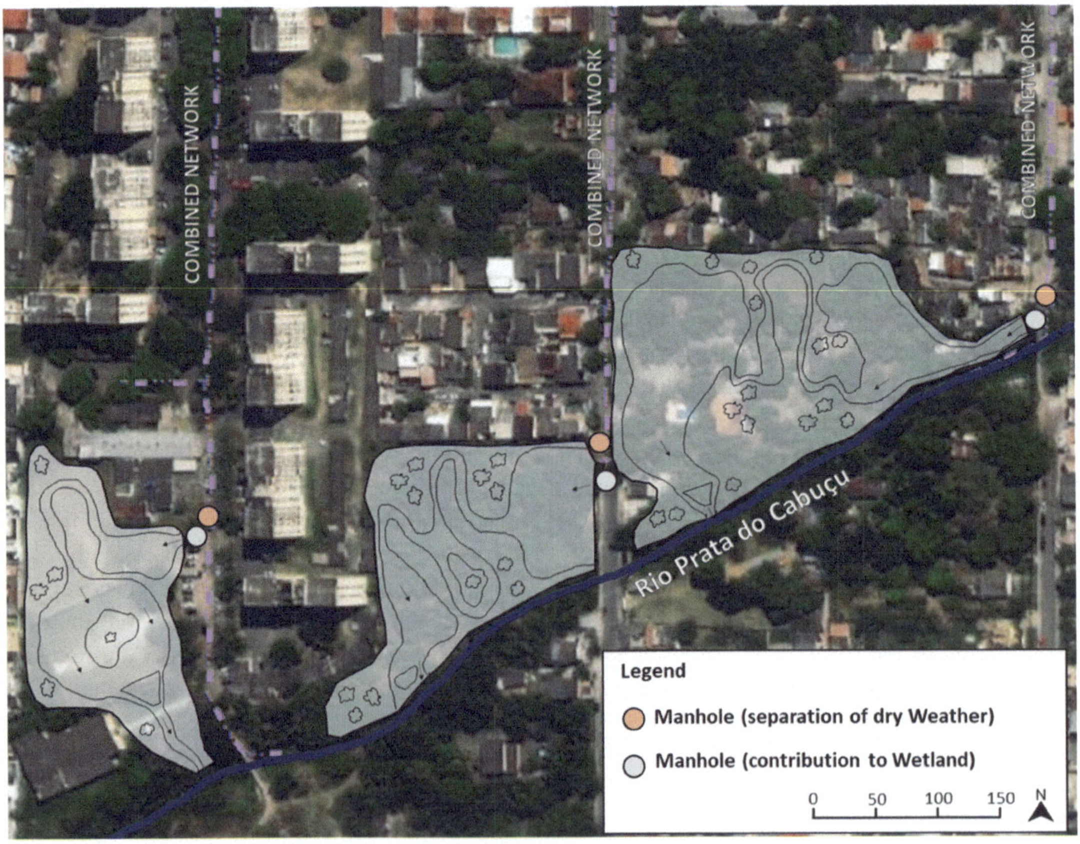

Fig. 6 Schematic design of the urban constructed wetlands to mitigate the interaction issues between the subsurface drainage and sanitary sewage networks

Table 2 Preliminary details of the three urban wetlands design

Factor	Value	Source
Category of constructed wetlands	Free surface flow	Bakhshoodeh et al. (2020)
Total volume to flood (10 years RP)	6040/7500/4640 m^3	Hydrodynamic simulation
Total volume to flood (25 years RP)	7300/9030/5600 m^3	Hydrodynamic simulation
Bottom slope, %	0	Hydrodynamic simulation
Manning's coefficient	0.1 $s/m^{1/3}$	Kadlec and Knight (1996)
Detention time (for soluble pollutants removal)	5 days	Hydrodynamic simulation
Detention time (for suspended pollutants removal)	0,5 days	Hydrodynamic simulation
Area	1 m^2 per served inhabitant	Lana et al. (2013)
Factor hydraulic efficiency	0.75	Persson et al. (1999)
BOD removal efficiency	80.6%	Bakhshoodeh et al. (2020)
Vegetation	Aquatic macrophytes	Rolon et al. (2010)

5 Conclusion

The methodology presented in this work for the integrated simulation of urban stormwater and wastewater drainage networks has proven to be a valuable tool for representing potential cases of cross-interference. The proposed multilayer simulation, using MODCEL-MHS, effectively simulated scenarios of interference between the three represented layers of flow: the surface layer of flow (including streets, lots, open spaces, and rivers), and the underground galleries of the stormwater and wastewater drainage systems.

This methodology holds practical applicability in interpreting and diagnosing cross-problems between drainage and sewer systems, as well as forecasting the operation of new compositions of these networks. It provides decision-making support for optimizing integrated urban water treatment. Integrated treatment here doesn't solely refer to a unified system, but rather to a holistic approach where these systems are considered together, aligning studies and projects more closely with real-world urban environments, enabling proactive problem anticipation.

The evaluation of the methodology's potential using the pilot basin suggests that this tool is not only innovative but also practical for simulating the interaction processes among different urban water components and their pollutant concentrations.

Furthermore, in the Piraquê-Cabuçu River basin, as well as in expanding urban areas, the application of this methodology emerges as a crucial tool for sizing drainage and sewage systems. It anticipates possible failures resulting from their interactions before urbanization or city expansion takes place. Consequently, it enables the proposal of more resilient systems against the adverse impacts of extreme rainfall events. Expanding regions offer open spaces that can accommodate solutions for both sectors. These solutions, while multifunctional, need to incorporate the concepts of nature-based solutions (NBS) and WSUDS (water-sensitive urban design solutions) presented here.

Notably, the implementation of constructed urban wetlands, as a key component of these multifunctional solutions, plays a pivotal role in mitigating floods and enhancing wastewater treatment capacity. These wetlands serve as reservoirs for temporary water accumulation, helping to retain runoff from the sub-basin until the main river's water level drops to manageable levels. This innovative approach not only addresses urban water management challenges but also contributes to biodiversity preservation within the urban environment, offering a range of ecosystem services to society.

Acknowledgment Special thanks are extended to the students and professors of the Universidade Federal do Rio de Janeiro (UFRJ) extension project titled "Multicriteria evaluation of flood resilience as support for territory planning and the social construction process of risk perception." Gratitude is also expressed to the UNESCO Chair on Urban Drainage in Regions of Coastal Lowlands, based at UFRJ. Their contributions and support have been invaluable in advancing this research and its applications.

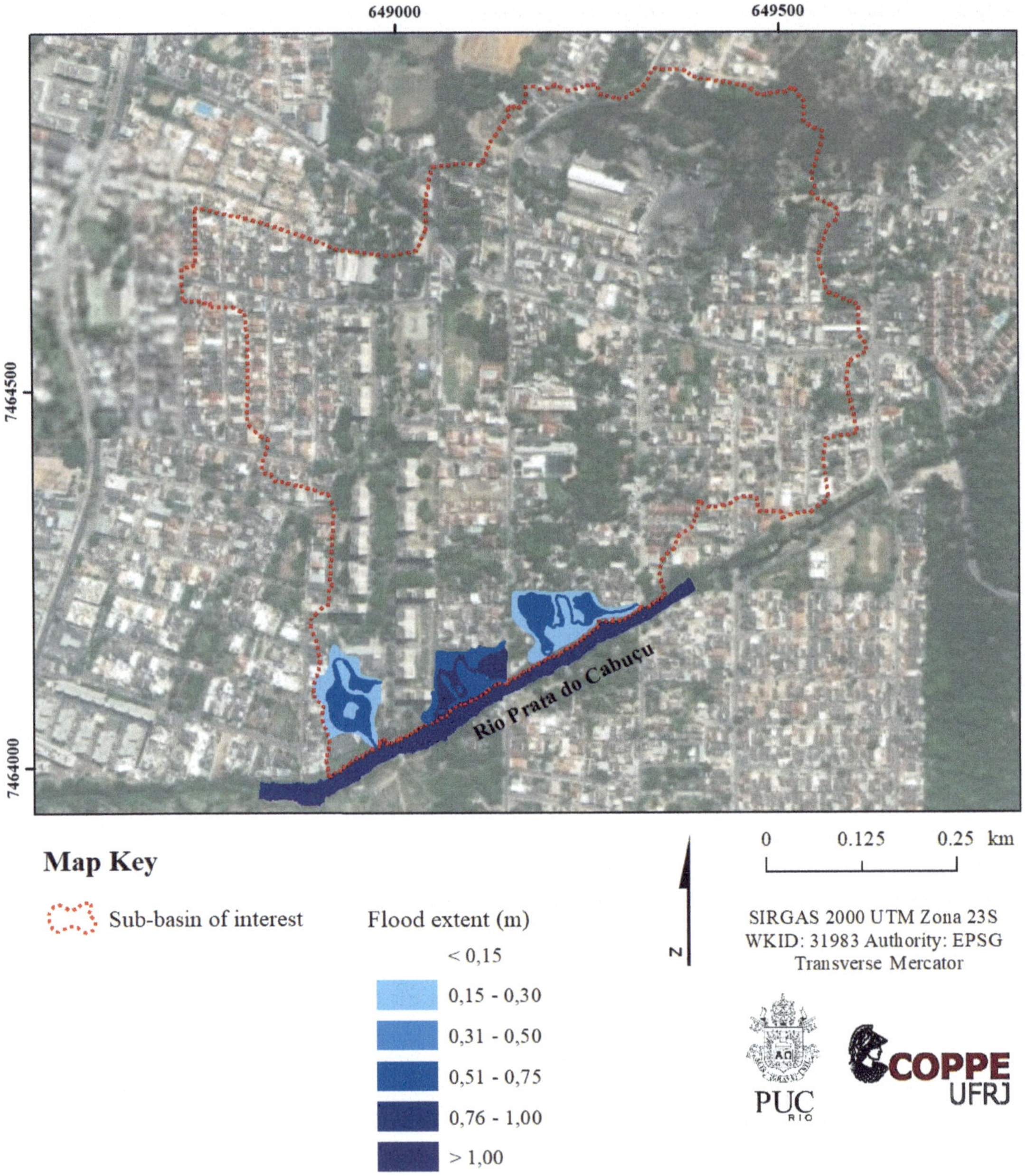

Fig. 7 Result of the floods considering a 25-year return period with the implementation of the constructed urban wetlands

References

ANA—Agência Nacional De Águas E Saneamento Básico. (2021). INDICADORES DE QUALIDADE - ÍNDICE DE QUALIDADE DAS ÁGUAS (IQA). Available at: http://pnqa.ana.gov.br/indicadores-indice-aguas.aspx

Bahia. (2010). Plano Estadual de Manejo de Águas Pluviais e Esgotamento Sanitário. TOMO VII—Concepção e Ações Difusas. Volume 1—Ações para o Estado da Bahia. Governo do Estado da Bahia. Secretaria de Desenvolvimento Urbano.

Bakhshoodeh, R., Alavi, N., Oldham, C., Santos, R. M., Babaei, A. A., Vymazal, J., & Paydary, P. (2020). Constructed wetlands for landfill leachate treatment: A review. Ecological Engineering, 146, 105725.

Bendoricchio, G., Cin, L. D., & Persson, J. (2000). Guidelines for free water surface wetland design. EcoSys Bd, 8(2000), 51–91.

Charles, B., Chase, M., Pociask, G., Bhattarai, R., & Matthews, J. W. (2023). Restored wetlands are greatly influenced by hydrology and non-native plant invasion. Wetlands Ecology and Management, 31(1), 129–142.

Confer, S. R., & Niering, W. A. (1992). Comparison of created and natural freshwater emergent wetlands in Connecticut (USA). Wetlands Ecology and Management, 2, 143–156.

Dias, A.P. (2003). Análise da interconexão dos sistemas de esgotos sanitário e pluvial da Cidade do Rio de Janeiro: valorização das coleções hídricas sob perspectiva sistêmica. Rio de Janeiro: Universidade do Estado do Rio de Janeiro.

Environmental Protection Agency—EPA. (1983). Results of the Nationwide Urban Runoff Program. Final Report.

Farooqi, I. H., Basheer, F., & Chaudhari, R. J. (2008). Constructed wetland system (CWS) for wastewater treatment. In Proceedings of Taal2007: The 12th World Lake Conference (Vol. 1004, p. 1009).

Fraga, J. P. R., Okumura, C. K., Guimarães, L. F., Arruda, R. N. D., Becker, B. R., Oliveira, A. K. B., ... & Miguez, M. G. Cost-benefit analysis of sustainable drainage systems considering ecosystems services benefits: case study of canal do mangue watershed in Rio de Janeiro city, Brazil. Clean Technologies and Environmental Policy, 1–18. 2021.

Gonsalves, J. F., & Mohan, P. (Eds.). (2011). Strengthening resilience in post-disaster situations: Stories, experience and lessons from South Asia. IDRC.

Guimarães, L. F., Battemarco, B. P., Oliveira, A. K. B., & Miguez, M. G. A new approach to assess cascading effects of urban floods. Energy Reports, 7, 8357–8367. 2021.

Gupta, Kapil; Saul, Adrian J. Specific relationships for the first flush load in combined sewer flows. Water research, v. 30, n. 5, p. 1244–1252, 1996.

Hey, D. L., & Philippi, N. S. (1995). Flood reduction through wetland restoration: the Upper Mississippi River Basin as a case history. Restoration Ecology, 3(1), 4–17.

IBGE. (2010). Censo Demográfico 2010. Brasília: Instituto Brasileiro de Geografia e Estatística.

IRM. Instituto Rio Metrópole. (2017). Strategic Plan for Integrated Urban Development for the Metropolitan Region of Rio de Janeiro. In Portuguese: Plano Estratégico de Desenvolvimento Urbano Integrado da Região Metropolitana do Rio de Janeiro (PEDUI/RMRJ).

Jacobson, R. B., Janke, T. P., & Skold, J. J. (2011). Hydrologic and geomorphic considerations in restoration of river-floodplain connectivity in a highly altered river system, Lower Missouri River, USA. Wetlands Ecology and Management, 19, 295–316.

Jean, M. et al. (2021). Optimization of Real-Time Control with Green and Gray Infrastructure Design for a Cost-Effective Mitigation of Combined Sewer Overflows. Water Resources Research, v. 57, n. 12, p. e2021WR030282.

Joshi, P. et al. (2021). Not all SuDS are created equal: Impact of different approaches on combined sewer overflows. Water Research, v. 191, p. 116780, 2021.

Kadlec, R. H., & Knight, R. L. (1996). Treatment Wetlands CRC Press, Boca Raton, FL. p. 893.

Kitanović, V., Greet, J., McKendrick, S. A., & Jones, C. S. (2023). Riparian and terrestrial grasses display unexpected tolerance to cool-season inundation. Wetlands Ecology and Management, 1–14.

Končalová, H., Květ, J., Pokorný, J., & Hauser, V. (1993). Effect of flooding with sewage water on three wetland sedges. Wetlands Ecology and Management, 2, 199–211.

Lana, L. C. O. et al. (2013). Performance of a single stage vertical flow constructed wetland system treating raw domestic sewage in Brazil. Water Science and Technology, v. 68, n. 7, p. 1599–1606.

Livingston, E. H. (2020). Use of wetlands for urban stormwater management. Constructed wetlands for wastewater treatment, 253–262.

Madeira, D. G. (2012). Implantação de sistema de esgoto do tipo unitário com posterior adequação ao sistema separador absoluto em Flores da Cunha-RS.

Miguez, M. G. (2001). Modelo Matemático de células de escoamento para bacia urbana. 2001. Tese (Doutorado). Programa de Engenharia Civil, COPPE, Universidade Federal do Rio de Janeiro, Rio de Janeiro.

Miguez, M. G.; Veról, A. P. (2017). A catchment scale Integrated Flood Resilience Index to support decision making in urban flood control design. Environment and Planning B: Urban Analytics and City Science, v. 44, n. 5, p. 925–946.

Montserrat, A. et al. (2014). Using data from monitoring combined sewer overflows to assess, improve, and maintain combined sewer systems. Science of the Total Environment, v. 505, p. 1053–1061.

Oliveira, A. K. B. (2018). O sistema de drenagem como eixo estruturante do planejamento urbano: caso da bacia hidrográfica do Rio Acari. Dissertação (Mestrado). Programa de Engenharia Civil, COPPE, Universidade Federal do Rio de Janeiro, Rio de Janeiro.

Oliveira, A. K. B., Rezende, O. M., de Sousa, M. M., Nardini, A., & Miguez, M. G. (2019). An alternative flood model calibration strategy for urban watersheds: the case study of Riohacha, Colombia. Water Science and Technology, 79(11), 2095-2105.

Oliveira, A. K. B et al. (2022). Evaluating the Role of Urban Drainage Flaws in Triggering Cascading Effects on Critical Infrastructure, Affecting Urban Resilience. Infrastructures, 7(11), 153.

Oliveira, A. K. B. (2022). Proposta de Metodologia de Simulação Integrada das Redes Urbanas de Drenagem Pluvial e Esgotamento Sanitário. Tese (Doutorado). Programa de Engenharia Civil, COPPE, Universidade Federal do Rio de Janeiro, Rio de Janeiro.

Oliveira, A. K. B., Alves, L. M. C., Carvalho, C. L., Haddad, A. N., de Magalhães, P. C., & Miguez, M. G. (2023). A framework for assessing flood risk responses of a densely urbanized watershed, to support urban planning decisions. Sustainable and Resilient Infrastructure, 1–19.

Oliveira, A. K. B., Magalhães, P. C. D., Rezende, O. M., Sousa, M. M. D., Marques, R. H. M., & Gomes Miguez, M. (2024). MODCEL-MHUS: a comprehensive multilayer hydrodynamic unified simulation for stormwater, sanitary sewer systems, and urban surface. *Water Science & Technology, 90*(1), 190–212. https://doi.org/10.2166/wst.2024.221

Pattison-Williams, J. K., Pomeroy, J. W., Badiou, P., & Gabor, S. (2018). Wetlands, flood control and ecosystem services in the Smith Creek Drainage Basin: A case study in Saskatchewan, Canada. Ecological economics, 147, 36–47.

Persson, J., Somes, N. L. G., & Wong, T. H. F. (1999). Hydraulics efficiency of constructed wetlands and ponds. *Water Science and Technology, 40*(3), 291–300. https://doi.org/10.1016/S0273-1223(99)00448-5

Radesca, F. D. (2018). Uso de Sistemas Unitários de Esgoto e Drenagem como estratégia alternativa e escalonada no tempo para o Saneamento Ambiental. Dissertação de Mestrado apresentada ao Programa de Pós-Graduação em Engenharia Civil- COPPE –Rio de Janeiro: UFRJ.

Rezende, O. M. et al. (2019) A framework to introduce urban flood resilience into the design of flood control alternatives. Journal of Hydrology, v. 576, p. 478–493.

Riechel, M. et al. (2016). Impacts of combined sewer overflows on a large urban river–Understanding the effect of different management strategies. Water research, v. 105, p. 264–273.

Rodrigues, E. B. et al. (2021). Utilização do biocarvão como meio filtrante de wetland construído vertical empregado no tratamento de esgoto sanitário.

Rolon, A. S., Homem, H. F., & Maltchik, L. (2010). Aquatic macrophytes in natural and managed wetlands of Rio Grande do Sul State, Southern Brazil. Acta Limnologica Brasiliensia, 22, 133–146.

Shinohara, P. Y. (2021). Implantação de um sistema francês de wetland construído como proposta para tratamento da água residuária do restaurante universitário. Trabalho de Conclusão de Curso. Universidade Tecnológica Federal do Paraná.

Sousa, M. M., Oliveira, A. K. B., Rezende, O. M., de Magalhães, P. M. C., Pitzer Jacob, A. C., de Magalhães, P. C., & Miguez, M. G. (2022). Highlighting the role of the model user and physical interpretation in urban flooding simulation. Journal of Hydroinformatics, 24(5), 976–991.

Souza, G. H. F. (2021). Projeto de um sistema wetland francês para tratamento de efluente em um campus universitário. 2021. Trabalho de Conclusão de Curso. Universidade Tecnológica Federal do Paraná.

Sztruhár, D. et al. (2002). Comprehensive assessment of combined sewer overflows in Slovakia. Urban Water, v. 4, n. 3, p. 237–243.

Tchobanoglous, G., Burton, F., & L. & Stensel, H. D. (1991). Wastewater engineering. Management, v. 7, n. 1, p. 4. Available at: https://d1wqtxts1xzle7.cloudfront.net/55688838/Wastewater_treatment_and_reuse-libre.pdf?1517478116=&response-contentdisposition=inline%3B+filename%3DWastewater_Engineering_Treatment_and_Reu.pdf&Expires-=1726068210&Signature=TIgWx7HsnOmeRNSeowTXp3XQIuQq0nP6HRV4JOaFYjbeZ~1TXl6523G9BXckx0S4ZQcX55XRWFKpqHoYhgjmwdhcHynz11sm5hweuliaZ6wAU76Tf0nRx5mOcE1XRti-stZZ19LExO~IutoW7A~1NEeliil5CexV61SYAotEY4Y4Q~fboAq96JSryht-wT47CAkCaQhmaE~WJqHuooRJYvei~ZcoJFED~senQtbjA__&Key-Pair-Id=APKAJLOHF5GGSLRBV4ZA

Tejaswini, V.; Sathian, K. K. (2018). Calibration and Validation of Swat Model for Kunthipuzha Basin Using SUFI-2 Algorithm. Int. J. Curr. Microbiol. App. Sci, 7(1), 2162–2172.

Thompson, J. R., Gavin, H., Refsgaard, A., Refstrup Sørenson, H., & Gowing, D. J. (2009). Modelling the hydrological impacts of climate change on UK lowland wet grassland. Wetlands Ecology and Management, 17, 503–523.

Todeschini, S.; Papiri, S.; Ciaponi, C. (2012). Performance of stormwater detention tanks for urban drainage systems in northern Italy. Journal of environmental management, v. 101, p. 33–45.

Von Sperling, M. (1996) Introdução à qualidade das águas e ao tratamento de esgotos. Editora UFMG.

Von Spering, M. (2014). Introdução à qualidade das águas e ao tratamento de esgoto. 4 ed. Belo Horizonte: Departamento de Engenharia Sanitária e Ambiental (SESA/ UFMG. 472 p.

Weibel, S. R.; Anderson, R. J.; Woodward, R. L. (1964). Urban land runoff as a factor in stream pollution. Journal (Water Pollution Control Federation), p. 914–924.

Yamanaka, S., Ishiyama, N., Senzaki, M., Morimoto, J., Kitazawa, M., Fuke, N., & Nakamura, F. (2020). Role of flood-control basins as summer habitat for wetland species-A multiple-taxon approach. Ecological engineering, 142, 105617.

Urban Wetlands as Resilient Landscape Infrastructure—The Case of Llanquihue Green Infrastructure Plan, Chile

Osvaldo Moreno

Contents

Abstract

The urban green infrastructure planning approach provides an innovative conceptual and operational framework to face the current challenges of conservation and rehabilitation of urban wetlands in the context of disturbances and vulnerabilities caused by urban expansion processes, climate change and disasters. In this sense, landscape units and their components can be conceived as a potential structuring network for the city and the territory, contributing to an integrated planning of natural and anthropic systems at both spatial and functional levels, considering the relationship of their ecologies with urban infrastructure systems through nature-based solutions.

The Llanquihue Green Infrastructure Plan—an applied research initiative presented in this chapter—addresses this conceptual framework to configure a landscape project based on the spatial and functional articulation of existing urban wetlands, transforming them into key elements of an infrastructure system designed to provide social benefits and ecosystem services to the city and its communities. Instead of "constructing" green spaces, with the high costs involved, the plan proposes the concept of "landscape activation" through the configuration of specific and delimited components designed to enable, equip and provide access to these areas, thereby promoting the efficiency of the public investment. The performance of these hybrid natural systems—related to the synergistic

O. Moreno (✉)
Escuela de Arquitectura, Pontificia Universidad Católica de Chile, Santiago, Chile
e-mail: omorenof@uc.cl

C. Rojas Quezada (ed.), *Urban Wetlands in Latin America*, Sustainable Development Goals Series,
https://doi.org/10.1007/978-3-031-69590-2_8

combination of ecological and anthropic components—contributes to the provision of socio-ecological functions and services related to risk reduction and adaptive capacity to climate change. At the same time, in the absence of public and green spaces, especially in vulnerable environments, it can contribute to the development of memorable places for urban living through the integration of ecologies, social programmes and multi-purpose infrastructures.

From a strategic approach, this initiative is proposed as a complementary and guiding platform to feed the current urban planning instruments, as well as to generate alternative mechanisms and tools for the integrated management of landscape and public spaces, becoming a potential model to be applied in other cases of regional cities also characterised by problems related to the recovery and enhancement of urban ecosystems.

Keywords

Urban wetlands · Green infrastructure · Nature-based solutions · Landscape activation · Urban resilience

1 Conceptual Framework: Linking Green Infrastructure Planning and Urban Wetland Management

Urban wetlands are important systems characterised not only by their scenic beauty and exceptional biodiversity, but also by their important role as ecological infrastructure that provides key ecosystem services for sustainable development, resilience and adaptability to climate change in vulnerable territories and communities (Rojas et al. 2022; Moreno and Arizaga 2022; Holland et al. 1991). At the same time, they are understood as the expression of broader and more complex ecological systems, related to regional hydrology and migratory movements of birds on a continental scale. From a socioecological perspective, wetlands are vital ecosystems for human survival, as one of the most productive environments in the world: they are cradles of biological diversity, sources of water and primary productivity on which countless plant and animal species depend for subsistence (Ramsar 2018). However, these systems are commonly exposed to serious threats and impacts on their components and functions, affecting not only the ecology of wetlands, but also the quality of life of communities. Although numerous research and studies have been generated in recent years in various fields of scientific and technical knowledge, there is a lack of an integrated vision to address the important challenges involved in the conservation, restoration, and rehabilitation of urban wetlands, especially in Latin American cities.

The urban green infrastructure planning approach provides an innovative conceptual and operational framework to face the current challenges of conservation, rehabilitation and enhancement of urban wetlands, caused by disturbances, crises and vulnerabilities related to urban expansion processes, climate change and disasters (Moreno 2020; Hung and Aquino 2013). As Benedict and Mcmahon (2006) argue, the landscape and its biophysical components can be conceived as a potential structuring network of the territory and the city, promoting integrated planning of natural and anthropic systems at both the spatial and functional levels, within the framework of urban and territorial development policies and instruments. Thus, the idea of landscape as infrastructure not only refers to the consideration of natural systems and their contribution to sustainability and resilience, but also contemplates the articulation of their ecologies with infrastructural, productive and urban systems, either as a product of strategies and design operations agreed upon from a given project, or derived from traditional and collective practices aimed at favouring the economic and cultural development of a community (Moreno 2020).

The influence of the green infrastructure approach in urban wetland conservation and restoration plans and projects is currently evident in

Fig. 1 On the left (**a**): Yanweizhou Wetland Park, in the Chinese city of Jinhua, developed by the Turenscape studio. On the right (**b**): designed by the same office, the Minghu Wetland Park in Liupanshui, China. In both cases, the design process has contemplated the articulation of ecosystems, infrastructures and use programs, transforming degraded former riverbanks into flexible systems for the provision of public spaces of high social value, ecological restoration and urban flood resilience management. (Source: Turenscape © https://www.turenscape.com)

several global initiatives, particularly in Asia, Europe and North America. Notable examples include the Minghu Wetland Park (Fig. 1a) and the Yanweizhou Wetland Park (Fig. 1b), in the Chinese cities of Liupanshui and Jinhua, respectively; the South Bay Sponge Plan in San Francisco; the Resilient Boston Harbor Plan (Fig. 2); and the Houston Coastal Roulette—Planning Resilient Communities for Galveston Bay (Fig. 3).

The contemporary panorama associated with this approach to landscape as infrastructure offers more and more references to plans and projects in the context of planning processes for resilient cities in post-disaster and climate change scenarios. The conservation and management of landscape units as part of an urban green infrastructure system, considering natural areas and also anthropic structures related to productive and urban areas, is a fundamental link to increase resilience in territories exposed to disasters, due to the adaptive capacities that these systems can provide to face geological risks—such as earthquakes, tsunamis, landslides and volcanic eruptions—and hydrometeorological risks, such as urban floods, landslides, tsunamis, storms, extreme rains and droughts. The study of the referential cases of plans and projects mentioned, which cover in temporal terms a decade[1] characterised by the increase in the frequency and intensity of disaster events, constitutes an important documentary source that exposes an emerging and innovative space of theoretical and practical development, whose conceptual basis in terms of design is informed both by ecological knowledge, as well as by the applied sciences of engineering, architecture and urbanism (Corner and Bick 2014; Reed and Lister 2014). The result of these research and professional practice experiences, which are framed within the guidelines established by the United Nations for disaster risk management through various intergovernmental agencies, opens a renewed scope of thought and action around the language of resilience, adaptive management and versatility, demonstrating the operability of these approaches applied to disaster vulnerability contexts, at variable scales of representation and project (Moreno 2020; Whaldheim 2016; Lister 2015; Mathur and Da Cunha 2001).

In Chile, it is possible to identify three important initiatives in recent years that are

[1]Including in this period important hydrometeorological events, such as hurricanes Katrina (2005), Ike (2008) and Sandy (2012), and the earthquakes and tsunamis in Indonesia (2004), Chile (2010) and Japan (2011).

Fig. 2 Resilient Boston Harbor, vista aérea. Iniciativa del gobierno de la ciudad, en desarrollo desde 2017, orientada a generar sistemas integrados de infraestructura verde en las franjas de borde costero y fluvial, mediante la conservación y recuperación gradientes ecotonales para la adaptabilidad y resiliencia ante fenómenos de inundabilidad, marejadas y tormentas. El proyecto amplifica el sistema de parques originalmente proyectado por Olmsted, otorgando una nueva escala de infraestructura verde a nivel urbano-territorial. (Fuente: City of Boston (2018) Resilient Boston Harbor)

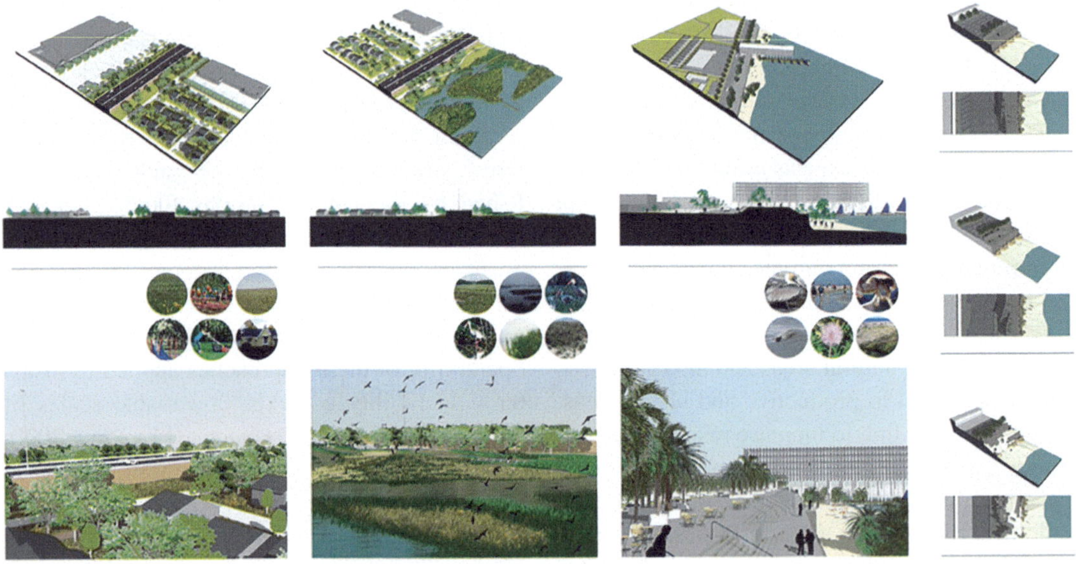

Fig. 3 Example of planning and design of green infrastructure for disaster impact mitigation in coastal areas, based on the operation of wetland systems, riparian vegetation and coastal forests, combined with agricultural areas and built infrastructure. These solutions are complemented by planning instruments to regulate the occupation of risk zones, while stimulating the conservation of coastal ecosystems. (Source: Own elaboration based on the Houston Coastal Roulette Planning Resilient Communities for Galveston Bay project (SWA Group and ASLA 2012))

directly related to the conservation of urban wetlands. Two of them respond to a regulatory dimension: the formulation of the National Wetlands Protection Plan (2018–2022) and the enactment of Law 21.202, known as the Urban Wetlands Law (2019). On the other hand, it is also worth highlighting the implementation experience of the Llanquihue Wetlands City Green Infrastructure Plan (2017–2019), a pioneering project at the national level that implements the urban green infrastructure planning approach and nature-based solutions for the conservation and sustainable management of urban wetlands.

2 Pilot Implementation: The Llanquihue Urban Green Infrastructure Plan

The city of Llanquihue[2] is located in southern Chile in Los Lagos region, on the shoreline of Llanquihue Lake, the second largest lake in the country. A quick look from the air over the urban area (Fig. 4) reveals the important presence of several aquatic ecosystems associated with the drainage area of the lake and the headwaters of Maullín River, the riparian ecotones of the floodplains on the lakeshore and the urban wetland system, where the Baquedano, Los Helechos, El Loto and Las Ranas stand out. These urban wetlands are the main habitat for more than 70 bird species, 100 plant species, 20 native fish species and nearly 50% of Chile's migratory birds. In addition, there are some remnant patches of native forest in the surrounding rural area, poplar groves and other exotic species typical of the productive landscape of this area of southern Chile. However, this promising landscape mosaic, which can be appreciated from a distance, reveals itself in a more complex way when it is perceived more closely.

Since its origins, as an urban and industrial settlement in the context of colonisation promoted by the Chilean government at the end of the nineteenth century, the city has progressively developed over a hydro-ecological network of wetlands and small streams with surface and underground connections to Llanquihue Lake and the Maullín River, resulting in highly altered ecosystems nowadays. In fact, due to the unplanned growth of the city in the past, several parts of this hydro-ecological system have been lost. Llanquihue's wetlands are threatened by the effects of climate change and various anthropogenic factors, including urban expansion, hydrological fragmentation, deforestation, extraction and burning of riparian vegetation, invasion of exotic species of flora and fauna, discharge of domestic and industrial wastewater, diffuse pollution from urban runoff and agricultural inputs (pesticides and fertilisers), among others (FLC 2020; FLC and MAPA UC 2017).

As Table 1 shows, these threats can be classified as: (I) physical threats, when they are capable of altering the system, changing its structure and functioning; (II) biological threats, when they can affect specific components of the system; and (III) chemical threats, when they alter specific components of the system (abiotic and/or biotic). However, if the concentration and/or properties of chemical compounds exceed the carrying capacity of the system—the natural processing capacity of the ecosystem—the whole system could be affected (MMA 2018; Fariña and Camaño 2012).

Despite these problems, in the case of Llanquihue there is a comparative advantage for the formulation of the plan: A community that is aware of the importance of valuing urban ecosystems. This awareness is complemented by municipal authorities and private actors committed to the sustainable management of the urban environment, which favours the understanding that Llanquihue's urban wetlands are key systems that provide the city with ecosystem services related to the management of the water cycle, flood control, the provision of clean water, biodiversity, scenic beauty and recreation (Moreno 2019; Mitsch and Gosselink 2000).

The proposal for the Llanquihue Urban Green Infrastructure Plan is the result of an

[2] The urban area of Llanquihue has 12.728 inhabitants, according to estimations of the National Statistics Institute (INE) and the Municipal Development Plan 2015–2018.

Fig. 4 Aerial view of Llanquihue. The image shows the riparian ecotones of the flood zones on the lakeshore, the urban wetlands system (Baquedano, Los Helechos, El Loto and Las Ranas) and the beginning of Maullín River, forming a green and blue network that coexists with the city. (Source: Own elaboration)

applied research initiative[3] developed jointly by the Master of Landscape Architecture of the Pontificia Universidad Católica de Chile (MAPA UC) and the Fundación Legado de Chile (FLC). Considering the various socio-ecological problems caused by urban pressure on the city's ecosystems, the "1st Forum for the Sustainability of Llanquihue" was launched in November 2016, a space of open participation that brought together more than 300 local residents, authorities, regional services, representatives of the local productive sector, educational institutions, scientists and planners. During this event, the team of researchers and professors from the Master of Landscape Architecture PUC and professionals from the Legado Chile Foundation conducted the "Green Infrastructure for Llanquihue" workshop, which established principles for the design of the Urban Green Infrastructure Plan. As a result, a preliminary roadmap and a first "target image" of the city in the medium- and long-term planning scenario were obtained, as well as a portfolio of related priority projects.

Starting in 2017, an "Environmental Action Plan" was implemented, applying an unprecedented methodology in Chile oriented towards the diagnosis, planning, implementation and participatory management of urban ecosystems, with transversal articulation with neighbourhood committees, educational institutions, municipal authorities and the regional ministerial secretariats of Housing and Planning (MINVU) and Public Works (MOP). Within the

[3]The Llanquihue Urban Green Infrastructure Plan is part of the research project "Plan de Acción Medioambiental de Llanquihue" (Environmental Action Plan for Llanquihue), financed by Innova CORFO through funds awarded in the call "Prototipos de Innovación Social Región de Los Lagos 2017".

Table 1 Main threats to wetland ecosystems

Physical threats	1. Expansion of urban areas (urban sprawl) 2. Alteration and fragmentation of natural watercourses due to urban infrastructure 3. Drainage of wetland areas for stabilisation of new urban land 4. Extraction of groundwater and surface water 5. Sedimentation of wetland ecosystems 6. Deforestation 7. Extraction and burning of riparian vegetation 8. Discharge of debris on wetland edges
Biological threats	9. Grazing and browsing in wetland areas 10. Eutrophication due to organic matter input (grazing) 11. Proliferation of toxic algae 12. Domestic predators (cats and dogs) that affect native fauna, especially birds 13. Monocultures, due to simplification and fragmentation of surrounding ecosystems and loss of biodiversity 14. Biomass extraction 15. Alteration of the biological capacity of soils 16. Discharge of domestic organic liquid waste 17. Deforestation of riparian vegetation 18. Introduction and propagation of exotic species
Chemical threats	19. Diffuse pollution from the use of agrochemicals (fertilisers, pesticides) in agriculture 20. Discharge of urban and industrial effluents with high concentrations of nutrients and salts 21. Discharge of heavy metals 22. Discharge of debris and household garbage on wetland riparian zones

Source: Own elaboration based on FLC (2020), RAMSAR (2018), FLC and MAPA UC (2017), Fariña and Camaño (2012) and MMA (2018)

framework of this agenda, the Urban Green Infrastructure Plan emerges as a priority strategy that acts as a catalyst for various local problems and demands that currently converge in urban ecosystems, such as flood prevention, household waste management, diversification of local economic activities, community empowerment and the improvement and reactivation of public space (Fig. 5).

3 Restoring, Programming and Linking: Landscape Activation Through Nature-Based Solutions

The Llanquihue Urban Green Infrastructure Plan addresses three strategic objectives: restoring, programming and linking. Firstly, it seeks to restore the riparian ecology of urban wetland ecosystems, understood as a key element for their enhancement and integration as part of the urban infrastructure matrix. In a complementary way, the Plan's initiating projects aim at inhabiting these spaces, integrating social programmes in the restoration and rehabilitation initiatives of urban wetlands, promoting their sustainable appropriation and use by the community. Finally, the action to link the different wetlands and other urban ecosystems through a continuous network of public spaces aims to create circuits and corridors that link the different landscape units socially and ecologically.

Instead of "building" green spaces, it is proposed to "activate" the remaining landscape units that persist in the emerging urbanisation, linking the diversity of ecosystems, public spaces and other valuable spaces of the city through simple and well-placed designed pieces. The project operations contemplated by the Plan are strategically positioned in the edge zones of the urban wetlands, because this space is understood as the critical area that must be recovered and consolidated as a buffer to conserve the qualities of its components and ecosystemic processes, regulating its accessibility and programmatic use. By optimising investments in strategic configurations and interventions, formulated in close consultation with the community and local governance actors, it is possible to encompass extensive green infrastructure systems based on the articulation of natural systems and public spaces, aimed at providing both memorable places for social gathering and recreation, as well as key ecological functions for urban sustainability and resilience (Fig. 6).

In order to achieve these objectives, the Plan considers three types of landscape planning and design operations: Strips, Nodes and Areas.

Fig. 5 "Green Infrastructure for Llanquihue" Workshop. Participation and co-design activities of the Llanquihue Urban Green Infrastructure Plan. (Source: FLC and MAPA UC 2017)

- The Strips help to promote human mobility in direct contact with different landscape units and their components, contribute to the conservation of water and ecosystem processes, and create ecological connectivity along the urban landscape mosaic. The strips are implemented through: (I) soil management and modelling operations; (II) the design of infiltration strips and pathways at the level of the natural terrain, with fine gravel soils to promote water infiltration; and (III) elevated wooden walkways to regulate accessibility, allow ecological recovery and water functionality of urban ecosystems.
- The Nodes are landmarks that define notable sites within the system, either because of the quality of the landscape elements they contain, their visibility from the surrounding landscape, or their suitability as a significant public space with specific programs and facilities such as playgrounds, shelters and viewpoints.
- The Areas are considered to be management spaces associated with larger zones that may contain various ecosystems of high ecological and/or cultural value, which require differentiated management measures to promote their recovery and conservation. This includes the restoration of marsh and riparian vegetation—on the edge zones of urban wetlands—and the planting strategies of native species to restore the habitat and refuge functions of flora and fauna.

4 A Prototype of Nature-Based Solutions: The Baquedano Urban Wetland Park

The plan proposes the implementation of a flagship project that will be a milestone and reference for the approach to urban green infrastructure and nature-based solutions: the Baquedano Urban Wetland Park[4]. The Baquedano wetland is located in the north side of the city and is classified as permanent wetland on inorganic soils by the Ramsar Convention, specifically the subtype of forest swamp. It is habitat for several species that are exclusive to wetlands, such as Piden (*Pardirallus sanguinolentus*), Chilean Frog (*Calytocephalla gayi*) and little remnant of swamp forest species such as Pitra (*Myrceugenia exsuca*), highlighting the value of these urban wetlands for biodiversity conservation. At the same time this wetland fulfils a role for the city, contributing to the mitigation of and protection against flooding. Despite its important role in the urban ecosystem, this wetland was highly fragmented and disconnected from the wetland network, covered in debris and garbage, affected by diffuse pollution from the streets runoff and also disturbed by stray dogs and parked vehicles.

[4]Project financed with complementary contributions provided by the Regional Government of Los Lagos to the Municipality, through the Regional Fund for Local Initiative (FRIL).

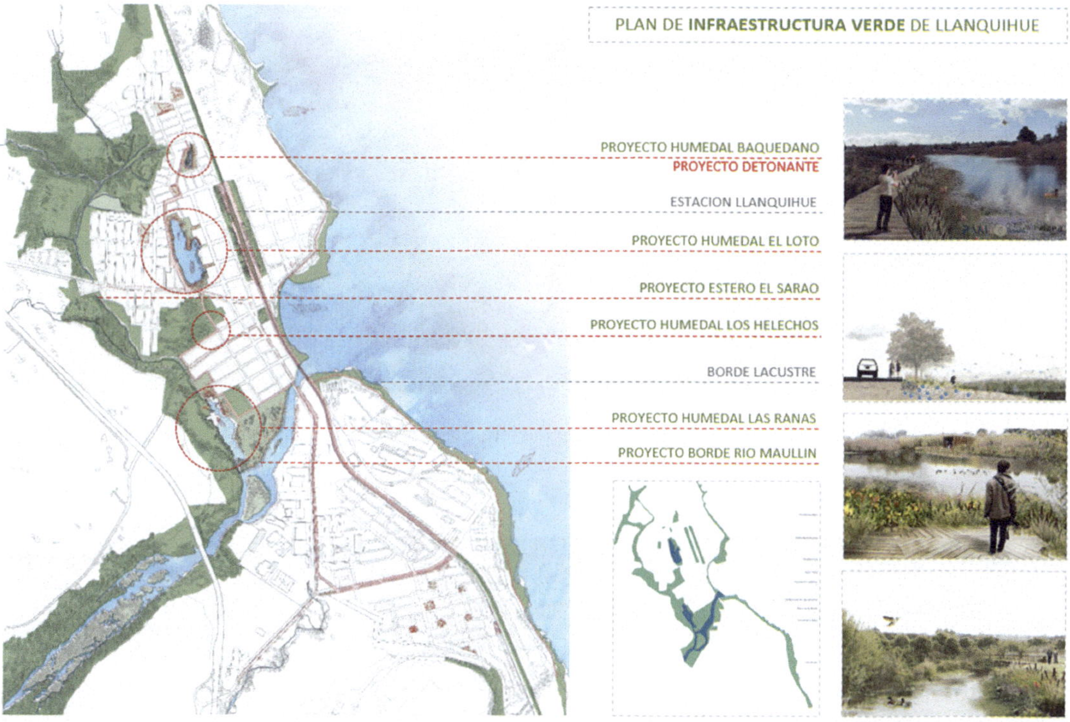

Fig. 6 Llanquihue Urban Green Infrastructure Plan, showing the landscape units that spatially structure the plan and the location of project actions to "restoring, programming and linking" the different urban ecosystem units to create a landscape network. (Source: Own elaboration)

With a current surface area of 10,000 m^2, this wetland has been declared in a critical state due to the systematic infilling of its edges, which has greatly reduced the presence of native vegetation and turned it into a focus of contamination and degradation. As a result of the deterioration of its edges and the eutrophication process that has affected it, the wetland's water surface has gradually disappeared. On the other hand, the urban area where this wetland is located—the Baquedano neighbourhood—is one of the sectors most affected by flooding, especially during the winter season, when rainfall easily exceeds 50 mm/day (Fig. 7).

In this context, planning strategies and landscape design interventions face the challenge of integrating the restoration of the wetland ecology, the improvement of the public realm through sustainable use, and the rehabilitation of the hydrological functions to retain and drain stormwater runoff. Based on this scenario and in accordance with the landscape planning objectives and actions described above, the project configures a system of strips, nodes and areas on the edge zone of the wetland, creating a space of ecological and social ecotone[5] (Figs. 8 and 9). This nature-based solutions strategy considers the restoration of the riparian ecosystems, the recovery of the hydraulic capacity through the enlargement of the wetland edges and the projection of a network of public paths and recreational areas. In this way, the Baquedano Wetland Park becomes a key link in the implementation process of the Llanquihue Urban Green Infrastructure Plan.

In the park, the strips are projected through a system of ground design on the perimeter of the wetland, defining different types of paths through wooden walkways and fine gravel surfaces. This system creates a border zone towards the built

[5]In ecology, the ecotone is understood as the transitional space between two bordering ecosystems, in which both interact generating synergies and complementarity of species, metabolic processes and ecosystem services.

Fig. 7 Target image of the Baquedano Wetland Park. On the left, the current scenario of the wetland. On the right, the target image of the Project. (Source: Own elaboration)

Fig. 8 Aerial view of the Baquedano Wetland Park, a prototype of nature-based solutions and flagship of the Llanquihue Urban Green Infrastructure Plan. (Source: Own elaboration.)

urban environment that integrates public space uses: playgrounds, multi-purpose areas, pedestrian paths and urban furniture. Towards the wetland itself, the strips generate an ecological gradient by restoring riparian vegetation in the ecotone zone.

The strips also include water flow management devices, configuring 'rain retention pockets' to collect streets runoff—to filter and decompose the water pollution—and 'slow-flow channels' to distribute and oxygenate water, from the rain retention pockets to the wetland (Fig. 10). In this process, the wetland's riparian vegetation also contributes to the phytoremediation of these water flows before they enter to the aquifer.

The nodes have been designed as landscape viewpoints and public gathering spaces in the park, creating platforms for observing and learning about the dynamics of the wetland ecosystem. At the same time, they define significant

Fig. 9 Plan view of the Baquedano Wetland Park. The project configures a system of strips, nodes and areas on the edge zone of the wetland, creating a socio-ecological ecotone. (Source: Own elaboration)

spaces within the city, that encourage social interaction, especially among children's groups, who have been actively involved in the formulation of this type of facility throughout the planning and design process of the project.

Finally, the areas are designed to the ecological restoration of the wetland's riparian ecotone. These areas consider soil modelling operations—terracing and cutting—to create a suitable slope that can allow the establishment of riparian vegetation to activate propagation and ecological succession processes. The species planted are individuals obtained from the wetland itself (*Juncus procerus*, *Schoenoplectus californicus* and *Typha angustifolia*), through relocation processes.

5 Conclusions and Projections

In the context of the challenges for conservation and management of urban wetlands in Latin American cities, the concept of resilience and the broader redefinition of the ecology of complex-adaptive systems opens up new opportunities to incorporate multipurpose design strategies based on the synergies resulting from the articulation between urban infrastructures and these natural systems, especially in territorial scenarios vulnerable to disasters. This multipurpose condition of design as an inherent quality of landscape architecture points directly to the articulation of risk reduction objectives with other parallel objectives—such as economic and productive development, ecological conservation and natural resource management, recreation and public space—which are no longer conceived as mere externalities of a given project, but are projected as a structuring part of it, from its very formulation and design.

Particularly in the case of the Green Infrastructure Plan in the city of Llanquihue, the performance of these hybrid systems—which synergistically combine the structure and functionality of ecological and anthropic components—favours the provision of regulating ecosystem services related to risk reduction and flood mitigation, as well as facilitating protocols for the evacuation or shelter of affected communities. At the same time, in the absence of public

Fig. 10 Nature-based solutions applied in the Baquedano Wetland Park. Above, slow-flow channels to distribute and oxygenate water, from the rain retention pockets to the wetland. Below, rain retention pockets to collect the street runoff. (Source: Own elaboration)

and green spaces, it can contribute to the development of memorable places for urban living through the integration of ecologies, social programmes and facilities. Medium-sized regional cities such as Llanquihue, due to their proximity to wilderness areas and rural areas that still preserve areas of high cultural and natural value, represent a favourable scenario for integrating the green infrastructure approach into urban planning and management. On the other hand, the presence of rivers, estuaries and wetlands within these cities is a relevant quality, given the potential to integrate these landscape components into the city's green infrastructure system, providing key ecosystem and cultural services for sustainability, resilience and improvement of urban environmental quality. As a result, it is possible to increase the competitiveness of these cities in terms of the attraction of human capital and investment, thus contributing to the decentralisation of the country.

Likewise, in terms of territorial governance, the fact that these cities have only one municipal government—instead of several municipalities as is the case in large cities—allows for better management and development of this type of initiative. Similarly, there are closer communication channels between different public agencies, such as ministerial secretariats and the regional government, which can provide an integrated planning and management framework, through the enhancement of urban ecosystems as part of a green infrastructure network. This proximity helps in the formulation of plans and projects to be socialised through technical and political levels, where it is possible to achieve the effective participation and involvement of key actors linked to decision-making in local government and in the different public sectors.

The Llanquihue Green Infrastructure Plan is an innovative initiative that aims to become a roadmap for integrated urban planning and environmental protection processes, focusing on sustainability and the conservation of urban ecosystems. Instead of constructing green areas, with the high costs that this entails, it proposes the idea of activating landscape areas through nature-based solutions and interventions related to the design of specific elements aimed at enabling, equipping and providing access to these areas, thereby favouring the efficiency of the public investment. On the one hand, the Plan configures a general strategy at the city level, articulated with municipal planning instruments (Municipal Development Plan and Regulatory Plan). On the other hand, it establishes a repertoire of specific interventions linked to portfolios of public investment projects associated with the regional government—such as the National Fund for Regional Development (FNDR), the Regional Fund for Local Initiative (FRIL) and sectoral programmes such as the Urban Improvement Programme (PMU)—through a complete technical dossier that specifies the different nature-based solutions for activating the urban green infrastructure.

In this sense, the Plan is proposed as a complementary and informative platform to feed the current urban planning instruments, as well as contributing to the generation of other toolkits that can promote the integrated management of landscape units, public spaces and urban infrastructure, becoming a potentially replicable model in other cases of regional cities also characterised by the problems related to the conservation and management of their urban ecosystems.

References

Benedict, M. & Mcmahon, E. (2006) Green Infrastructure. Linking Landscapes and Communities. Island Press. Washington DC.

Corner, J. & Bick, A. (2014). The Landscape Imagination. The Collected Essays of James Corner 1990–2010. Princeton Architectural Press. New York.

Fariña, J.M., y Camaño, A. (2012) Humedales costeros de Chile. Aportes científicos a su gestión sustentable. Santiago: Ediciones Universidad Católica de Chile.

FLC—Fundación legado Chile (2020) Plan de Conservación de los Humedales Urbanos de Llanquihue. Ediciones FLC. Puerto Varas, Chile.

FLC—Fundación legado Chile & MAPA UC—Magíster en Arquitectura del Paisaje Pontificia Universidad Católica de Chile (2017) Plan de Acción medioambiental para la ciudad de Llanquihue. Informe Final. Proyecto CORFO Prototipos de Innovación Social Región de Los Lagos.

Holland, M.; Risser, P.; Naiman, R. (1991) Ecotones. The role of landscape boundaries in the management and restoration of changing environments. Routledge, Chapman & Hall.

Hung, Y. & Aquino, G. (2013) Landscape Infrastructure: Case Studies by SWA. Berlin, Boston: Birkhäuser.

Lister, N. M. (2015) Resilience: Designing the New Sustainability The need for an evidence-based approach to adaptivity. En, Topos The international Review of Landscape Architecture and Urban Design, 90: 14–21.

Mathur, A.; Da Cunha, D. (2001) Mississipi Floods: Designing a Shifting Landscape. Yale University Press. USA

Mitsch, W.; Gosselink, J. (2000) The value of wetlands: Importance of scale and landscape setting. Special Issue-The value of wetlands: Landscape and institutional perspectives. Ecological Economics; 35: pp. 24–33.

MMA, Ministerio de Medio Ambiente (2018). Plan Nacional de Protección de Humedales 2018–2022.

Moreno, O. (2020). El paisaje como infraestructura para la adaptabilidad y resiliencia de territorios en riesgo de desastres. El caso de los Parques de Mitigación en localidades de borde costero del centro-sur de Chile post tsunami 2010. Tesis doctoralpara acceder al Grado de Académico de Doctor en Arquitectura y Urbanismo en la Universidad Nacional de La Plata, Argentina.

Moreno, O., & Arizaga, X. (2022). Rehabilitación y conservación de humedales urbanos costeros en contextos de vulnerabilidad hídrica. Caso de estudio: Humedal de Huasco. Cuadernos de Investigación Urbanística, 0(143), 138-158. https://doi.org/10.20868/ciur.2022.143.5002

Moreno, O. (2019). Infraestructura verde urbana. Estrategias de planificación y diseño del paisaje para la resiliencia y adaptabilidad socioecológica de ciudades regionales en Chile. El caso de Llanquihue. En From the South. Global Perspectives on Landscape and Territory. Santiago de Chile: Ediciones UDD, pp. 82–91.

RAMSAR Convention Report (2018). 13th Meeting of the Conference of the Contracting Parties to the Ramsar Convention on Wetlands. COP13, Dubai.

Reed, C. & Lister N.M. (2014). Projective Ecologies. Harvard University Graduate School of Design. Actar.

Rojas, C.; Jorquera, Felipe, & Steiniger, Stefan. (2022). Acceder caminando a los humedales urbanos, una oportunidad de recreación y bienestar. Urbano (Concepción), 25(46), 56–67.

Waldheim, C. (2016) Landscape as Urbanism. Princeton Architectural Press. New York.

Restoration of the Rocuant Andalién Coastal Wetland (Metropolitan Concepción, Chile): A Transdisciplinary Proposal

Octavio Rojas, Carolina Rojas Quezada, Juan Munizaga, Evelyn Soto, Ivonne Rueda, Grace Mallea, Loreto Alvarez, and Francisco de la Barrera

Contents

Abstract

After securing wetland protection, it is critical to implement restoration actions aimed at restoring their ecological integrity by stopping the negative pressures and reestablishing the ecological functions, including ecosystem services and connectivity. This study proposes a set of restoring areas for the Rocuant-Andalién coastal wetland (2000 ha; Metropolitan Concepción, Chile), which has been highly degraded by the rapid development of urban, port, road, and agricultural lands. To identify

O. Rojas (✉) · J. Munizaga · E. Soto
F. de la Barrera
Facultad de Ciencias Ambientales, Centro EULA-Chile, Departamento de Planificación Territorial y Sistemas Urbanos, Universidad de Concepción, Concepción, Chile
e-mail: ocrojas@udec.cl; juanmunizaga@udec.cl; fdelabarrera@udec.cl

C. Rojas Quezada
Instituto de Estudios Urbanos y Territoriales, Centro de Desarrollo Urbano Sustentable CEDEUS, Insti tuto Milenio SECOS, Santiago, Chile
e-mail: carolina.rojas@uc.cl

I. Rueda
Centro de Desarrollo Urbano Sustentable (CEDEUS), Universidad de Concepción, Concepción, Chile

G. Mallea
Facultad de Arquitectura y Urbanismo, Universidad de Chile, Santiago, Chile

L. Alvarez
Conservación de humedales costeros de la zona centro sur de Chile, Ministerio del Medio Ambiente, Santiago, Chile
e-mail: lalvarez@mma.gob.cl

C. Rojas Quezada (ed.), *Urban Wetlands in Latin America*, Sustainable Development Goals Series,
https://doi.org/10.1007/978-3-031-69590-2_9

the priority potential restoration areas (PRA), we adapted the IUCN-ROAM methodology, including: GIS and ecological connectivity analyses, ecosystem services evaluation, and stakeholder consultation at two spatial scales. For the watershed scale, we identified 45 PRA covering 6053 ha and 17 PRA covering 418 ha for the wetland scale, mainly located in the bottom section of the watershed. Stakeholders prioritized five potential restoration areas (PRA) primarily due to their heightened vulnerability to urbanization and potential landfill exposure. Priority restoration areas have the potential to incorporate water-sensitive design measures. The existing robust collaboration between the public sector and citizen organizations will play a crucial and supportive role in implementing restoration actions within these prioritized zones. However, the tensions between the private sector and citizen organizations can weaken and impede the achievement of restoration in critical areas at both scales. Although all sectors were involved in identifying restoration priorities, there is a need to strengthen and increase the relations between all stakeholders to support the recovery of susceptible and complex natural systems like wetlands.

Keywords

Restoration · Transdisciplinary science · Urban wetlands · Chile

1 Introduction

Mediterranean regions are among the most complex and richest in biodiversity on a global scale (Fierro et al. 2019; Blondel et al. 2010). However, the harmful impact of urbanization, farming and ranching practices, and climate change has resulted in a gradual decline in its ecological health (Rojas et al. 2019; Hernández et al. 2016; Moraga et al. 2022). Urban wetlands have been especially vulnerable, with numerous studies in recent years citing reduced surface area and increased fragmentation and deterioration (Ramsar 2018; Davidson et al. 2018; Rojas et al. 2019; Darrah et al. 2019)

This situation is particularly worrisome in developing countries, where ecosystems have little legal protection and fewer financial resources to reduce their impacts. In addition, these regions are expected to be strongly impacted by climate change, and wetlands are a natural tool to reduce hydroclimatic disasters and increase the population's well-being through the ecosystem services they provide (Da Cruz and Lopes 2018; Rojas et al. 2022). However, the quality of these services is heavily reliant on the health of the ecosystems, which multiple stressors have compromised (Arriagada et al. 2019). This is precisely why undertaking ecological restoration actions is crucial to preserve biodiversity and promote human well-being (Law No. 21,202).

Ecological restoration is the process that helps the reestablishment and rational use of an ecosystem that has been degraded, damaged, or destroyed (SER 2004) to recover the environmental conditions that prevailed in a given site and that, for some reason, were negatively affected (Sol Sánchez et al. 2002). Restoration can have different objectives and levels of depth depending on technical and economic capabilities. Thus, it is possible to go from a more restricted scope, such as reclamation, to intermediate rehabilitation approaches to ecological restoration, focusing on transitioning from a degraded ecosystem to a historical reference ecosystem.

In order to effectively preserve and restore natural areas, it is crucial to collaborate with local experts who possess valuable knowledge and experience. We can establish sustainable and impactful conservation strategies by joining forces with professionals and leveraging scientific data (Borrini-Feyerabend 2019) to advance toward a transdisciplinary science (Guimarães et al. 2013). The study of the dynamic between humans and nature has evolved considerably over time. While previous perspectives viewed humans as dangerous to the environment, contemporary thinking recognizes the significance of fostering a harmonious relationship between society and nature (Anderson et al. 2015) and planning accordingly.

According to a recent study by Dayer et al. (2020), the effectiveness of conservation programs is heavily influenced by cultural, political, and economic traditions. While this may be less significant in areas with minimal human intervention or sparse populations, it becomes crucial in urban wetlands that still maintain natural habitats, where society has direct contact with nature (Shao et al. 2021). As such, involving local professionals with expertise in ecosystem conservation programs is imperative since it enables the prompt execution of actions to safeguard or restore wetlands, ensuring their preservation for future generations.

In Chile, wetlands have been facing challenges as a result of urbanization. However, there is hope for their restoration thanks to the Strategic Plan for Biological Diversity 2011–2020, a framework developed within the Convention on Biological Diversity. This plan focuses on water-linked ecosystems, contributing to both climate change adaptation and mitigation efforts (MMA 2016). Additionally, the Urban Wetlands Law (Law No. 21,202) has been approved and implemented, offering an opportunity to protect historically degraded and threatened ecosystems (Rojas et al. 2022; De Urresti 2020). The Ministry of Environment is responsible for ensuring its implementation. It is important to note that ecological restoration alone is not enough; protection efforts are also necessary to ensure that the landscape reaches its optimal ecological condition and provides essential ecosystem services for people's well-being.

In the present study, the Rocuant-Andalién coastal wetland system and its associated Andalién river basin in the Mediterranean region of Chile were selected as a case study. The Andalién watershed has a degraded and eroded landscape (Jaque 2010). Urbanization has negatively impacted the water system in its lower section due to environmental factors (Vidal and Romero 2010; Rojas et al. 2019; Rojas et al. 2017a, b), evidenced in its fluvial system. Therefore, Espinosa et al. (2015) devised exercises to enhance the connection between the city and the Andalién River. Building upon this, Espinosa et al. (2020) conducted a comprehensive analysis of fluvial restoration measures, encompassing both geomorphological and ecological actions. They recommended strategies to reinstate the fluvial system, ensuring water-sensitive urban design. This is a crucial consideration given the region's susceptibility to natural disasters like river floods, tsunamis, and urban floods).

This particular research is a component of the GEF Coastal Wetlands initiative, which has designated the wetland system as a pilot ecosystem. Unfortunately, the wetlands have experienced a marked decrease in surface area due to urbanization, with the remaining areas becoming degraded and no longer ecologically connected (Rojas et al. 2013b, 2017a, b, 2019). The wetland system comprises two distinct types of wetlands: those that have been filled or extensively modified and those that remain but need restoration due to their degraded state.

The wetland system is under great strain and is deteriorating rapidly due to the swift growth of urban zones, ports, agricultural areas, and roads. Additionally, nearby communities have been impacted by natural disasters such as floods, flooding, and tsunamis. The GEF Project's goal is to implement Law 21.202 and restore areas to unify the natural and semi-natural network of the landscape. In order to enhance the restoration of hydrological functions, this research proposes potential restoration areas and measures. The methodology employed follows a transdisciplinary approach adapted from the IUCN-ROAM methodology.

2 Study Area

The research area is situated in Concepción Metropolitano (36°S), the second largest urban settlement in Chile (more than 1 million inhabitants), is located in an area where tectonic processes and marine cycles modeled palaeobays that conditioned the existence of coastal wetlands and relicts now called urban wetlands, surrounded by the urbanization process (Isla et al. 2012). The entire wetland system includes many other little wetlands such as Vasco Da Gama,

Paicavi, or Tucapel Bajo, reaching 2000 ha, and it is located in the Andalién river basin (71,000 ha), being part of this metropolitan area. The basin's plains and low slopes of <0.5° are home to numerous people, but this poses a challenge for preserving the ecosystem, given the impact of human activities like urbanization and industrialization. Additionally, as Arriagada et al. (2019) and Rojas et al. (2013a) noted, this region is subject to multiple stressors.

The development of this area's urban and economic landscape has always been heavily influenced by water. Unfortunately, this has come at the expense of vital ecosystems such as lagoons, wetlands, and native vegetation, as Smith and Romero (2009) and Rojas et al. (2013b) noted. Prior to the 2000s, it was already apparent that the wetlands had shrunk by 40% since 1975, with their place being taken by scattered vegetation and eventually urbanization (Smith and Romero 2009). Since 2000, an additional 500 ha of wetlands have been lost due to changes in land use (Rojas et al. 2013c). The impact of this loss extends beyond surface area, as it has also severely impacted the biodiversity and ecosystem services these areas provide, as highlighted by Villagrán et al. (2006) and Sepúlveda et al. (2012).

This study considers two scales: the watershed and the wetland system, as seen in Fig. 1.

3 Methodology

Identifying potential areas for restoration (PRA) involved a comprehensive approach that engaged various stakeholders. The Local Technical Committee (LTC) of the GEF coastal wetlands project was assembled to facilitate this process, with a total of 56 stakeholders taking part. The participants included 42% from the public sector, 29% from civil society, 18% from the private sector, and 11% from academia. Despite the challenges posed by COVID-19, five interactive online workshops were conducted, with an average attendance of 41 participants.

Experts from the International Union for Conservation of Nature (IUCN) and the World Resources Institute (WRI) created the Restoration Opportunity Assessment Methodology (ROAM) in 2014 to identify areas of the wetland and watershed system that needed restoration. In order to apply ROAM to this specific situation, various stages were employed, including preparation and planning, data collection and analysis, and recommendations based on results. The tools used included prioritizing restoration interventions among stakeholders, identifying areas that could benefit from restoration, and selecting the most effective restoration methods for each area. The process began with initial steps such as:

Defining the Problem and Restoration Objectives The Urban Wetlands Law framework was considered, which states implicitly not only to protect the declared urban wetlands but also to define minimum sustainability criteria to safeguard their ecological characteristics and functioning, as well as to maintain the hydrological regime, both surface, and subsurface. On this basis, the restoration objective was defined as: "To recover the hydro-ecological functionality of the Rocuant-Andalién-Vasco Da Gama-Paicaví-Tucapel Bajo Wetlands System to ensure its rational use through the conservation of biodiversity (especially nesting birds and aquatic fauna) and the assurance of the provision of essential ecosystem services, through an approach at the scale of the contributing hydrological basin." The priority ecosystem services to restore were:

- Protection against river floods and waterlogging
- Maintaining Populations and Habitats (refuge)
- Flora and Fauna Appreciation
- Climate regulation and local air quality
- Opportunities to enhance tourism (hiking, walking, photography).

For the purpose of restoration, two reference ecosystems have been identified within the same eco-region. The first ecosystem is "Tubul-Raqui" marsh, located in a coastal basin affected by high tides up to 6 km. It is surrounded by forest and scrubland uses and contains wetland species,

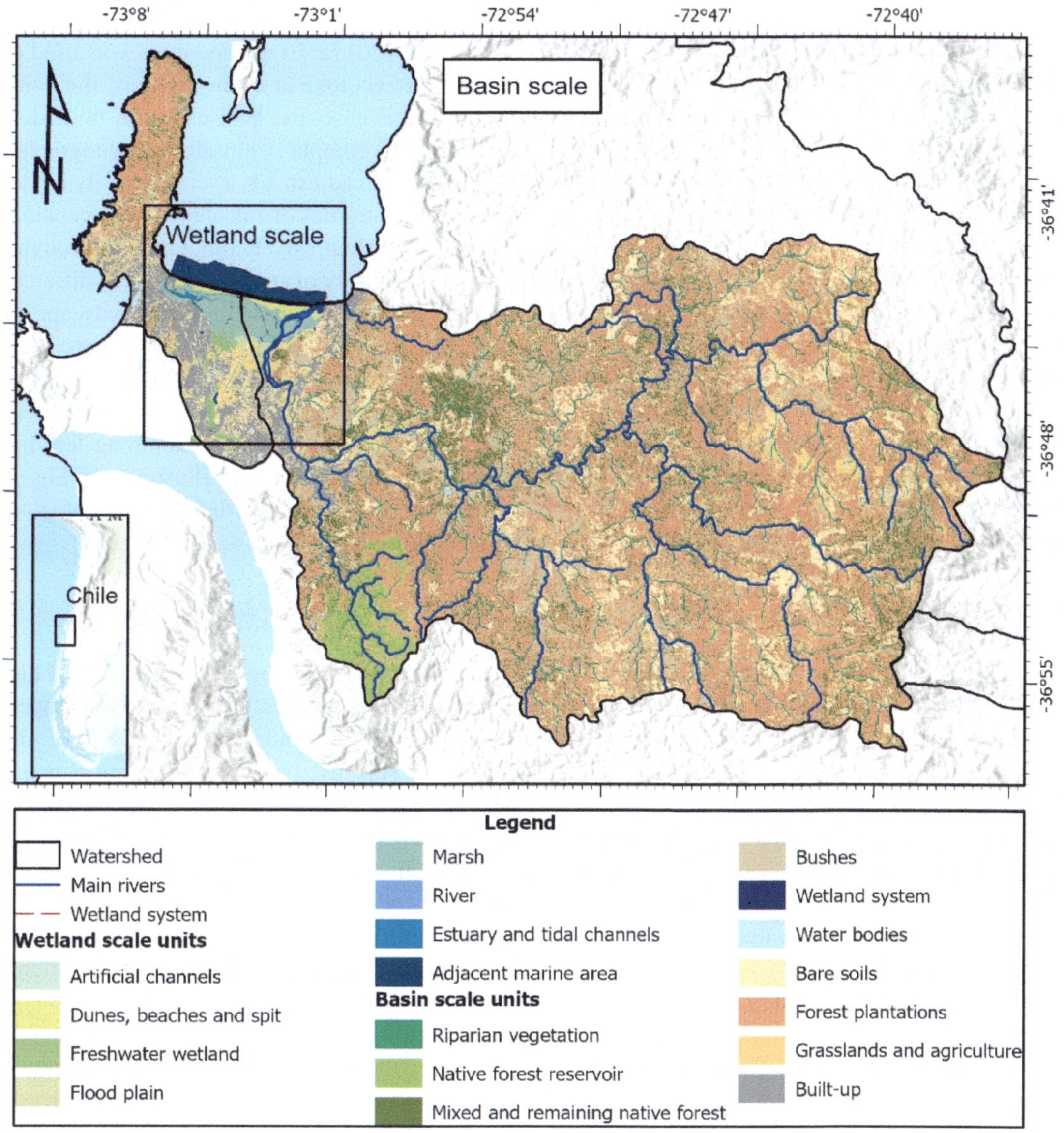

Fig. 1 Study area. Stratification of the area at wetland and watershed scale. (Source: Own elaboration)

including *Spartina densiflora*. This ecosystem is crucial for the preservation of threatened bird species. The second ecosystem is the "Laguna Verde" marsh wetland, situated in the Nature Sanctuary "Peninsula de Hualpén." A natural matrix and mixed forests surround it, and it is home to halophyte plants, wetland species such as *Juncus procerus*, and a diverse range of insects.

Stakeholder Involvement The GEF wetlands initiative placed significant emphasis on identifying and involving stakeholders in the region, as determined by the Local Technical Committee (LTC) overseeing the study. Furthermore, the initiative conducted thorough analyses of stakeholder networks and relationships, surpassing the LTC's original scope.

Scope and Product Definition The main product consisted of the preparation of a map with the priority areas to be restored (traffic light type), which was accompanied by an explanatory document in index card format with a set of priority restoration initiatives/interventions, for each vulnerable area with restoration priority.

Area Stratification The study was conducted on a large-scale involving basins and wetland systems (refer to Fig. 1). Stratifications were created for this purpose. The wetland area included temporary and permanent wetlands, beaches, dunes, associated vegetation, adjacent marine zones, and artificial wetlands. For the watershed level, the following were considered: riverside vegetation, native and mixed forests, bushes, water bodies, forest plantations, grasslands, agriculture, and urbanization.

Two complementary analyses were developed: ecological connectivity of the ecosystems of interest and evaluation of ecosystem services.

Participation of Local Stakeholders
In order to gather valuable insights from key stakeholders, we carried out two exercises. Firstly, we examined various options for restoring each unit at both the watershed and wetland system levels (as illustrated in Fig. 1). To kick-start the conversation and generate ideas, we shared an initial catalog (Annex 1) of restoration actions linked to each unit. The previously mentioned enabled us to pinpoint areas that required urgent attention and determine the most effective course of action.

During the second activity, stakeholders prioritize restoring the wetland units, selecting areas in 3 priority levels. The prioritization process was determined by the number of mentions each unit received, with areas receiving a maximum weighting (value = 1) if they were considered a first priority, a medium weighting (value = 0.5) for second priority, and a low weighting (value = 0.1) for third priority. As a result, six priority levels were established, which were then plotted on a map, along with the opportunities and units that required restoration.

Ecological Connectivity Index Calculation
Rojas et al. (2017a, b) methodology was used to evaluate the ecological connectivity of the landscape. In this case, the landscape can be understood as a complex mosaic of geographic elements that adjust to a complex dynamic, depicted by the cities of Talcahuano, Concepción, Penco, and Hualpén. Ecological connectivity refers to the ability of species to traverse different land covers or habitats within the landscape. It quantifies the obstacles that organisms may encounter when moving between natural habitats due to fragmentation and other landscape features. Our analysis involved examining ten distinct landscape coverages, as illustrated in Fig. 1.

The following equation defines the ecological connectivity index (ICE):

$$\mathrm{ICE} = \sum_{i=1}^{n} \mathrm{CD}_i$$

The ICE is defined by a cost function of distances between patches of land use coverages; therefore, all the landscape coverages of the basins are selected because they require maintaining and/or improving their connectivity for the upkeep of the wetland system. These are reserve native forests, bushes, grassland and agriculture, wetlands, riverside vegetation, mixed and remaining native forests, water bodies, bare and sparsely vegetated soils, forest plantations, and urban land.

A friction or impedance matrix was created to determine the barrier effect of the ten natural ground covers in the basins. The resulting friction matrix was used as an input for the Cost Distance (CD) function, which calculates the shortest weighted distance from each cell on a raster surface. The friction matrix was developed using both an affinity matrix and an impedance matrix.

The affinity matrix (Table 1) gauges the potential for species to prosper in different habitats and reveals resemblances between them. This research measures affinity using a scale that ranges from no affinity (0.001) to low (0.01–0.1), moderate (0.25–0.5), and high (1), according to the catalog of species.

Using the affinity matrix (Table 2), we could determine the habitat's resistance or impedance

Table 1 Affinity matrix

N°	Coverage		Affinity	1	2	3	4	5	6	7	8	9	10
1	Reserve native forest	M	**1**	**1**	0,5	0,001	0,25	0,001	0,25	0,25	0,1	0,001	0,5
2	Remaining mixed and native forests	A T	**2**	0,5	**1**	0,1	0,5	0,5	0,25	0,25	0,1	0,1	0,25
3	Water bodies	R	**3**	0,001	0,1	**1**	0,1	0,1	0,1	0,5	0,001	0,001	0,5
4	Bushes	I	**4**	0,25	0,5	0,1	**1**	0,5	0,5	0,25	0,25	0,25	0,25
5	Forestry plantations	X	**5**	0,001	0,5	0,1	0,5	**1**	0,25	0,25	0,25	0,25	0,25
6	Meadows and agriculture		**6**	0,25	0,25	0,1	0,5	0,25	**1**	0,5	0,25	0,25	0,5
7	Wetlands System		**7**	0,25	0,25	0,5	0,25	0,25	0,5	**1**	0,25	0,25	0,5
8	Bare soils and sparse vegetation		**8**	0,1	0,1	0,001	0,25	0,25	0,25	0,25	**1**	0,1	0,25
9	Urban		**9**	0,001	0,1	0,001	0,25	0,25	0,25	0,25	0,1	**1**	0,25
10	Riverside Vegetation		**10**	0,5	0,25	0,5	0,25	0,25	0,5	0,5	0,25	0,25	**1**

Source: Own elaboration

Table 2 Impedance matrix

N°	Coverage		Impedance	1	2	3	4	5	6	7	8	9	10
1	Reserve native forest	M	**1**	**1**	2	1000	4	1000	4	4	10	1000	2
2	Remaining mixed and native forests	A	**2**	2	**1**	10	2	2	4	4	10	10	4
3	Water bodies	T	**3**	1000	10	**1**	10	10	10	2	1000	1000	2
4	Bushes	R	**4**	4	2	10	**1**	2	2	4	4	4	4
5	Forestry plantations	I	**5**	1000	2	10	2	**1**	4	4	4	4	4
6	Meadows and agriculture	X	**6**	4	4	10	2	4	**1**	2	4	4	2
7	Wetlands system		**7**	4	4	2	4	4	2	**1**	4	4	2
8	Bare soils and sparse vegetation		**8**	10	10	1000	4	4	4	4	**1**	10	4
9	Urban		**9**	1000	10	1000	4	4	4	4	10	**1**	4
10	Riverside vegetation		**10**	2	4	2	4	4	2	2	4	4	**1**

Source: Own elaboration

to the movement of organisms across the landscape. This is shown as the inverse of the affinity (Pino and Marull 2012), which means it is obtained by dividing one by the affinity. As a result, the impedance values are 1, 2, 4, 10, 100, and 1000.

For example, urban areas and transportation infrastructures can create barriers that cause fragmentation, negatively impacting the size, number, shape, and connectivity of natural and semi-natural landscapes. This fragmentation can also disrupt various ecological processes (Forman 1995).

In order to create a final map that shows the ease or difficulty for organisms to move between ecosystems, both matrices are used in the Cost Distance function. This results in continuous distance or accumulated cost maps. These maps show how easily, or complex organisms can move between different ecosystems. The natural coverages are then summed up, taking into account the ones that impede movement. These coverages act as barriers, limiting movement and causing greater friction. The cost values are classified into five different levels of ecological connectivity, ranging from "Very High" to "Very Low." This classification is done using the natural break classification.

Ecosystem Services Evaluation

Using the Burkhard et al. (2009, 2013) methodology, we could value areas of significant importance in terms of ecosystem services. This approach involves a comprehensive biophysical evaluation system (non-monetary) that draws upon surveys conducted by both academic and local experts. Our assessment divided the Wetland System into eight distinct geomorphological-ecological units. In comparison, the watershed was segmented into nine geo-biophysical units based on specific criteria such as land use, ecosystem, and geomorphology. Our analysis factored in the optimal condition of each ecosystem to estimate its capacity to provide valuable ecosystem services.

Multi-criteria Evaluation

A GIS Multi-criteria Evaluation was applied using map algebra. Its objective was to identify areas to restore degraded areas to improve ecological connectivity and biodiversity and the ecosystem services it provides. Three criteria were used, each of which was assigned a weight and whose sum should total 1.

Criterion 1 – Stakeholder Perception – 0.5 The units of the wetland system and the basin were prioritized for restoration based on stakeholder perception, with a ranking system of 1 to 5. Units with a rank of 5 have the highest priority, while those with a rank of 1 have low priority (Table 3).

Criterion 2 – Ecological Connectivity – 0.25 The connectivity map above measures the connection level between different habitats. The highest connectivity values with better conditions will have a value of 1, and low connectivity will have a value of 5 (Table 4).

Criterion 3 – Ecosystem Services – 0.25 The map shows the direct and indirect benefits people would gain from ecosystem services as a result of a restoration process (Table 4).

In summary, restoration areas should (1) have a high level of support from the Local Technical Committee, (2) have low connectivity to increase the structural and functional connectivity of the landscape, and (3) safeguard ecosystem services for the population's well-being. In other words, this process will positively value those areas with a high perception of restoration by local stakeholders, low ecosystem connectivity, and good capacity to provide ecosystem services under a restoration process.

Final Review with Local Stakeholders

A final workshop, known as N°5, was organized to review the outcomes of restoration efforts that were identified using weighted maps. The workshop employed a participatory approach on the Zoom platform, with participants split into five groups comprising a moderator and 9–10 members. During the initial session, attendees discussed restoration areas on a digital Miro

Table 3 Restoration priorities as perceived by stakeholders

Criterion	Score	Descriptor	Units
Priority of restoration for watershed and wetland units	5	Very high priority	Marshy wetlands
	4	High priority	Marshes, estuaries, and canals
	3	Priority	Urbanized areas, dunes, beaches, uncovered soils
	2	Regular	Headwaters, Andalién River, riverside vegetation, creeks
	1	Low priority	Concentration areas, forestry plantations, artificial canals, bushes, agricultural land, mixed forest, native forest
	0	No priority	Other areas

Source: Own elaboration

Table 4 Restoration needs according to ecological connectivity and ecosystem services

Criterion	Score	Descriptor	Criterion	Score	Descriptor
Restoration needs by ecological connectivity	1	Very high connectivity	Restoration needs based on ecosystem services.	5	Very high restoration capacity
	2	High connectivity		4	High restoration capacity
	3	High connectivity		3	Regular capacity
	4	Regular connectivity		2	Low capacity
	5	Low connectivity		1	Very low capacity
	0	No connectivity		0	No capacity

Source: Own elaboration

platform (Fig. 2), providing feedback that was documented using sticky notes and later analyzed by the team. Based on the feedback, restoration areas were revised. The second session involved prioritizing the restoration areas, with each team selecting five areas as a top priority, five as a secondary priority, and ten as a third priority, culminating in the final prioritization. The viewpoints of territorial stakeholders and local government were also considered to validate the findings.

The restoration process had relevant stakeholders identified, and a matrix of their linkages was prepared. The LTC was tasked with completing the required data to identify the connections between the actors. The workshop-provided relationships were also considered. A circular sociogram was created using yEd Graph Editor software to visualize the relationships.

4 Results

(a) *Identifying Restoration Priorities According to Territorial Stakeholders*

Table 5 shows the units and the number of mentions by each stakeholder team. Based on the weighted value, 6 priority levels have been defined on a scale of 0–5. The areas of marsh wetlands, marshes, and estuary/natural channel systems have been identified as having a high and very high priority for restoration based on the opinions of the territorial stakeholders.

When it comes to prioritizing restoration efforts, the most pressing actions involve removing debris and micro-landfills that threaten biodiversity, as well as restoring water flow. Renaturalizing and converting degraded soil use are considered a medium priority, as is limiting access to people and automobiles to allow natural

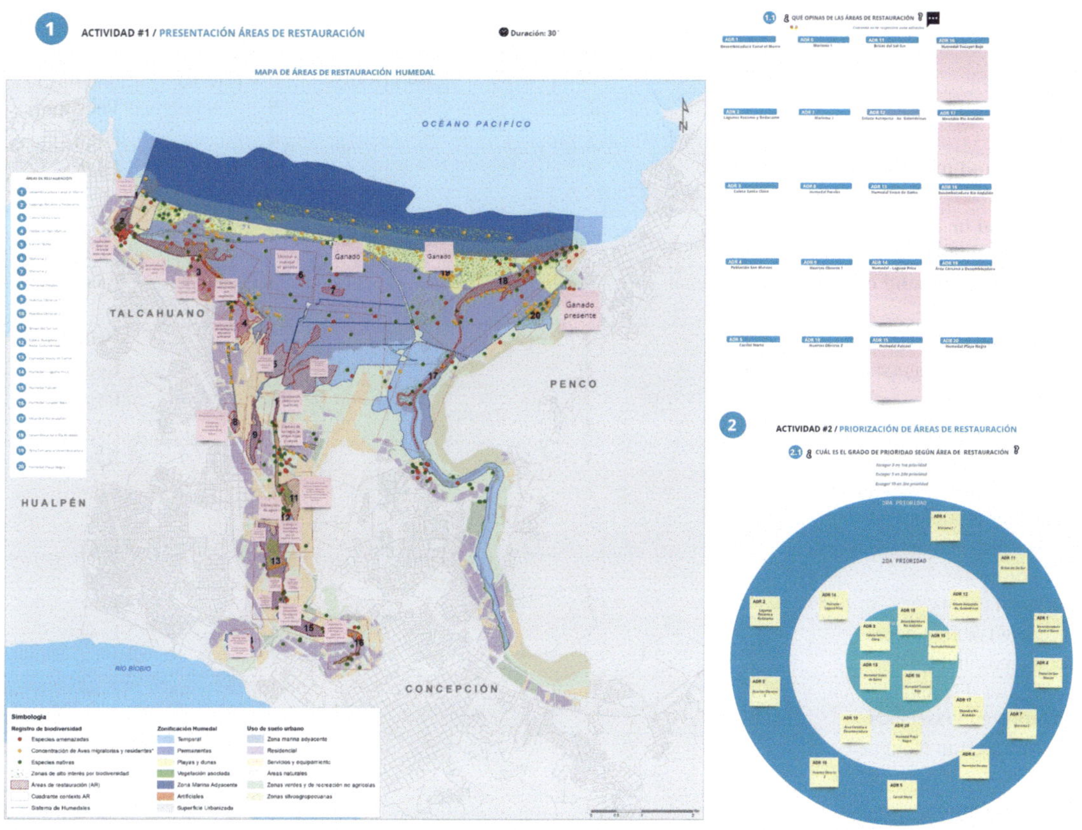

Fig. 2 Example of a panel work on the Miro platform for the two activities of workshop 5. (Source: Own elaboration)

processes to occur. Table 6 provides a comprehensive list of these priorities.

(b) *Ecological Connectivity Index*

Impedance represented the friction of the movement. It was observed that wetlands require water bodies and riparian vegetation to maintain and connect better in the landscape. Regarding connectivity, the watershed landscape presented high connectivity in 48% of its territory, mainly due to the abundance of forestry plantations (Fig. 3a). Within this landscape, a notable 37% of highly connected regions are found in areas such as forests, bushes, and marshes that can link up with bodies of water. The upper section of the basin is under the most substantial strain, with just 7.5% of the region consisting of mixed forests and remnants of native forests that have regular connectivity. Conditions in the lower part of the basin are the most challenging, with meager (5.15%) and insufficient connectivity (1.82%) found in mixed forests, native forests, and wetland systems like Vasco de Gama.

(c) *Ecosystem Services*

In total, 16 ecosystem services surveys were collected from both the basin and wetland levels. The majority of those who participated were from civil society and academia, accounting for 50% of respondents. Interestingly, 57% of participants did not reside within the basin yet utilized its services regularly (86%). Figure 3b illustrates the potential for ecosystem services adjustment to support restoration efforts, with five different capacities ranging from very low to very high. The wetland system, riparian vegetation, shrublands, and native forests were all iden-

Table 5 Units weighting based on input at stakeholder's workshop

Units	Mentions in 1st priority	Mentions in 2nd priority	Mentions in 3rd priority	Mentions in 3rd priority	Priority ranking
Marshy wetland	4	1		**5**	**Very high**
Marsh	2	3		**4**	**High**
Estuary and natural channel system	2	4		**4**	**High**
Urbanized or industrial area on wetland edge	2	1	2	**3**	**Priority**
Dunes, beaches, and coastal bar	1	4		**3**	**Priority**
Floodplains	1	4		**3**	**Priority**
Headwaters of watersheds	1	1	1	**2**	Regular
Andalién River	1	2	1	**2**	Regular
Riverside vegetation		4		**2**	Regular
High slope streams		3	1	**2**	Regular
Species concentration areas (e.g., migratory and nesting)	0	2		**1**	Low
Active and unmanaged forestry plantations	1		4	**1**	Low
Artificial canals		2	3	**1**	Low
Degraded or fragmented scrublands and forests		2	1	**1**	Low
Agricultural land and anthropized grasslands		1	3	**1**	Low
Native forest in Nonguén Reserve			5	**1**	Low
Lagoon edges			1	**0**	No priority

Note: Priority value = n mentions 1st priority * 1.0 + n mentions 2nd priority * 0.5 + n mentions 3rd priority * 0.1
Source: Own elaboration

Table 6 Prioritized restoration actions

Priority	Potential restoration actions
Very high priority	Removal of debris and micro-landfills that threaten biodiversity
High priority	Recovery of natural water flows
Priority	Renaturalization and reconversion of degraded soil use, entrances closing to facilitate natural processes
Regular	Native species enrichment, elimination of exotic species, implementation of efficient irrigation systems, and soil improvement for agricultural purposes
Low priority	Development of green spaces and urban trees, Naturalization of artificial channels, Reinforcement of mitigation measures on the Interportuaria highway, Creation of firebreaks, and removal of dead or dry vegetation fuel
No priority	Other actions

tified as having high to very high capacities for providing ecosystem services.

(d) *Areas to be Restored Following the Application of Multi-criteria Assessment*

The result of the multi-criteria evaluation was a prioritization at the basin level that integrates the degree of detail of the scale of the wetland system. The extension of the basin analyzed had a value ranging from 0 to 5. Out of this scale, 73% had a value equal to or less than 2, while 23% had a value between 2 and 3. Only 4% had a value greater than 3.

Subsequently, the threshold above 1% was selected as areas to be restored, which on a scale

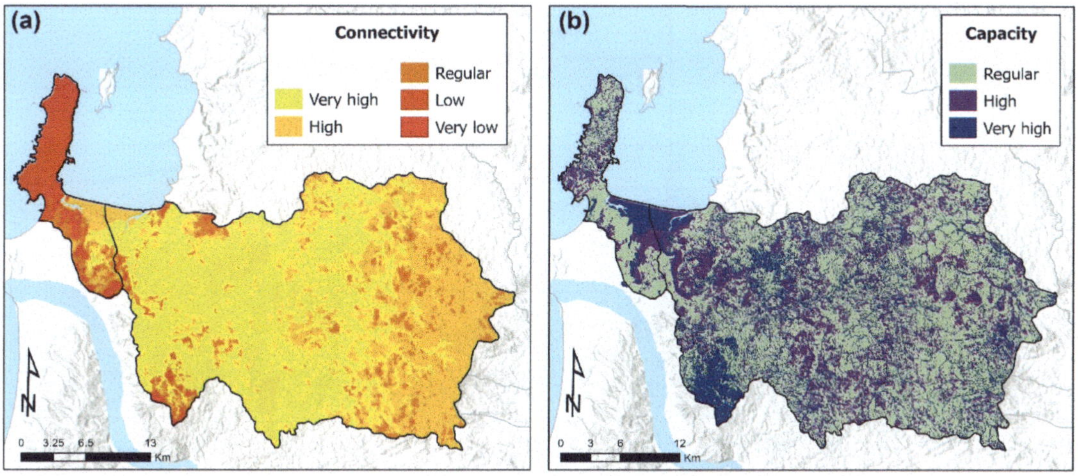

Fig. 3 (**a**) Ecological connectivity; (**b**) Capacity of providing ecosystem services. (Source: Own elaboration)

of 0–5 is represented by the areas that obtained a value higher than 3.75. These areas were identified as a high and very high priority for restoration at the watershed level and cover a total area of 418 ha. All of these areas were located in the Wetland System and its immediate surroundings (Fig. 4a).

It should be noted that the outcome of the multi-criteria evaluation prioritizes the implementation of restoration actions in and around the Wetland System. Suppose we utilized an average restoration value as a systematic criterion to identify other areas in the basin, we would necessitate the restoration of the entire Wetland System and a vast surrounding area, totaling 6053 ha (PRA).

In addition to the areas of interest for biodiversity that were identified earlier in the Wetlands System and watershed, there are also complementary areas. One of these areas is the Nonguén National Park, which is more significant in the watershed. The protected area is renowned for its native forests and a remarkable variety of species. Nonetheless, it is vulnerable to the effects of upper watershed activities, including rock erosion, landslides, and substandard forestry plantations that exacerbate erosion. Municipalities (local governments) have taken the lead in promoting participatory and educational programs aimed at mitigating these challenges, which can complement ongoing efforts to revive the Wetland System further downstream.

Finally, implementing Rueda's (2020) green belt proposal can further elevate the ecological planning vision for the metropolitan area of Concepción. Creating a continuous ecosystem near the Wetlands System will enable the realization of Fig. 5b at the watershed scale. This will serve to foster sustainable development and safeguard the environment for generations to come.

During Workshop No. 5 (Fig. 2), all areas with high and very high restoration priority were discussed. Each polygon was analyzed in the context of its surrounding quadrant, which includes elements outside of the polygon determined through multi-criteria evaluation. To define these quadrants, we gathered information about land use and cover, threatened species, migratory bird concentration points, areas of high biodiversity interest, wetland typologies, and boundaries. As a result, 20 polygons were identified for restoration along with their corresponding quadrants and information (Fig. 6).

(e) *Final Prioritization of Areas to be Restored*

Figure 7 shows the final prioritized restoration areas in three levels. The sectors with the highest priority correspond mainly to marsh wetlands (Carriel Norte Wetland, Vasco de Gama or San

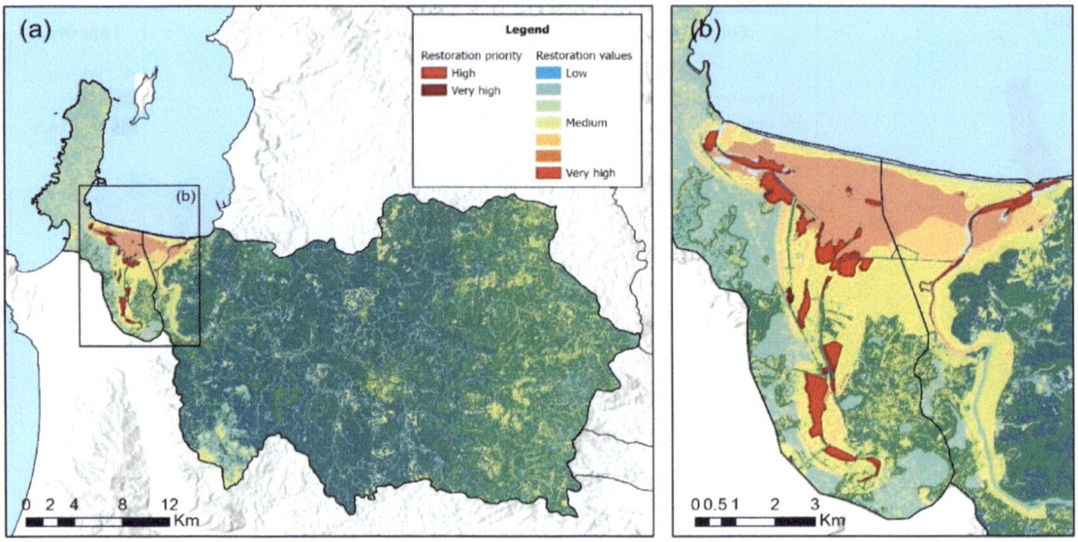

Fig. 4 (**a**) Identification of areas with high and very high restoration priority at basin level. (**b**) Detail in the Wetland System of identifying areas with high and very high restoration priority. (Source: Own elaboration)

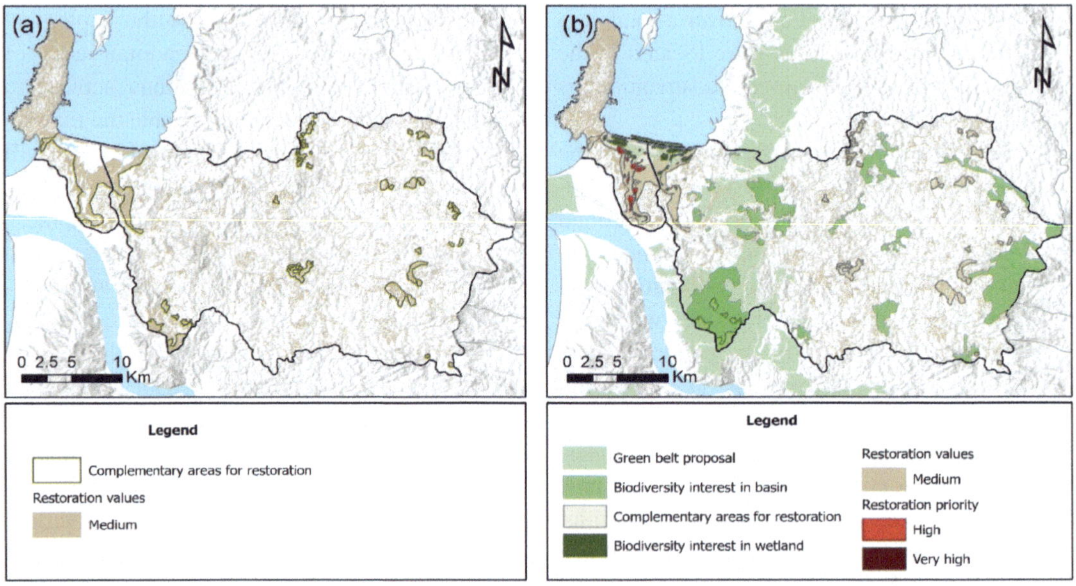

Fig. 5 (**a**) Scenario of incorporating more areas for restoration at the watershed level based on average restoration values for the entire study area. (**b**) Integration of areas of biodiversity interest at the Wetland System and watershed level, complementary areas for restoration, and areas with high and very high restoration priority at the watershed level and proposed green belt for the Metropolitan Area of Concepción. (Source: Own elaboration)

Andrés/Chimalfe Wetland, Paicaví-Tucapel Bajo Wetland) pressured by the advance of urbanization and areas adjacent to river systems (meander and mouth of Andalién River) that canalizations have intervened. It is important to emphasize that distinct sheets were crafted for every region, complete with pertinent information such as priority ranking, present land coverage, ecosystem benefits, territorial planning instrument zoning, and possible hazards. Furthermore, each sheet

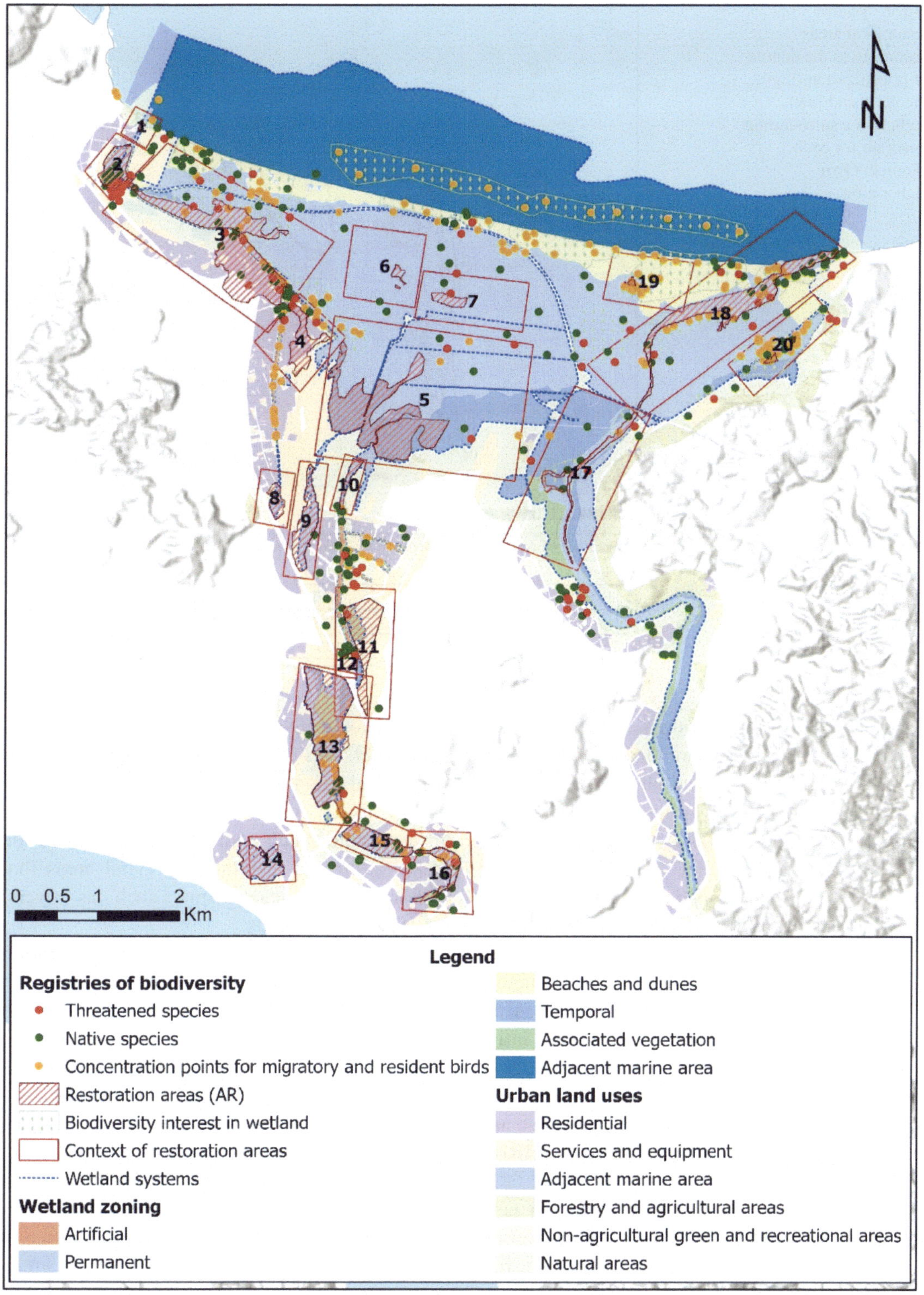

Fig. 6 Initial restoration areas identified after the multi-criteria assessment, including a surrounding quadrant for each and contextual information (biodiversity records, wetland zoning, and land uses). (Source: Own elaboration)

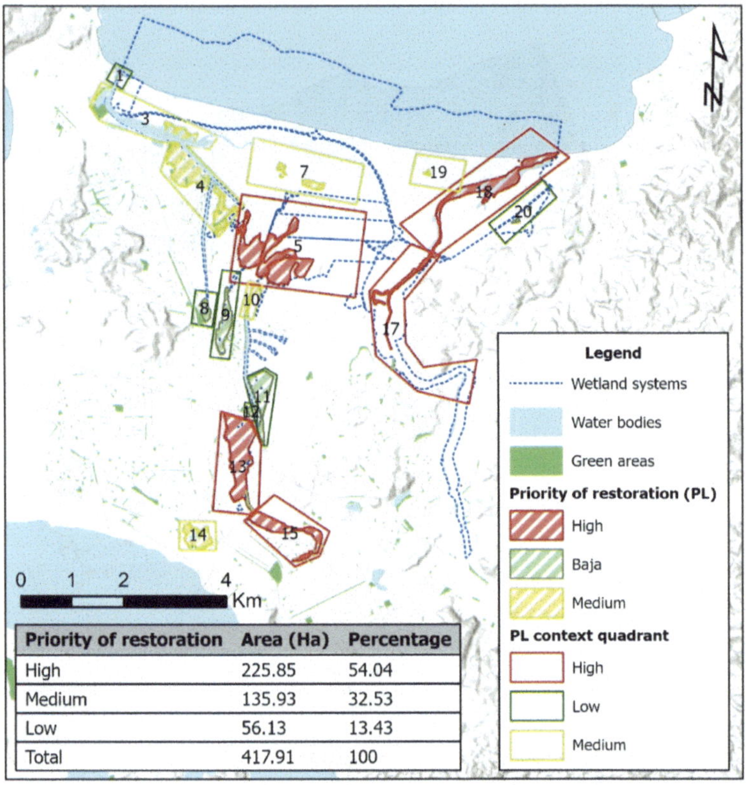

Fig. 7 Twenty restoration areas according to the degree of prioritization (high, medium, and low), including a surrounding quadrant for each. (Source: Own elaboration)

incorporates a restoration map, suggested remedial measures by the research team, and the individuals taking part in the initiative.

To tackle the issue of climate change and bolster our ability to withstand its effects, a promising approach is to adopt green infrastructure incorporating water-sensitive design in critical locations. By carefully selecting relevant criteria, we can develop effective strategies and practical recommendations to guide infrastructure initiatives. These priority areas can then be seamlessly integrated into a broader green infrastructure network that spans seven communes of Concepción, utilizing an innovative watershed approach that prioritizes ecological planning and responsible territory management (AMRBB 2021).

Ecological planning in watersheds can be a complex task, particularly when it comes to coordinating with territorial planning instruments that focus solely on land use and overlook landscape planning. Rueda's (2020) work offers a solution in the form of green infrastructure, which can enhance connectivity between natural areas at the basin level. This approach would facilitate metropolitan and communal level initiatives, ultimately leading to a comprehensive strategy for green infrastructure. Such a strategy would aid in the preservation and restoration of vital areas that sustain diverse ecosystems, such as rivers, lagoons, and wetlands.

Indeed, investments located in the restoration areas must be feasible through intervention with sustainable urban drainage and landscape architecture projects, especially on the edges of the transition to consolidated urban areas. In this sense, one of the limitations of the proposed methodology is that hydrological connectivity needs to be evaluated.

The proposal aligns with the views expressed by local stakeholders. Toledo (2020) highlights the importance of recognizing the significance of the water system and improving urban green spaces in the metropolitan area of Concepción. The Strategic Action Plan aims to transform Greater Concepción into a "Sustainable and

Resilient Metropolis" by 2030. As part of this effort, the "Water Route" has been established to identify and enhance the value of the water system of the AMC. Its goal is to protect and promote wetlands as urban spaces that have ecological and social value.

Relevant Stakeholders for Restoration Areas

The sociogram in Fig. 8 presents the various actors involved in restoration efforts, with a total of 38 individuals hailing from the public sector, civil society, academia, and private sector. The data reveals that the municipalities of Concepción and Penco are actively engaged and connected with other public sector organizations. Meanwhile, civil society groups display strong interconnectivity with each other, as well as links to both public and private sectors. The private sector is positioned on the circle's periphery, with only small white representations, except for Forestal Arauco, which is represented in a light blue color and has more significant connections. Notably, three out of four academic representatives and two private sector representatives are

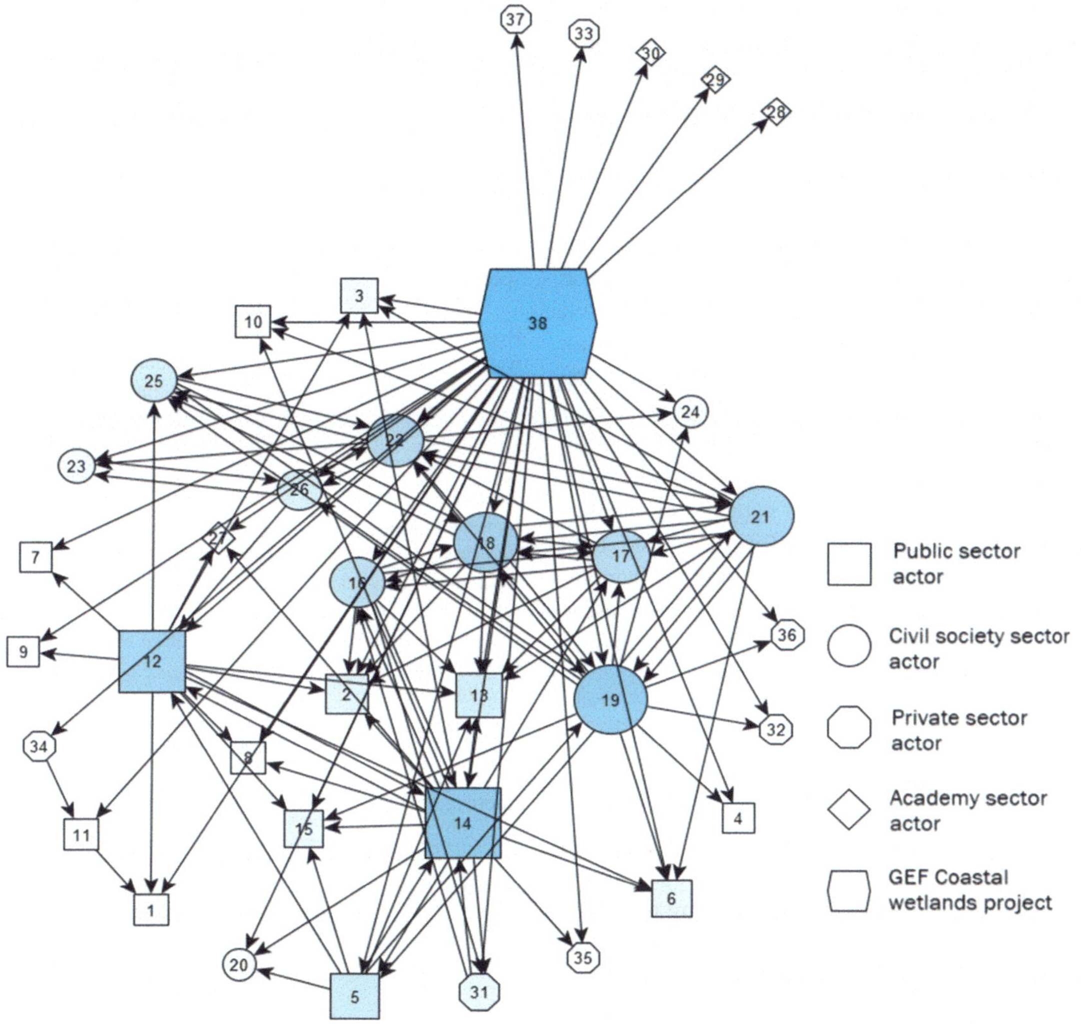

Fig. 8 Links sociogram of the Local Technical Committee*Note*: The figures' size corresponds to the number of identified links, with larger sizes indicating more links. The intensity of colors indicates the number of relationships. Actors represented in colors closer to the system's color have fewer links associated with other members and the wetland system. (Source: Own Elaboration)

situated outside of the central circle, signaling a dearth of connections with other organizations.

For restoration projects, detailed sheets are prepared for each area to be restored. These sheets carefully outline the actors involved, taking into account their specific roles and geographic areas of influence. Strengthening the partnerships between civil society and the private sector is vital to securing community support and the backing of landowners for proposed restoration efforts. By fostering mutually beneficial relationships within ecosystems, we can ensure the success of these crucial initiatives.

In 2017, Navarro studied wetlands protection in the metropolitan area of Concepcion and discovered that urban governance poses both opportunities and challenges. The research revealed that although territorial planning measures limit land use on wetlands, landowners often prioritize economic gain over preservation. This is attributed to public-private partnerships that restrict urban development within areas owned by these actors, neglecting conservation efforts. Moreover, decision-making does not mandate citizen involvement, resulting in exclusion from the process.

In fact, urban wetlands located near cities can serve as natural classrooms for education, research, and as examples of nature-society interaction. Citizen science platforms, guided field trips, and natural areas for studying human-nature interactions are essential for increasing knowledge about the conservation of natural areas (Estévez et al. 2010).

5 Conclusion

By identifying priority restoration areas (PRA), it is possible to enhance the provision of ecosystem services, particularly for those residing near wetlands. PRAs can be utilized for water-sensitive design actions or incorporated into green infrastructure systems, improving water storage areas' functionality. This contributes to urban resilience against floods or other water-related disasters.

Although all sectors have collaborated to identify restoration priorities, it is crucial to strengthen further and enhance the relationships among the public sector, private industries, and citizen organizations. By working together, they can facilitate the recovery of delicate and vulnerable natural systems, such as wetlands, and promote a mutually beneficial relationship between human communities and the environment.

Annex 1 Restoration Actions List

Type of action	Description
Renaturalization and reconversion of degraded soil use	Transform degraded or developed land into a natural area by replacing species and modifying land use
Native species enrichment	Planting more native species in areas with existing vegetation
Elimination of exotic species	Mechanical removal of invasive plants such as blackberry or aromo
Recovery of natural water flows	Physical measures to allow flooding of undeveloped areas (e.g., channel breaching, barriers, and fills). Removal of artificial channels
Closing accesses to facilitate natural processes	Physical measures to block access and destruction of roads
Removal of debris and micro-dump sites	Threatening biodiversity
Creation of firebreaks and removal of dead or dry plant fuels	Elimination of forest fire threats
Implementation of efficient irrigation systems and soil improvement for agricultural purposes	Elimination of canalizations and incorporation of drip irrigation systems
Reforestation for productive purposes	Planting of forest trees or exotic fruit trees in areas previously destined for these purposes and currently degraded or unmanaged (e.g., affected by fires)
Development of green areas and urban trees	Increase vegetation in urbanized areas

Type of action	Description
Stakeholder proposals	
Biomass management focusing on fire	Reducing biomass and avoiding reforestation denotes a fire risk, increasing heterogeneity
Livestock control and management	Avoiding soil erosion, including the prohibition of slurry
Improvement of watershed conditions	Mainly in steep creeks, as they were fundamental during the earthquake, and neighbors have protected and improved them with public funds
Sustainable agriculture	Promotion of responsible agricultural practices
Urban and stormwater management	Rainwater re-direction and preventing the water that reaches wetlands from carrying pollutants, considering filtering systems

Source: Own elaboration

References

AMRBB, Asociación de Municipios de la Región del Biobío (2021). Propuesta de red de infraestructura verde de 7 comunas de Concepción, desde un enfoque de cuencas. In. Chile.

Anderson, C. B., Pizarro, J. C., Estévez, R., Sapoznikow, A., Pauchard, A., Barbosa, O., Moreira-Muñoz, A., & Valenzuela, A. E. J. (2015). ≤Estamos avanzando hacía una socio-ecología? Reflexiones sobre la integración de las dimensiones "humanas" en la ecología en el sur de América. *Ecología austral, 25*. http://www.scielo.org.ar/scielo.php?script=sci_arttext&pid=S1667-782X2015000300012&lng=es&tlng=es

Arriagada, L., Rojas, O., Arumi, J., Munizaga, J., Rojas, C., Farías, L., & C., V. (2019). A new method to evaluate watershed vulnerability facing several stressors: A case study of the Chilean Mediterranean. *Science of the Total Environment, 651*, 1517–1533.

Blondel, J., Aronson, L., Bodiou, J., & Boeuf, G. (2010). The Mediterranean Region: Biological Diversity in Space and Time. *The Quarterly Review of Biology, 85*(4), 497–497. https://doi.org/10.1086/656852

Borrini-Feyerabend, G. y. H., R. (2019). Gobernanza para la conservación de la naturaleza. In M. L. G.L. Worboys, A. Kothari, S. Feary e I. Pulsford (Ed.), *Gobernanza y gestión de áreas protegidas* (pp. 175–214.). Editorial Universidad El Bosque y ANU Press.

Burkhard, B., Crossman, N., Nedkov, S., Petz, K., & Alkemade, R. (2013). Mapping and modelling ecosystem services for science, policy and practice. *Ecosystem Services, 4*, 1–3. https://doi.org/10.1016/j.ecoser.2013.04.005

Burkhard, B., Kroll, F., Müller, F., & Windhorst, W. (2009). Landscapes' capacities to provide ecosystem services—A concept for land-cover based assessments. *Landscape Online*, 15. https://doi.org/10.3097/LO.200915

Darrah, S. E., Shennan-Farpon, Y., Loh, J., Davidson, N. C., Finlayson, C. M., Gardner, R. C., & Walpole, M. J. (2019). Improvements to the Wetland Extent Trends (WET) index as a tool for monitoring natural and human-made wetlands [Article]. *Ecological Indicators, 99*, 294–298. https://doi.org/10.1016/j.ecolind.2018.12.032

Davidson, N. C., Fluet-Chouinard, E., & Finlayson, C. M. (2018). Global extent and distribution of wetlands: trends and issues [Article]. *Marine and Freshwater Research, 69*(4), 620–627. https://doi.org/10.1071/mf17019

Dayer, A. A., Barnes, J. C., Dietsch, A. M., Keating, J. M., & Naves, L. C. (2020). Advancing scientific knowledge and conservation of birds through inclusion of conservation social sciences in the American Ornithological Society [Article]. *Condor, 122*(4), 6, Article duaa047. https://doi.org/10.1093/condor/duaa047

Da Cruz, R. & Lopes, O. (2018). Incorporating wetlands in hydrologic and hydraulic models for flood zone delineation: An application to Durán, Ecuador. *International Journal of Disaster Risk Reduction*, 28, 375–383. https://doi.org/10.1016/j.ijdrr.2018.03.023

De Urresti, A. (2020). *Humedales Urbanos. Historia de una ley pionera y ciudadana.*

Espinosa, P., De Meulder, B., Alarcón, M., & Pérez, L. (2015). Interacciones de agua y ciudad / Una investigación de Urbanismo del Paisaje aplicado al caso del río Andalién, Concepción. *Revista De Urbanismo, 33*, 52–72. https://doi.org/10.5354/0717-5051.2015.36569

Espinosa, P., De Meulder, B., & Ollero, A. (2020). Restauración fluvial como estrategia de diseño urbano. Un diálogo entre investigación y diseño. Concurso río Andalién, Concepción, Chile. *AREA: Agenda de Reflexión en Arquitectura, Diseño y Urbanismo, 26*, 1–25.

Estévez, R. A., Sotomayor, D. A., Poole, A. K., & Pizarro, J. C. (2010). Formando una nueva generación de investigadores capaces de integrar los aspectos socio-ecológicos en conservación biológica. *Revista chilena de historia natural, 83*(1). https://doi.org/10.4067/s0716-078x2010000100003

Fierro, P., Valdovinos, C., Arismendi, I., Diaz, G., De Gamboa, M. R., & Arriagada, L. (2019). Assessment of anthropogenic threats to Chilean Mediterranean freshwater ecosystems: Literature review and expert opinions [Review]. *Environmental Impact Assessment Review, 77*, 114–121. https://doi.org/10.1016/j.eiar.2019.02.010

Forman, R. (1995). *Land Mosaics: The ecology of landscapes and regions*. Cambridge University.

Guimarães M.H, Ballé-Béganton, J., Bailly, D., Newton, A., Boski, T., & Dentinho, T. (2013). Transdisciplinary conceptual modeling of a social-ecological system—A case study application in Terceira Island, Azores [Article]. *Ecosystem Services*, 3, e22–e31. https://doi.org/10.1016/j.ecoser.2012.12.007

Hernandez, A., Miranda, M. D., Arellano, E. C., & Dobbs, C. (2016). Landscape trajectories and their effect on fragmentation for a Mediterranean semi-arid ecosystem in Central Chile [Article]. *Journal of Arid Environments*, *127*, 74–81. https://doi.org/10.1016/j.jaridenv.2015.10.004

Isla, F. I., Flory, J. Q., Martinez, C., Fernandez, A., & Jaque, E. (2012). The Evolution of the Bio Bio Delta and the Coastal Plains of the Arauco Gulf, Bio Bio Region: the Holocene Sea-Level Curve of Chile [Article]. *Journal of Coastal Research*, *28*(1), 102–111. https://doi.org/10.2112/jcoastres-d-10-00035.1

Jaque, E. (2010). Diagnóstico de los paisajes mediterráneos costeros. Cuenca del río Andalién, Chile. *Boletín de la Asociación de Geógrafos Españoles*.

Law No. 21,202. (2020). Modifica diversos cuerpos legales con el objetivo de proteger los humedales urbanos. In. Chile: Ministerio del Medio Ambiente.

MMA, Ministerio del Medio Ambiente (2016). Metodología base para la revisión de sitios prioritarios y otras áreas de valor ecológico, y propuestas de áreas con potencial de restauración para el desarrollo de una infraestrucutura ecológica con aplicación en una región piloto. Realizado por la Facultad de Ciencias Forestales de la Universidad de Concepción. In. Chile.

Moraga, D., Vivancos, A., Ruiz, V. H., Rojas, O., Diaz, G., Manosalva, A., Vega, P., & Habit, E. (2022). A century of anthropogenic river alterations in a highly diverse river coastal basin: Effects on fish assemblages [Article]. *Frontiers in Environmental Science*, *10*, 12, Article 943586. https://doi.org/10.3389/fenvs.2022.943586

Navarro, V. (2017). *Oportunidades y desafíos para la protección de los humedales rocuant-andalién y los batros en el área metropolitana de concepción: una mirada desde la gobernanza urbana.* [Memoria para optar al título de Geógrafa, Universidad de Chile].

Pino, J., & Marull, J. (2012). Ecological networks: Are they enough for connectivity conservation? A case study in the Barcelona Metropolitan Region (NE Spain). *Land Use Policy*, *29*(3), 684–690. https://doi.org/10.1016/j.landusepol.2011.11.004

Ramsar Conventioon on Wetlands. (2018). *Global Wetland Outlook: State of the World's Wetlands and Their Services to People.*

Rojas, C., De la Barrera, F., Aránguiz, T., Pino, J., & Munizaga, J. (2017a). Efectos de la Urbanización en la Conectividad ecológica de paisajes metropolitanos. *Revista Universitaria de Geografía*, *26*, 155–182.

Rojas, C., Munizaga, J., Rojas, O., Martinez, C., & Pino, J. (2019). Urban development versus wetland loss in a coastal Latin American city: Lessons for sustainable land use planning [Article]. *Land Use Policy*, *80*, 47–56. https://doi.org/10.1016/j.landusepol.2018.09.036

Rojas, C., Muñiz, I., & Pino, J. (2013a). Understanding the Urban Sprawl in the Mid-Size Latin American Cities through the Urban Form: Analysis of the Concepción Metropolitan Area (Chile). *Journal of Geographic Information System*, *5*, 222–234. https://doi.org/10.4236/jgis.2013.53021

Rojas, C., Pino, J., Basnou, C., & Vivanco, M. (2013c). Assessing land-use and -cover changes in relation to geographic factors and urban planning in the metropolitan area of Concepción (Chile). Implications for biodiversity conservation. *Applied Geography*, *39*, 93–103. https://doi.org/10.1016/j.apgeog.2012.12.007

Rojas, C., Pino, J., & Jaque, E. (2013b). Strategic Environmental Assessment in Latin America: A methodological proposal for urban planning in the Metropolitan Area of Concepcion (Chile) [Article]. *Land Use Policy*, *30*(1), 519–527. https://doi.org/10.1016/j.landusepol.2012.04.018

Rojas, O., Mardones, M., Rojas, C., Martinez, C., & Flores, L. (2017b). Urban Growth and Flood Disasters in the Coastal River Basin of South-Central Chile (1943-2011) [Article]. *Sustainability*, *9*(2), 21, Article 195. https://doi.org/10.3390/su9020195

Rojas, O., Soto, E., Rojas, C., & Lopez, J. J. (2022). Assessment of the flood mitigation ecosystem service in a coastal wetland and potential impact of future urban development in Chile [Article]. *Habitat International*, *123*, 10, Article 102554. https://doi.org/10.1016/j.habitatint.2022.102554

Rueda, I. (2020). *Un Anillo Verde para el Área Metropolitana de Concepción. Propuesta de Infraestructura verde en base a criterios geográficos* [Tesis para optar al título de Magíster en Análisis Geográfico, Universidad de Concepción.]. Chile.

Sepúlveda, E., Parra, L. E., Benítez, H. A., & Rojas, C. (2012). Estados de naturalidad y heterogeneidad vegetacional de humedales palustres y su efecto sobre la diversidad de Macrolepidoptera (Insecta: Lepidoptera). *SHILAP Revista de lepidopterología*, *40*(158), 155–170. https://www.redalyc.org/articulo.oa?id=45524831004 (IN FILE)

Shao, Y. H., Xu, X. Y., & Yuan, J. (2021). The Intension and Values of Urban Wildscapes [Article]. *Landscape Architecture Frontiers*, *9*(1), 14–25. https://doi.org/10.15302/j-laf-1-020039

Smith, P., & Romero, H. (2009). Efectos del crecimiento urbano del Área Metropolitana de Concepción sobre los humedales de Rocuant-Andalién, Los Batros y Lenga. *Revista De Geografia Norte Grande*, *43*, 81–93.

Society for Ecological Restoration (SER) International Grupo de trabajo sobre ciencia y políticas. (2004). Principios de SER International sobre la restauración ecológica. www.ser.org In. Tucson: Society for Ecological Restoration International.

Sol, A., Zenteno, C., Zamora, L., & Torres, E. (2002). Modelo para la restauración ecológica de áreas alteradas. *Kuxulkab'. Revista de Divulgación, VII*, 48–60.

Toledo, C. (2020). Construyendo metrópolis en Chile: el caso ruta del agua como piloto de gestión del territorio metropolitano en el Gran Concepción. *Territorios Y Regionalismos.*, *2*, 15–23.

Vidal, C., & Romero, H. (2010). Efectos ambientales de la urbanización de las cuencas de los ríos Bíobío y Andalién sobre los riesgos de inundación y anegamiento de la ciudad de Concepción. In L. Pérez & R. Hidalgo (Eds.), *Concepción Metropolitano (AMC). Planes, procesos y proyectos*. Serie GEOlibros, Instituto de Geografía, Pontificia Universidad Católica de Chile.

Villagrán, R., Aguayo, M., Parra, L., & González, A. (2006). Relación entre características del hábitat y estructura del ensamble de insectos en humedales palustres urbanos del centro-sur de Chile. *Revista chilena de historia natural*, *79*, 195–211. https://doi.org/10.4067/S0716-078X2006000200006.

Quebrada Parque: Social and Ecological Restoration in Puerto Varas, Chile

Camila Teutsch, Rodrigo Garrido, Pedro Bonacic, Bárbara Cid, Alejandra Díaz, Jorge Gallardo, and Jaime Rau

Contents

Abstract

Quebrada Honda drains a small basin with a strong urbanization pattern in Puerto Varas, southern Chile. Urban growth and unregulated interventions—including deforestation and wetland filling—have severely compromised the ecosystem services that this area used to provide, particularly in terms of flood control and water quality management. Since 2016, a multidisciplinary team led by Patagua has been working on a collaborative initiative, called Quebrada Park, that aims to restore the creek and wetland, as well as their deteriorated surroundings, so as to contribute to a more sustainable and resilient urban environment. The project is based on four key pillars: sustainable urban drainage, urban nature, public space, and community engagement. Initial governmental funding was directed toward the latter, allowing for an extensive process of participatory activities and environmental education. Through this process, several tons of waste were removed from the area by the local community and hundreds of native trees and bushes were planted to control soil erosion, as well as to contribute to a greener and more biodiverse city. Subsequent public and private alliances have led to the design of a nature-based solution to recover the wetlands' rainwater storage capacity, and to the development of a master plan for an urban park. The park is conceived as an integrated restoration initiative that aims for the ongoing recovery and protection of this urban nature hotspot, as for social integration

C. Teutsch (✉) · R. Garrido · P. Bonacic · B. Cid · A. Díaz
Patagua, Puerto Varas, Chile
e-mail: camila@patagua.cl; rodrigo@patagua.cl; pibonacic@uc.cl; barbara@patagua.cl

J. Gallardo · J. Rau
Ecology Laboratory, University of Los Lagos, Los Lagos, Chile
e-mail: jrau@ulagos.cl

C. Rojas Quezada (ed.), *Urban Wetlands in Latin America*, Sustainable Development Goals Series,
https://doi.org/10.1007/978-3-031-69590-2_10

through the habilitation of high-quality public spaces.

Keywords

Socio-ecological restoration · Community engagement · Urban drainage

1 Introduction

Ravines and wetlands are spaces that pose a challenge for urban management in Latin America. On one hand, they are often perceived as marginal and low-value sites, so practices such as the consolidation of micro-dumps, filling, deforestation, discharge of wastewater, informal settlements, and crime, among others, are perpetuated in them (Ciudad Emergente 2018). On the other hand, they constitute highly functional systems that can be a key piece to improving the climate adaptation and resilience of cities (Vásquez 2016).

In this context, the Guide for sustainable management of urban ravines (Patagua 2018) proposes the figure of *"Quebrada Parque"* ("Ravine Park") as a replicable model to contribute to the development of sustainable cities, recognizing and enhancing the ecosystem services that these water systems offer. This guide was developed based on the initial learnings obtained from the case of Quebrada Honda, in the city of Puerto Varas, southern Chile. This document summarizes the process and the results obtained after more than five years of work led by Patagua in the ravine (2017 to date), including adjustments to the original methodological approach.

Quebrada Honda—a ravine-wetland system of around 3.5 ha—is one of the main components of the natural drainage system of Puerto Varas, a predominantly touristic city located in the Llanquihue lake basin (Fig. 1). Like several other small and intermediate cities in southern Chile, Puerto Varas has experienced an accelerated demographic growth. Specifically, the population of the commune increased by 204% between 1992 and 2017 (INE 2017). This has brought light to the challenge of managing and enhancing urban water and nature spaces in order to contribute to resilience and quality of life in the city. From the collaborative work experience in Quebrada Honda, we seek to identify learnings that contribute to the socio-ecological restoration and sustainable management of urban creek and wetland systems in similar contexts.

2 Methodological Approach

The applied methodological approach consists of 6 steps that are shown in Fig. 2 and are briefly described below.

The situation analysis included a review of secondary information, with an emphasis on urban hydrology; field surveys of environmental variables such as flora, fauna, and water quality; systematization and processing of primary and secondary territorial information through QGIS software; and an urban analysis of variables such as density, housing types and access to green areas. Along with these technical studies, participatory methodologies (i.e., collective mappings, focus groups, surveys, and semi-structured interviews) were used to identify uses, perceptions, and expectations around this space. The results of this diagnosis were discussed within a Technical Committee integrated by representatives from public, private, academic, and civil society institutions, allowing for the conceptualization of an integrated system around four key components that reflect both current and potential functionalities of the ravine. This conceptual framework laid the foundation for a strategic planning process, whose priority actions have been implemented in a participatory manner. The last step—monitoring and evaluation—provides insight into the effectiveness of the actions taken, and feeds the process of strategic planning with a continuous improvement approach. Participation and collaboration are conceived as cross-cutting elements of the methodological approach, which allows involving key actors throughout the process and decision-making, while ensuring the feasibility of implementing the plan.

Fig. 1 Quebrada Honda is a creek-wetland system of about 3.5 ha that drains into Lake Llanquihue (the second largest in Chile) in the heart of the city of Puerto Varas

Fig. 2 Methodological approach based on 6 steps, where participation and collaboration are conceived as cross-cutting elements

3 Key Results

Next, we present the main results of the situational analysis and the conceptual framework that derives from it, and which is proposed as a replicable model for managing similar systems. Then, the results associated with strategic planning and implementation are integrated, and findings from the incipient process of monitoring and evaluation are presented. The most important strategic alliances are indicated along with the results of each of these steps.

3.1 Situation Analysis

Quebrada Honda is part of a 350-ha micro-basin, with a growing urbanization pattern that has resulted in an increase in soil sealing and greater pressure on the natural drainage system. The analysis of the city's rainwater master plan (DOH 2017) shows that the existing urbanization in the lower part of the basin is located in a flood zone (Fig. 3), which implies risk to people, damage to infrastructure, and deterioration of the quality of life in the city. Along with this, the soil movements produced by the ongoing urbanization process in the upper part of the basin have caused an increase in the sediment load and an accelerated filling of the wetland located in the lower section. Hence the need to recover the system's capacity to contain and control floods, in order to minimize the risk of flooding.

In 2017 and 2018, two non-systematic **flora surveys** were conducted within the *Quebrada*, where a total of 69 species belonging to 39 families were identified. The results (Fig. 4) show a highly intervened space, with about 67% of species of exotic origin—some of them invasive, such *Ulex europaeus* and *Rubus ulmifolius*—and the presence of micro-dumps on the west slope. Despite the marked degradation of the slopes, the deepest part of the ravine proved to maintain natural conditions typical of wetland environments, and to be the refuge of native species not present in other parts of the city. A relevant finding was the identification of a *hualve* or flooded forest, formed by representatives of the myrtle family such as *Myrceugenia exsucca* and *Luma apiculata.* In other areas, the presence of *Schoenoplectus californicus* and *Blechnum chilense* was noted. The data collected in these studies were shared with the local government to support its initiative of declaring Quebrada Honda as an urban wetland.

Other relevant results come from the **bird survey** carried out in collaboration with the Ecology

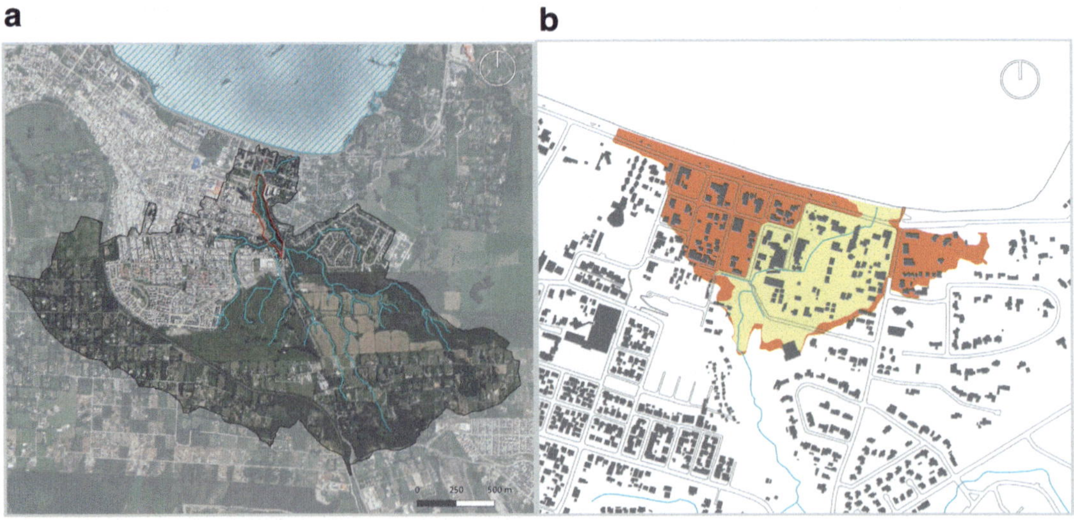

Fig. 3 (**a**) The basin contributing to the Quebrada Honda system presents a pattern of growing urbanization. (**b**) The lowest part of the basin is located in a flood zone (yellow area corresponds to T = 5; orange area corresponds to T = 100)

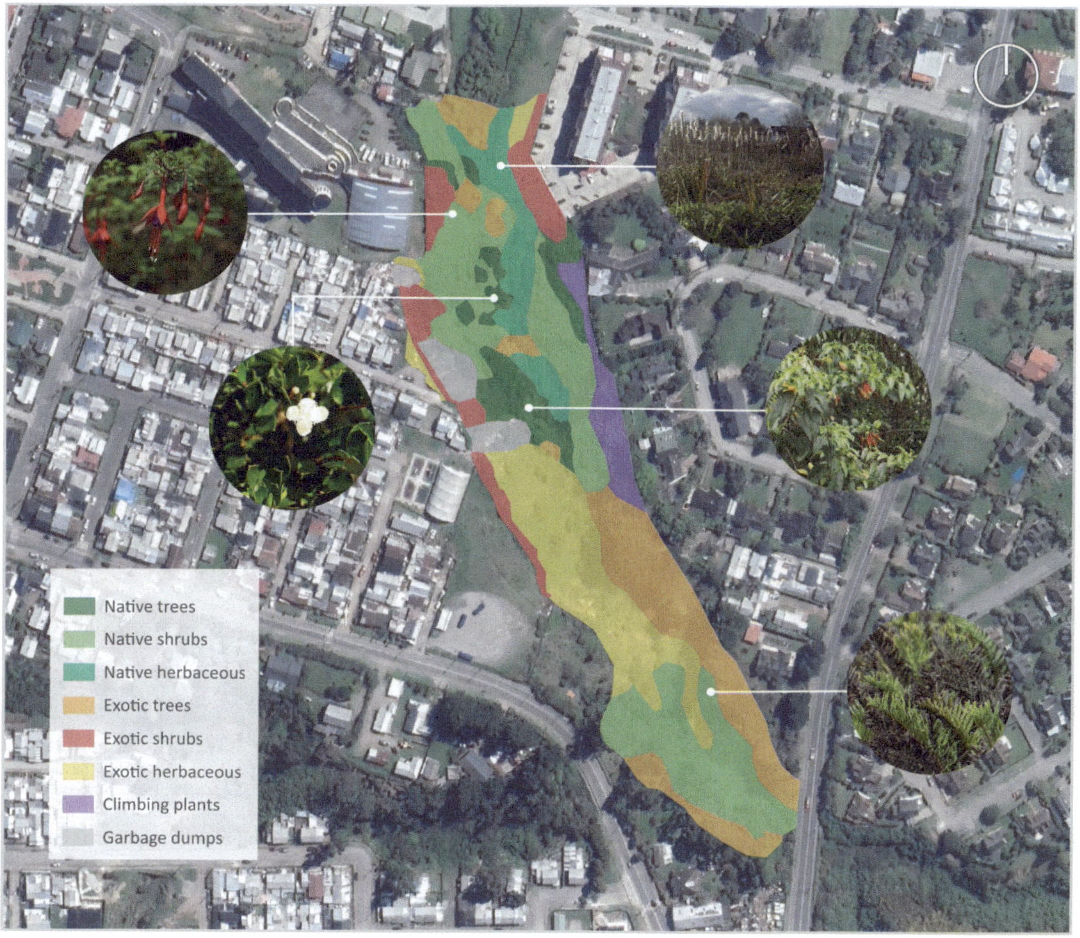

Fig. 4 Results of the vegetation surveys in Quebrada Honda

Laboratory of the University of Los Lagos, between April 2018 and April 2019. The survey shows a total richness of 25 species, 92% of which are native. Also, 2 species are total austral migrants, that is, species that nest in the southern and austral zone of Chile and migrate to the north-central part of the country in autumn; 9 partial austral migrants, that is, they present the same behavior as the total austral migrants, but depending on the climatic conditions and the species, part of the population does not return to the south and nests in the north-central zone of the country; and 14 are residents (Fig. 5). The survey was carried out using three sampling transects, two located in the central area of the ravine and another in the surrounding urban area, which allowed to compare the presence of birds in natural and urban conditions. Out of the 25 species that were observed in total, 15 were identified only in the *Quebrada* transects, reflecting the role that this area plays as a refuge for local fauna.

In terms of **water quality**, a microbiological study led by researchers of San Sebastian University was carried out in September 2021, as part of the project "Implementation of molecular strategies to characterize the microbiological condition of urban wetlands". The results of this study indicate that the ravine's waters are subject to significant pollution from anthropogenic sources, and low occurrence of natural processes and cycles (Opitz and Medina 2021).

From a **sociocultural** point of view, some traditional uses in the ravine were identified through a participatory appraisal approach. One that

Fig. 5 (**a**) *Sephanoides sephaniodes*, partial austral migrant bird with higher relative observation frequency in *Quebrada* transect; (**b**) *Anairetes parulus*, resident bird; (**c**) *Elaenia albiceps*, total austral migrant bird

Fig. 6 In the heart of the ravine, Pedro Mansilla plants and harvests *Salix* sp. with which he makes traditional baskets

stands out is the handcrafting of wicker baskets using raw material that is planted in and harvested from the heart of the *Quebrada* (Fig. 6). The performance of ceremonies associated with the Mapuche culture was also identified. Despite these uses, most of the neighboring community that participated in the process declared not to have a connection with the ravine. This lack of valuation and social appropriation is reflected, for example, in the micro-dumps that were mentioned above.

The **urban analysis** of the immediate environment of the *Quebrada* reveals significant contrasts. On the western edge, a high density of homes on small lots is observed (90–200 m^2; mostly social housing), while on the eastern edge, larger lots with lower construction densities predominate (300–2000 m^2) (Rammsy 2018). Thus, the ravine becomes not only a physical barrier, but a socioeconomic one, which segregates these contrasting realities (Fig. 7).

On a **city scale**, the analysis carried out shows that large green areas (over 5000 m^2) are concentrated in the western area of Puerto Varas and that less than 20% of the population has a good level of accessibility to them (up to 5 min on foot; Steiniger et al. 2021). This constitutes a gap in terms of social equity and environmental justice (Steiniger et al. 2021). As shown in Fig. 8, the enabling of the *Quebrada* as a multipurpose green area offers an opportunity to significantly shorten this gap.

3.2 Conceptual Framework

Based on the results of the situation analysis, the ravine is conceptualized as an integrated system that considers four pillars: urban drainage, urban nature, public space, and community. Each of them seeks to enhance the key functionalities that were recognized in the system, shaping the figure

Fig. 7 The satellite image (left) and the Nolli map (right; Rammsy 2018) show the ravine as a physical barrier that segregates two areas of the city with contrasting socioeconomic realities

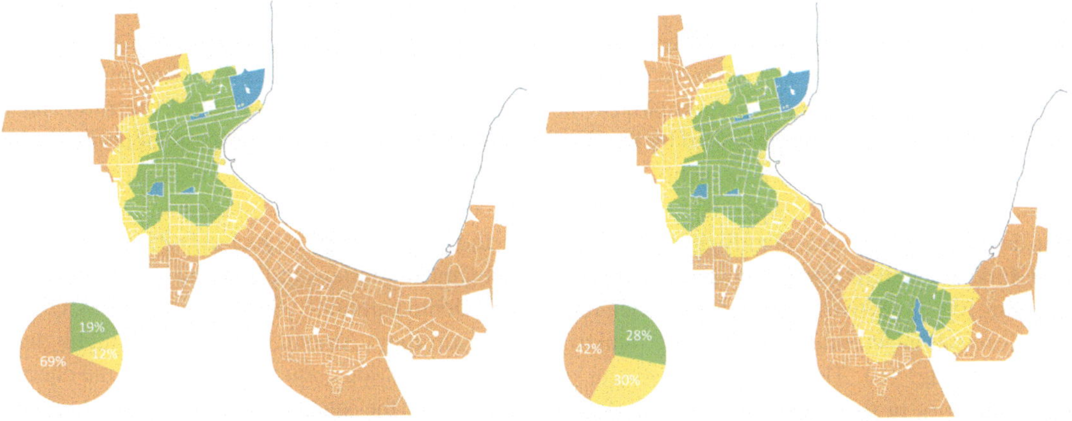

Fig. 8 Accessibility to large green areas (surface equal to or greater than 5000 m^2; in blue) in Puerto Varas, without *Quebrada Parque* (left) and with *Quebrada Parque* (right). In green, areas with good accessibility (up to 5 min on foot); in yellow, areas with medium accessibility (between 5 and 10 min on foot); in orange, areas with low accessibility (more than 10 min on foot). Sustainability indicator adapted from Steiniger et al. (2021)

of *Quebrada Parque* as a replicable model for ecological and social restoration (Fig. 9).

- **Urban drainage:** refers to the functional role of the *Quebrada* in terms of regulation, purification, and storage of urban runoff. These functions help to prevent floods and potential problems associated with water quality that could put the wellbeing of the community at risk.
- **Urban nature:** recognizes the value of the *Quebrada* as a biodiversity core within the city, which offers habitat to various species of native flora and fauna and contributes to the balance between the natural environment and the built space.
- **Public space:** refers to the value of the *Quebrada* as a green area with public use potential, which can be achieved by the creation of a linear park that would add landscape, educational, and recreational value.
- **Community:** recognizes the identity value of the *Quebrada* and its potential to become a place that fosters gathering, sense of belonging, and social integration.

Fig. 9 The figure of *Quebrada Parque* is proposed as a replicable management model for ecological and social restoration, based on four pillars: urban drainage, urban nature, public space, and community. The key functionalities that the model seeks to enhance are indicated for each of them. (Adapted from Patagua 2018)

3.3 Strategic Planning and Implementation

The strategic objectives proposed for each pillar, as well as the main results and strategic alliances for their implementation, are summarized in Table 1. These alliances are promoted by Patagua as the leading organization, and involve public, private, and civil society organizations. Their contributions (which are mainly in-kind) have been fundamental to carry out actions toward achieving the objectives. Participatory cleanups and reforestation activities have been particularly successful (Fig. 10). The active involvement of the community in these activities contributes to the restoration of the system, both from an environmental and a social point of view (Gómez-Ruiz and Bonfil 2021).

Along with this, the participatory design process of the public space allowed to develop and validate a master plan with the zoning and architectural program of the park (Fig. 11), whose infrastructure will soon be designed in detail and implemented through public funding. This master plan is based on the recognition of the hydrological and ecological functions of the space and, therefore, promotes a respectful use and limited to specific areas of the Ravine (Rammsy et al. 2019).

In order to also contribute to biological connectivity at an urban scale, a green corridor was projected that connects the Quebrada with other remnants of urban nature and large green areas of the city (Fig. 12). The layout of the corridor was proposed based on the analysis of available green spaces around the road system, to intervene them through the design of ecosystemic gardens in replacement of the grass that predominates in them. The first sections of the corridor have already been implemented in collaboration with the Municipality of Puerto Varas, neighborhood associations, and private companies.

In December 2022, Quebrada Park became the first Chilean initiative to join the Global Network of Demonstration Sites of UNESCO-PHI Ecohydrology. Among the attributes highlighted by the evaluation committee regarding the characteristics of the site, the approach used, and the actions implemented to date, are the following: the urban relationship with the stream-wetland-lake system, the design of sustainable urban drainage strategies, reforestation as an erosion control strategy, the interdisciplinary team, community involvement, and public-private collaboration.

3.4 Monitoring and Evaluation

The formalization of a program that allows monitoring and evaluating the results of the socio-ecological restoration process in an integral way is one of the central challenges that has been posed in the planning of the next steps of this initiative. To date, some progress has been made in monitoring the areas of urban drainage and urban nature, which are summarized below.

Table 1 Threats and opportunities identified around the four pillars of the Quebrada Parque framework

Pillar	Strategic objectives	Implementation and alliances
Urban drainage (UD)	*Main*: Recover the flood control capacity through the design of a sustainable urban drainage solution (SUDS) involving the wetland *Secondary*: Purify urban runoff before its incorporation into the natural drainage system (stream-wetland-lake) through SUDS	SUDS has been designed and approved by the sectorial authority *Alliance*: Real estate company
Urban nature (UN)	*Main*: Foster native species, control invasive species, and mitigate erosion *Secondary*: Connect the Ravine with other cores of nature through the design of a green corridor at an urban scale	*Ulex europaeus* (invasive) removed in 63.5% of its initial coverage *Alliance*: Construction company Over 400 native trees and shrubs planted *Alliance*: NGOs, Municipality of Puerto Varas, San Sebastian University More than 4.5 tons of garbage removed *Alliance*: Neighborhood associations, NGOs, Municipality of Puerto Varas Green corridor designed and pilot sections implemented *Alliance*: Municipality of Puerto Varas, Local companies
Public space (PS)	*Main*: Consolidate public use spaces that promote social integration and outdoor learning, through the design and implementation of low environmental impact infrastructure	Zoning and master plan of the park designed and validated through a participatory process Detailed design and construction of infrastructure soon to be tendered. *Alliance*: Municipality of Puerto Varas, Ministry of Housing and Urbanism.
Community (CO)	*Main*: Promote social appropriation of the *Quebrada* as well as local integration through citizen involvement strategies *Secondary*: Contribute to social restoration on a city scale	Participatory activities of cleaning, reforestation, and maintenance *Alliance*: Neighborhood Associations, NGOs, Municipality of Puerto Varas Outdoor environmental education activities *Alliance*: Mirador del Lago School and other educational establishments

- *Urban Drainage*: The Quebrada was recently incorporated by the General Water Directorate into its national water quality monitoring program, committing to the semi-annual analysis of 28 physicochemical parameters at 3 points of the natural drainage system. The first results will constitute the baseline against which the effectiveness of future actions aimed at purifying the urban runoff that enters the system will be evaluated.
- *Urban Nature*: The participatory maintenance days of the reforested areas have allowed monitoring the state of the native plants and controlling outbreaks of invasive plants. From this follow-up, it was identified that the main threat to the establishment of the planted trees is the careless grass cutting, which causes wounds in the trunks and has completely cut many of the smaller plants. This observation led to the installation of protections for all the trees; along with this, the Municipality was asked to train the staff in charge of maintaining green areas to avoid damage to the plants.

4 Lessons Learnt

After more than five years of implementing the management model associated with the figure of *Quebrada Parque* in Puerto Varas, a series of lessons have been learned that could be of interest for the application of this model in similar contexts. The most important ones are summarized below:

Fig. 10 Citizen involvement in activities like participatory cleanups and reforestation contributes to the restoration of the environmental and social system

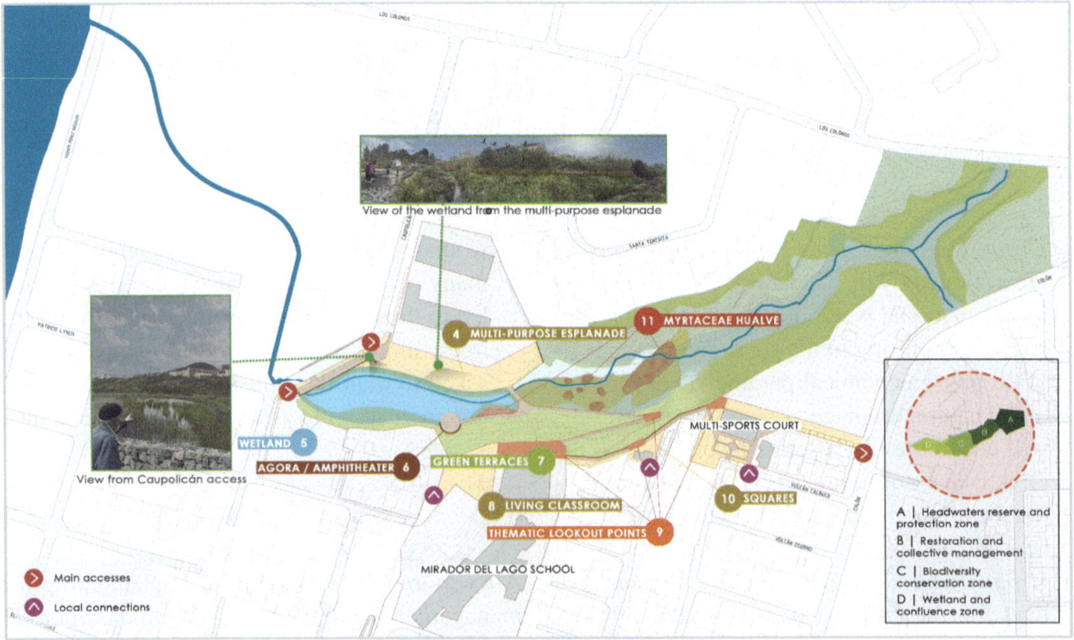

Fig. 11 Zoning and master plan of low-impact public use infrastructure. (Adapted from Rammsy et al. 2019). The detailed design and implementation of this infrastructure will be developed with already committed public funds

Fig. 12 A green corridor was projected to connect the Quebrada with other parks and remnants of urban nature through the implementation of native sidewalks

- The conceptualization of the quebrada from a systemic perspective—around the axes of urban drainage, urban nature, public space, and community—facilitated a process of integrated planning and action for socio-environmental restoration and urban resilience. This, in line with what is proposed with goals 11.3, 11.4, and 11.7 of Sustainable Development Goal 11 of the 2030 Agenda, referred to Sustainable communities and cities (UN 2018). For work with this approach, the formation of interdisciplinary teams is crucial.
- The principles of participation and collaboration stand out as central and strategic. The involvement of key actors from early stages of analysis and planning is fundamental to generate ownership and add wills that translate into the materialization of results.
- Having the support of the local government is an important contributing factor to give viability and sustainability to initiatives of this type. In this case, the Municipality of Puerto Varas has contributed from logistical facilities for the removal and safe disposal of large volumes of garbage, to the leveraging of public funds for the design and enablement of infrastructure.
- Partnerships with local universities have been critical to generate reliable information that guides decision-making. Likewise, the alliance with the public body in charge of water management in Chile has allowed to ensure the long-term data collection, in a particularly costly area such as water quality.
- The implementation of restoration activities (mainly, cleaning, reforestation, and mainte-

nance) through citizen involvement strategies has allowed to advance in the implementation of the action plan, while it has allowed to consolidate a group of "ambassadors" of this space. In this, the articulation with civil society organizations and nearby educational establishments has been key.

- The role of the management team is fundamental to carry forward the process, articulate the key actors, establish alliances, coordinate the implementation of actions, communicate, and ensure the fulfillment of the proposed objectives. In this case, the management team fulfills this role pro bono and is housed in a single organization (Patagua), but it has set the goal of forming a broader management and governance system in the medium term to ensure the sustainability of the initiative over time.

References

Ciudad Emergente. (2018). *Manual de Barranqueros: Plan de Habilitación Participativa de Barrancos*. 2a ed. https://static1.squarespace.com/static/5a657d43bff200b621e34a72/t/5b5b514303ce64489cc8232a/1533561843682/Manual+de+Barranqueros_2nd-ilovepdf-compressed.pdf

Dirección de Obras Hidráulicas, DOH. (2017). *Plan Maestro de Evacuación y Drenaje de Aguas Lluvias de Puerto Varas, Región de Los Lagos: Resumen Ejecutivo*. Ayala, Cabrera y Asociados Ltda. & AC Ingenieros Consultores Ltda. Informe final.

Gómez-Ruiz, P. A. y Bonfil, C. (2021). La participación comunitaria en la restauración ecológica: el reto de integrar diversos actores. *Clavigero*, (18), 11. https://clavigero.iteso.mx/wp-content/uploads/sites/88/2021/01/Clavigero-18-11.pdf

Instituto Nacional de Estadísticas, INE. (2017). Censo nacional de población y vivienda.

Opitz, C. y Medina, D. (2021). *Composición microbiológica de humedales urbanos en Puerto Varas (work in progress)*. Proyecto colaborativo de vinculación con el medio: Implementación de estrategias moleculares para caracterizar la condición microbiológica de humedales urbanos. Universidad San Sebastián.

Organización de las Naciones Unidas, ONU. (2018). La Agenda 2030 y los Objetivos de Desarrollo Sostenible: una oportunidad para América Latina y el Caribe (LC/G.2681-P/Rev.3), Santiago de Chile.

Patagua. (2018). *Quebrada Parque: Guía para la Gestión de Quebradas Urbanas*. https://drive.google.com/file/d/1Lzn8JLa9jw8QxzydvYsnzgeFkzo7uZwu/view

Rammsy, T. (2018). *Estudio análisis urbano, Quebrada Parque, Puerto Varas*.

Rammsy, T., Teutsch, C., Corrales, B. & Infante, J. (2019). *Propuesta de plan maestro Quebrada Parque, parque del agua y naturaleza urbana*. Puerto Varas, Chile.

Steiniger, S., Villegas, R., De la Fuente, H., Rueda, I. & Truffello, R. (2021). *Indicadores de sustentabilidad urbana—Aplicación en capitales chilenas*. Informe ejecutivo. Centro de Desarrollo Urbano Sustentable, Santiago de Chile.

Vásquez, A. (2016). Infraestructura verde, servicios ecosistémicos y sus aportes para enfrentar el cambio climático en ciudades: el caso del corredor ribereño del río Mapocho en Santiago de Chile. *Revista de Geografía Norte Grande*, 62, 63–86. https://doi.org/10.4067/S0718-34022016000100005

The "Re-naturalization" of Green Public Spaces: A Multi-scale Landscape Architecture Proposal for the Integration of Natural Systems in Urban Contexts. The Case of Bosque Londres, Valdivia, Chile

Paulina Ibieta Illanes, Camila Wirsching Fuentes, and Angela Mimica Godoy

Contents

Abstract

Bosque Londres is a landscape architecture project part of the recovery plan for green areas, parks, and areas of natural value of the Municipality of Valdivia, which aims to recover green areas of natural and social value as an action against climate change. For the latter purpose, a design was commissioned to protect and enable Bosque Londres as a relic valdivian rainforest located on the edges of an urban wetland. This forest is recognized for its environmental and community value, however, with socioenvironmental problems associated with its enclosed edges such as illegal occupations and micro dumps, among others.

Thus, the project was confronted with the task of effectively balancing ecological restoration principles with community needs and expectations at both the neighborhood and

P. I. Illanes (✉) · C. W. Fuentes · A. M. Godoy
Estudio Tepa, Valdivia, Chile
e-mail: mpibieta@uc.cl

C. Rojas Quezada (ed.), *Urban Wetlands in Latin America*, Sustainable Development Goals Series,
https://doi.org/10.1007/978-3-031-69590-2_11

communal levels. Extensive technical studies were conducted, alongside participatory processes, to gain insight into the space's inherent identity and utilization patterns while safeguarding its intrinsic forested character. Subsequently, the project adopted a multi-faceted approach, addressing three distinct scales of intervention: (a) A system scale, the project focused on the broader wetland ecosystem encompassing the forest, envisioning a cohesive neighborhood vision centered around the restoration of natural elements; (b) An edge scale, attention was directed toward the immediate boundaries of the forest and its internal components. A comprehensive master plan was developed, delineating various zoning regulations tailored to different levels of protection and usage; (c) An enablement scale, detailed design interventions were proposed for a specific portion of Bosque Londres. These interventions were guided by the principle of minimal intervention, aiming to facilitate the forest's human use while ensuring its long-term preservation.

In this manner, the project proposes a set of design guidelines aimed at fostering a new understanding of natural systems within urban contexts. Departing from the traditional concept of the urban park, the project seeks to establish re-naturalized spaces that serve as platforms for environmental education and stewardship. Ultimately, these spaces contribute to mitigating the adverse effects of climate change in urban environments, fostering a deeper connection between communities and their natural surroundings.

Keywords

Urban nature · Urban wetlands · Urban design · Landscape design · Climate change

1 Introduction

The city of Valdivia is distinguished by its vast green coverage that extends throughout its diverse neighborhoods, encompassing approximately 213 hectares. According to data from the local municipality, these green areas include various spaces such as wetlands, beaches, squares, parks, walks, and other natural environments. However, this quality is compromised when noting that only 40.4% of this territory is maintained by the municipality.

Based on this evidence, and in response to the pressing need to face climate change, in 2022 the municipality of Valdivia formulated the Recovery Plan of Green Areas, Parks, and Areas of Natural Value, main goal of which is "to improve the quality of life of people, reduce greenhouse gas emissions and increase access to the city's green infrastructure" (Ilustre Municipalidad de Valdivia, 2022a, b), and proposes as one of its actions to formalize and incorporate new surfaces of natural green areas into the network of urban parks managed by local municipality.

In this context, Bosque Londres design project emerges as a public municipal contract awarded to TEPA design company at the end of 2022, with the goal to transform the forest into a public urban park. An urban forest that stands out as one of the few relics valdivian rainforest with biodiversity left within the urban fabric of the city, with more than 5.2 hectares located on the edges of the protected urban wetland called "El Bosque – Miraflores – Las Mulatas – Guacamayo [1]", in Huachocopihue neighborhood. It is actively used by the community for well-being and engagement activities related to environmental education and scientific research, despite facing challenges associated with its enclosed edges, such as illegal occupations, inappropriate consumption of alcohol and drugs, and micro dumps, among other issues.

For the latter, the project's foundation aimed to achieve as general objective the development of a Master Plan for Bosque Londres (5.2 hectares), followed by a landscape architecture design for the development of a first phase (1.6 hectares), which is a space currently occupied by the community. The goal was to ensure the inclusion of citizen opinion at all stages of the project through effective and binding participation.

The project's methodology was developed on three design scales: (a) System scale; which covers the study of the urban wetlands system in relation to the neighborhood-city scale through the development of an *Urban Imaginary*, (b) Edge scale; which studies the intermediate scale of the 5.2 hectares of Bosque Londres through landscape planning defining zones and land use intensities, and last, (c) Enablement scale; which focused on the architecture and landscape design of 1.6 hectares of Bosque Londres, which is currently occupied by the community, called phase 1.

Each stage included technical and participatory processes, considering both quantitative and qualitative methods for gathering and processing information, as well as their validation, which will be detailed later. Among these are fundamental studies such as soil mechanics, topographic, phytosanitary surveys, and qualitative studies such as exploratory walks with neighbors, guided tours with experts, memory workshops with the elderly, and focus groups with strategic stakeholders.

As a result, the design process has sparked a paradigm shift, evolving from a traditional urban park approach toward one more aligned with natural conservation and the integration of existing natural systems within the city. Furthermore, it has proposed design guidelines with the potential to be replicated in similar projects aimed at consolidating an urban wetland cityscape.

2 From Urban Park to Natural Forest: Redefinitions from a Participatory Design Process

Throughout the various instances of citizen participation that shaped the design process of the project, a recurring idea emerged that permeated all development: Bosque Londres is a scarce natural heritage in Chilean cities, being a remnant of a highly biodiverse primary forest nestled within the city. It is considered "what remains of what once was," or a fragment of "Valdivia before Valdivia," and that this patch of forest is an example of resilience that must be protected. Resilience defined as the ability of a system to absorb changes and reorganize during the pressure of a new situation, essentially retaining its function, structure, and identity (Walker and Salt 2006) (Fig. 1).

This attitude is also largely due to the origin of the settlement of the Huachocopihue neighborhood with habitants relocated from the 1960 earthquake and its urbanization model typology of the Garden City by Ebenezer Howard, where dwellers inherit a system of solidarity organization alongside a context of human-scale design and privileged contact with nature. In this context, Bosque Londres is the central focus of their social projects (Sanhueza 2018).

This notion has significant precedents. In 2008, the local organization called Biosphere, through the acquisition of an environmental protection fund from the Ministry of the Environment, identified more than ten natural spaces for public use as Urban Natural Reserves in Valdivia. The latter is defined as fragments of forests and wetlands that maintain the structure and characteristic dynamics of natural ecosystems. Among them is Bosque Londres, classified as a "Forest," achieving recognition not only at a neighborhood scale but also at the communal level. While this designation still lacks legal support in the city's planning instruments, there is consensus among social, ecological, and institutional organizations in Valdivia about the importance of preserving these natural spaces as essential ecosystem services to regulate vital processes such as climate, air, and water quality, as well as providing habitats for biodiversity and pollination, among others (Figs. 2 and 3).

Consequently, the project acknowledges this view by transitioning from a design mandated for an urban park to a natural forest. This involved incorporating design criteria for its ecological recovery and preservation, characteristic of an ancestral forest, while also enabling its community use. The main challenge was to find a balance between respectful enabling and preservation of its ecosystem function within the city. The project addressed this challenge with a multiscale approach, through 3 working scales: a system scale, a neighborhood scale, and an edge scale.

Fig. 1 Exploratory walk conducted on January 25th, 2023 the first citizen participation workshop of the project. (Source: Municipality of Valdivia, 2023)

Fig. 2 Bosque Londres in the context of declared urban wetlands, (blue border) urban wetlands surveyed in the national survey (shaded blue), and green areas. (Source: own elaboration with data from Municipality of Valdivia and Ministerio del Medio Ambiente geoportal, 2024)

3 System Scale: A Multiscale Approach to Landscape Design

The city of Valdivia is strongly influenced by its topographic and hydrological characteristics. Situated at the confluence of the Calle-Calle, Cruces, and Valdivia rivers, the city is part of a hydrological network of estuaries, rivers, streams, and wetlands that shape four main natural macro-systems: Wetlands, Floodplains, Forests, and Plains (Ilustre Municipalidad de Valdivia 2011). These macrosystems form an interconnected ecological continuity, and any interruption of this, such as road crossings or fillings, can affect it. Therefore, urban wetlands cannot be understood as isolated units within the urban area but rather constitute a continuous hydrological system with groundwater and sediment flows, serving as ecological corridors for various species of native flora, fauna, and fungi. In this context, Bosque Londres corresponds to a type of temperate-rainforest and is

Fig. 3 Neighbors in the installation of a sign with the title of "Urban Natural Reserve", erected by the local community in 2011. (Source: Organización Biósfera 2011)

part of the system of native forest patches and floodplains surrounding the wetlands in Valdivia. It is located at the intersection of an urban area composed of multiple neighborhoods and a protected urban wetland system called: El Bosque—Miraflores—Las Mulatas Guacamayo. Thus, by analyzing this edge condition, it was possible to reveal its vocation as an interface between urban and natural elements through the development of an urban-wetland imaginary, from both an ecological and social perspective. For this purpose, El Bosque wetland and all its edges were defined as the study area, considering aspects of its natural, built, historical, and social environment for analysis, diagnosis, and development of the urban imaginary (Figs. 4 and 5).

The natural environment is defined by its geomorphology, characterized by three distinct elements: (1) Upper terrace, where the urban fabric has consolidated; (2) Transitional slopes, which vary in steepness depending on the location; and (3) Low-lying wetland area, which sank during the 1960 earthquake, giving rise to El Bosque Wetland as well as other wetlands in the city (Figs. 6 and 7).

Its hydrological system is strongly determined by the presence of El Bosque wetland and the surrounding forest system. This wetland serves as a collector and recipient of rainwater from the city and is part of the city's stormwater master plan prepared by the Ministry of public works. The forests, on the other hand, help regulate the hydrological cycle and cool the environment by reducing evaporation (Peña 2016).

Regarding flora, the study area is in the Valdivian Laurifolio Forest region (or Valdivian Rainforest), as defined by R. Gajardo in 1994 (Gajardo 1994). This region is characterized by a rainy climate throughout the year with temperatures showing little variation, remaining constant throughout all seasons. The geographic area where this type of forest is found is of limited extent and in many cases fragmented, with most of its elements considered phytogeographical relic. In this sense, the value of the Valdivian rainforest present in the study area goes beyond its urban condition, as it is part of the ecoregion of the southernmost temperate rainforest in the world, with a high level of endemism and one of the most biodiverse areas on the planet (Oyarzún et al. 2005) (Fig. 8).

The concurrence of the following types of vegetation coverages were identified:

- Communities of Valdivian Laurifolio Forests consisting of roble-laurel (*Nothofagus obliqua-Laurelia sempervirens*) and roble-

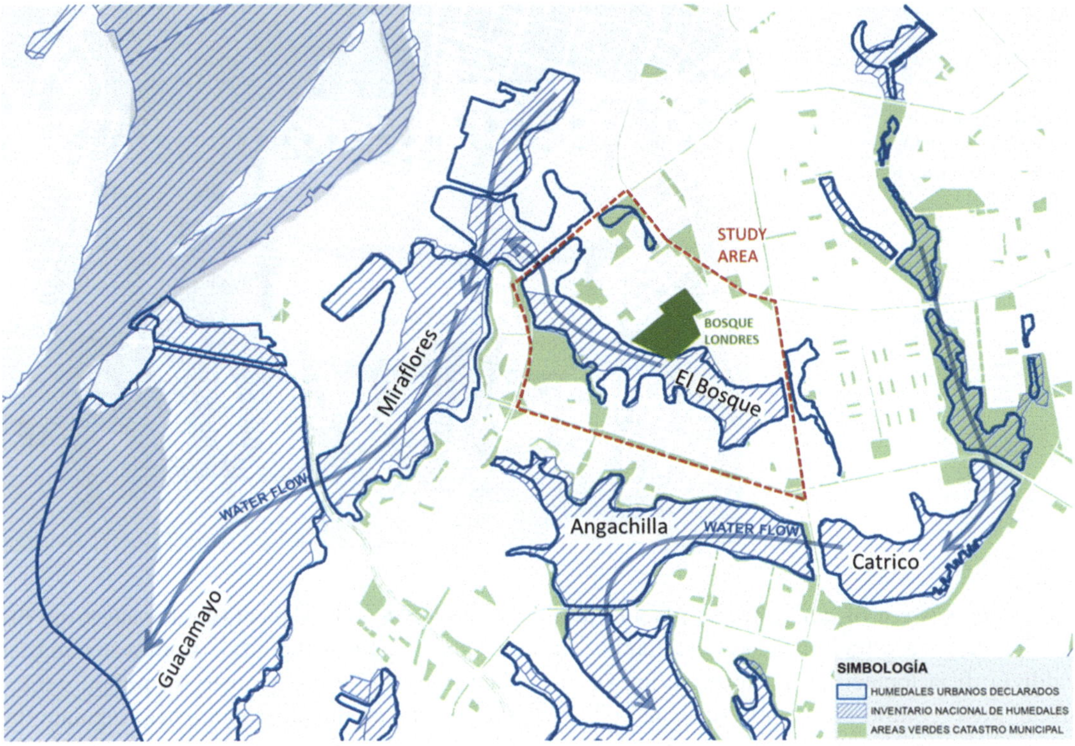

Fig. 4 Study area and Bosque Londres, in relation to the Urban Wetlands surrounding system. (Source: Own elaboration with data from the local Municipality of Valdivia and map from the geoportal of the Ministry of Environment of Chile, 2024)

coihue (*Nothofagus obliqua-Nothofagus dombeyi*) communities.

- Swamp forest communities of myrtle known as "hualves."
- Wetlands of the reed or rush type formed by various species of Typha known as "totora," "vatro," and "cortadera," among others.
- Communities of invasive exotic species.
- Floodable plains.

The first three are native formations, while the latter are of anthropogenic and introduced origin. In this context, it is relevant to highlight that there are two very important remnants of native forest for the city on the edges of the El Bosque wetland: El Bosque Urban Park (a privately owned park located at the southwest of the study area) and Bosque Londres (the project, located north of the study area) (Fig. 9).

Regarding birdlife, according to data from FORECOS - a local Foundation- and the Citizens' Network for Wetlands of Valdivia, this sector of the city is home to different species of amphibians such as the masked frog, spotted frog, and Chilean big frog. Also, land shrimp live here on the edges of the wetland. On the other hand, records from the El Bosque Urban Park administration indicate the existence of 46 bird species, including birds from reed areas such as the siete colores, trile, run run, and carpintero, ducks such as the jergón, herons, and huairavo, as well as forest birds such as the churrín, cometocino, and chuncho. In swamp forests or floodplains, it is possible to find species with conservation issues, including the Huillín or River Otter (*Lontra provocax*), the Güiña (*Oncifelis guigna*), the Chilean Pigeon (*Columba araucana*), and the Quique (*Galictis cuja*) (Hauenstein et al. 2002) (Fig. 10).

In city of Valdivia humid environment and dark undergrowth allow for the development of many fungi in the forests and inland wetlands (Ladera Sur, 2022). The fungi of these ecosys-

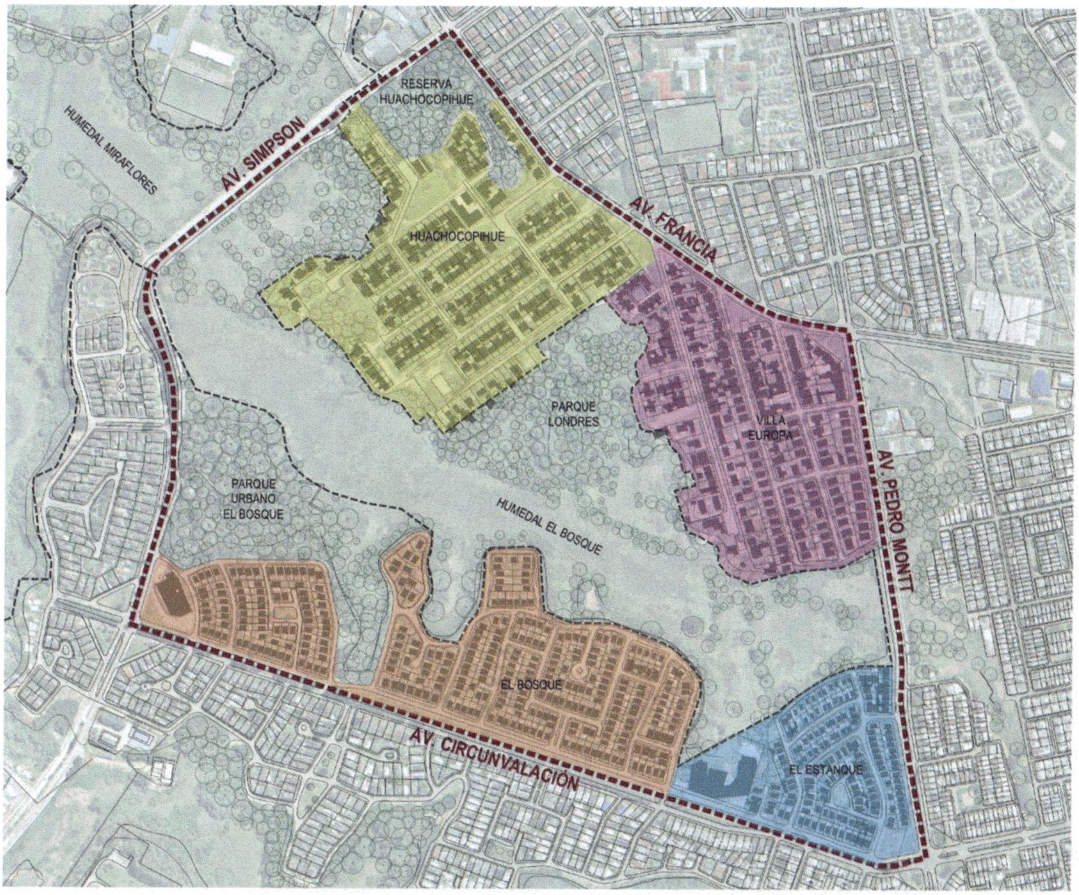

Fig. 5 Definition of the study area and surrounding neighborhoods. (Source: Own elaboration on Google Earth's base map, 2024)

tems consist of ectomycorrhizal fungi during the autumn (small Mycena among the mosses and various mushrooms in shapes, sizes, and colors). Then in spring, cup-shaped and gummy fungi, as well as digüeñes, emerge in great abundance. In summer, stick fungi, which persist throughout the year, can be found (Fig. 11).

From an urban perspective, the area has not been planned as a unit around the wetland and its natural systems. Each neighborhood is embedded in the urban fabric connecting to the main roads (Av. Francia, Av. Pedro Montt, Av. Simpson, Av. Circunvalación Sur), and the wetland has been treated as "the back part." The perimeter roads of the study area have the positive quality of having central medians of varied width and urban tree planting, connecting with the rest of the city. On the other hand, the relationship between urban fabric and wetland occurs differently north and south of the wetland: in the north (where Huachocopihue neighborhood resides), the urban fabric ends at the wetland, with only specific public roads balconing over the slope, which are identified as segments with potential. On the other hand, on the south side, urban design has included a street bordering open to the wetland. Lastly, urban green areas are distributed as partially connected patches with diverse identities and vocations (Figs. 12 and 13).

Fig. 6 Aerial photograph 2023. (Source: Google Earth, 2024)

In summary, it is concluded that the main attributes and potentials of the studied area are associated with the ecological value of the existing natural systems, while the identified issues and threats mainly relate to the relationship between urban life and the natural environment, both morphologically and in terms of informal uses and antisocial practices. In recent years, a new paradigm of recognizing the value of urban wetlands has been established, both among the population and in institutions. This implies recognizing that the wetland is a central attribute, linked and connected to valuable ecosystems: forests and floodplains and also inexorably connected to a larger water system of rivers and wetlands (Table 1).

4 Wetland Forest Neighborhood: An Urban Imaginary Proposal for a Wetland City

The proposed urban imaginary for the whole study area encompasses the section bordering the El Bosque wetland. However, it is envisioned as a future scenario for a wetland city, capable of being applied in other contexts of protected urban wetlands, as an integrated system of neighborhoods that coexist harmoniously with the natural environment in which they are located. In this way, the main objective of the proposal is to highlight the natural attributes of the place, transforming the wetland into the central and unifying element for the surrounding neighbor-

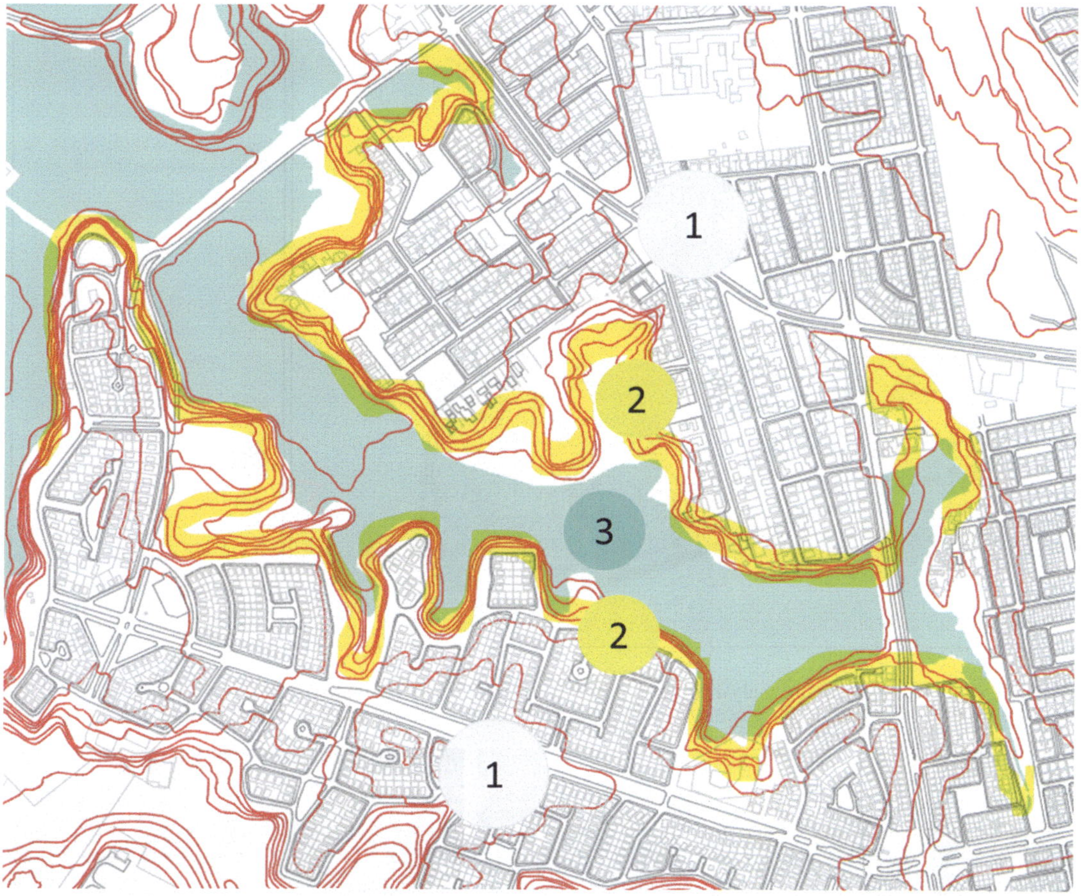

Fig. 7 Topography diagram (contour lines) and three morphological elements: (1) Upper terrace, (2) Transitional slopes, and (3) Low-lying wetland area (Source: Own elaboration, 2024)

Fig. 8 Vegetation present in Bosque Londres. (Source: photos by the authors, 2024)

Fig. 9 Vegetation cover map for analysis of the study area. (Source: Own elaboration, 2024)

Fig. 10 Fauna present in the study area. (Source: Laderasur.com website)

hoods. Three project guidelines are proposed to shape the proposed wetland city image, each of them composed by a series of strategies and actions commented below (Fig. 14).

4.1 Consolidation of an Integrated System of Natural Value Areas

(a) Wetland Core: The wetland is proposed as the central core of the imaginary, aiming to integrate it into the urban fabric as the structuring element of the proposed natural areas system, consolidating a natural buffer zone. This buffer will not only protect it from threats but also consolidate natural spaces for public use.

(b) Urban Natural Reserves: These correspond to the existing patches of relict native forests, in this case, Parque Urbano El Bosque, Reserva Huachocopihue, and Bosque Londres. Their condition as border forests allows them to serve as public spaces of natural transition between the city and the surrounding wetlands. They are proposed as

Fig. 11 Fungi present in the study area. (Source: Chilebosque.cl website)

Fig. 12 Road connectivity map. (Source: Own elaboration, 2023)

delimited areas with controlled access, with a vocation for environmental learning and community outreach.

(c) Seasonal Floodplains: The floodable areas are proposed as spaces for approaching the wetland with seasonal use. These areas flood in winter and during summer they transform into natural meadows, compatible with non-invasive recreational uses such as hiking, birdwatching, or similar activities.

(d) Forest Squares: These correspond to native groves inserted into the urban fabric, composed of old trees and grassy meadows used as recreational areas, with low presence of undergrowth. This recognizes their current condition and integrates them into the natural areas system as free-access forest enclaves, where plaza uses and facilities are allowed, serving as a prelude and access to natural reserves.

(e) Neighborhood Squares: These are squares with neighborhood use programs and facilities such as playgrounds, community centers, sports infrastructure, etc. The objective of these squares is to provide recreation and leisure areas for the community, aiming to preserve natural value areas as natural spaces, with minimal intervention, loading programs, and greater intensity of use to green areas within the neighborhoods.

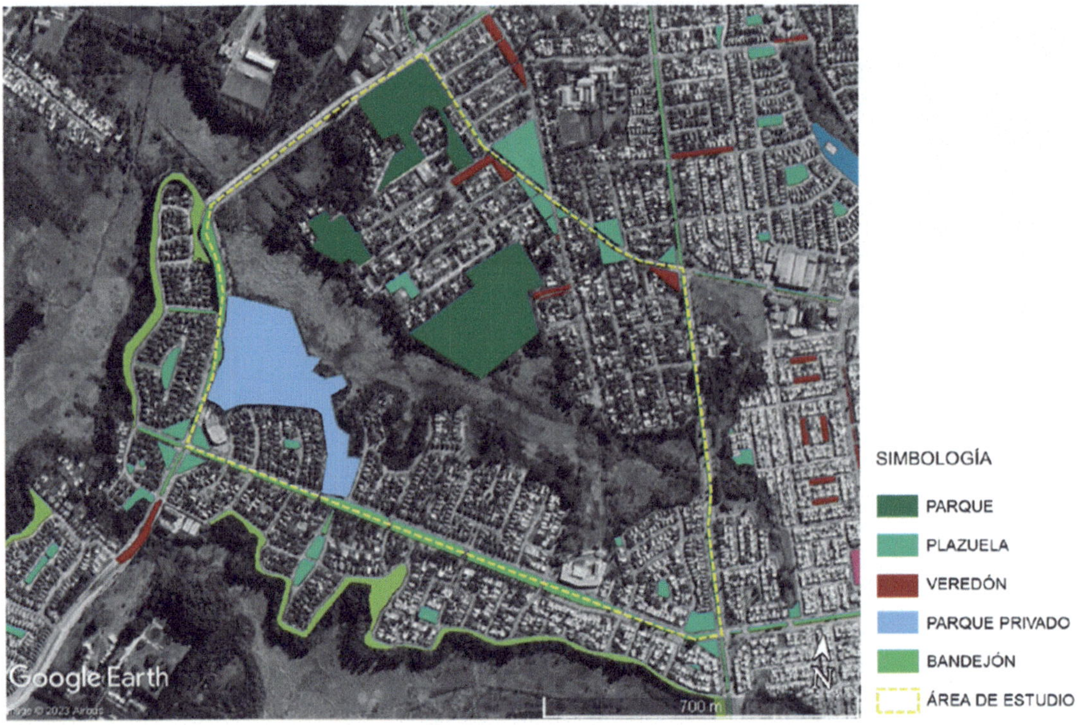

Fig. 13 Green areas of Valdivia by type, and study area boundaries. (Source: Own elaboration, 2023, based on Google Earth and KMZ Green Areas Classification, Municipality of Valdivia)

Table 1 Summary table of attributes, potentials, problems, and threats identified around El Bosque Londres site (Source: Own elaboration, 2023)

Attributes and potentialities	Problems and threats
El Bosque wetland: Neighborhood's core	Disruption of El Bosque wetland's continuity
Remnant forest patches	Sediment accumulation
Hualves or flooded myrtle forests patches	Edge slope—risk zone identified in the local regulatory plan
Perimeter green avenues	Irregular land seizures
Neighborhood green areas	Some closed and inaccesible edges
Edge streets towards the natural system	Micro-landfills
Part of the city's stormwater system network	Area perceived as unsafe
Existing schools	Sectors dominated by invasive plants
Viewpoints or wide-view points	

(f) Native Gardens: These correspond to all those green medians, sidewalks, small squares, and urban patches with the potential to be native gardens that provide ecological continuity to neighborhood routes and allow direct contact with native flora as a learning space.

4.2 Restoration of Hydrological Systems Through the Implementation of Nature-Based Solutions for Stormwater Management

(a) Restoration of Wetland Hydrological Connectivity. It is proposed to increase the flow capacity of the water passing underneath the streets, historically built as land-

Fig. 14 Wetland forest neighborhood's proposal: synthesis image. (Source: Own elaboration, 2023)

fills, by increasing connections or existing pipes and replacing pavements with draining surfaces in low-lying areas of the avenues.

(b) Improvement of Watercourses. A landscape redesign of these watercourses is proposed as an interconnected network, allowing for greater storage, flooding, and water infiltration capacity before being discharged into larger water bodies.

(c) Ecological Restoration Ponds. To restore the wetland's dampness condition, the creation of stationary ponds is proposed to receive rainwater from watercourses before being discharged into the wetland. This increases the flood-prone area and water retention capacity, creating new habitats to enrich its biodiversity.

(d) Rain Gardens. A network of rain gardens is proposed to depurate water from streets and sidewalks before entering the sewer system that discharges into the wetlands, as well as to serve as learning spaces, with the implementation of signage explaining their functioning.

4.3 Implementation of a Network of Pathways at Different Scales and with Different Approaches to the Natural Environment and Green Areas

(a) Perimeter Greenway. It consists of transforming the perimeter ring formed by the main avenues surrounding the wetland into an urban green corridor. This includes incorporating bike lanes, improving sidewalks, and consolidating urban tree planting in central medians with native species.

(b) Inner Border Route. Corresponds to the streets surrounding the wetland and existing forests, where visual connections and openings to the surrounding landscape are generated. It is proposed to maintain its function as a residential street for smaller vehicles, complemented by an inclusive pedestrian promenade along the edge, allowing for a safe and consolidated route for all people.

(c) Natural Pedestrian Trails. These are trails that penetrate the various natural spaces of the neighborhood (wetlands and forests). Existing footprints and routes are recognized, as well as viewpoints, bridges, docks, or other infrastructures that highlight landmarks in the landscape.

5 Edge Scale: Landscape Planning for Bosque Londres

In accordance with the urban imaginary developed at the macro scale of the wetland system, landscape planning for Bosque Londres begins with its ecosystem function as a forested wetland edge under the scope of an Urban Natural Reserve. Thus, the concept of an Urban Park is reimagined as "Bosque Londres Urban Natural Reserve," defining zones with different protection statuses and intensity of use. The area considered for landscape planning encompasses a total of 5.2 hectares of natural surroundings composed of forest formations and wetland areas. Nestled within the urban fabric, the site exhibits various landscape units primarily defined by its geomorphology, vegetation cover, and edge typology, detailed as follows.

5.1 Geomorphological Description

The topography of the site features two predominant levels: a terrace or plateau at a higher and a lower part at the wetland level. These levels are abruptly separated by a slope that creates a disconnection between both sectors, with predominant use and pathways in the high terrace sector, directly connected to the urban fabric, while the lower sector lacks formal access points (Figs. 15 and 16).

5.2 Existing Landscape Units

Eight groups of vegetation coverage were identified to characterize and understand the various ecosystems present in the study area.

1. Valdivian evergreen native forest
2. "Hualve" or Swampy Forest of myrtaceous
3. Clearings in floodplains
4. "Quilantos" or *Chusquea quila* forests
5. Sedges of hydrophytic vegetation
6. Floodable groves of exotic origin
7. Edge groves of exotic origin
8. Scrubland of invasive exotic vegetation (Fig. 17)

In this manner, the previously described units were classified into two groups to propose appropriate guidelines for their management and conservation. On one hand, there are the native ecosystems as objects to be preserved and regenerated, and on the other hand, there are the vegetation communities with invasive behavior, which represent a threat to the conservation of native ecosystems (Fig. 18).

5.3 Characterization of Edges

Different edge typologies were defined for the forest boundaries, to understand the relationship and exchange with the surroundings and their role as urban-natural transition spaces.

5.3.1 Identification of Edge Typologies

The boundary defining the area of Bosque Londres covers a total of 5.2 hectares and its perimeter a total of 1090 linear meters, of which only 16% currently allows access or connection to the urban

Fig. 15 Topography map and Bosque Londres' boundary. (Source: Own elaboration, 2023)

Fig. 16 Levels scheme in Bosque Londres. (Source: Photo by the authors, own elaboration, 2023)

Fig. 17 Vegetation cover map. (Source: Own elaboration, 2023)

Fig. 18 Existing landscape units in Bosque Londres study area (native ecosystems on the right, invasive vegetation to the left). (Source: Own elaboration, 2023)

Fig. 19 Bosque Londres edge typologies (yellow for private edge, blue for public edge, green for wetland edge). (Source: Own elaboration, 2023)

fabric (Fig. 19). The fact that a high percentage of its edges are privatized, hinders visual control and citizen security over large areas of the forest, and the walls of the houses produce blind spots that encourage the appearance of hidden sites where various illicit activities occur. On the other hand, there is inappropriate occupation of the space that poses a threat to the environmental conservation of the forest, which derives in landfills, micro dumps, proliferation of invasive species, and discharge of housing sewage.

5.4 Edge Scale Strategy and Proposal

The area features a variety of landscape units that must be understood as the main attributes of the forest. In this way, existing ecosystems were worked on, promoting environmental care and protection, as well as the provision of a natural good for public use. This approach links the community with nature and serves as a space for learning and valuing all natural systems present in the city. In this sense, the project has an imprint that seeks to "re-naturalize" public spaces, to respond to the demands that the Valdivian society currently has. Subsequently, the main objectives of the project are:

(a) Protect and regenerate the natural value of Bosque Londres.
(b) Foster nature learning and conservation experiences.
(c) Provide safe and inclusive well-being spaces for all audiences.

To accomplish this, the proposal is to emphasize the transitional nature of this natural area, showcasing wetlands as an integral part of the urban landscape. This area features a natural vantage point that enables observation of the wetlands without the necessity of direct intervention (Fig. 20).

(a) Landscape Zoning

A strategic zoning approach is proposed for the landscape of Bosque Londres, aiming to recognize and respect the existing ecosystems while

Landscape Zoning	***Enablement Zones:***
	Restricted use: habitat conservation (Low forest).
	Moderate use: regenerative landscaping (High forest).
	Public use: green areas and native gardens.
	Transition Zones:
	Slope edge: regeneration and visibility management.
	Private edges: firebreak landscaping
	Public edges: regeneration and exotic vegetation management
Edges consolidation	***Public Spaces Integrated System:***
	Access squares
	Urban viewpoints
	Native pathways and rain gardens
	Enclosure and Delimitation of Private Boundaries:
	Boundaries along Londres Street
	Boundaries along Italia Street
Enablement of pathways and community programs	***A Route Network as a Learning Experience***
	Central inclusive route
	Sensory edge route
	Hualves route
	Stay Areas and Programs
	Access pavilions
	Outdoor classrooms
	Viewpoints
	Forest clearing
	Organic waste management and composting Areas
Complementary actions	***SInformative, interactive, and inclusive signage***
	Flora and fauna friendly lighting design

Fig. 20 Design strategies for Bosque Londres at the Edge Scale (Source: Own elaboration, 2023)

integrating human use of the area highlighting its status as a relic forest. Six zones are identified to regulate the use and purpose of the forest, gradually transitioning from its urban and public condition to its natural and conservation-oriented state. Two types of zones are defined: enablement zones and transition zones. Each of these zones entails specific landscape management practices and regulates the intensities of use allowed within them (Fig. 21).

(b) Edge Consolidation

The proposal involves reclaiming illegally occupied spaces at the forest edges to establish a network of squares, viewpoints, and public green areas. These spaces will facilitate access to and visibility of Bosque Londres and the wetland. Furthermore, it suggests installing boundary fencing along private properties to deter further unauthorized occupations and enhance the sense of security.

(c) Enablement of Pathways and Community Programs

The zoning proposal arises from the experience and recognition of the ways in which the place is currently used by the local community. In this sense, the proposed pathways and social

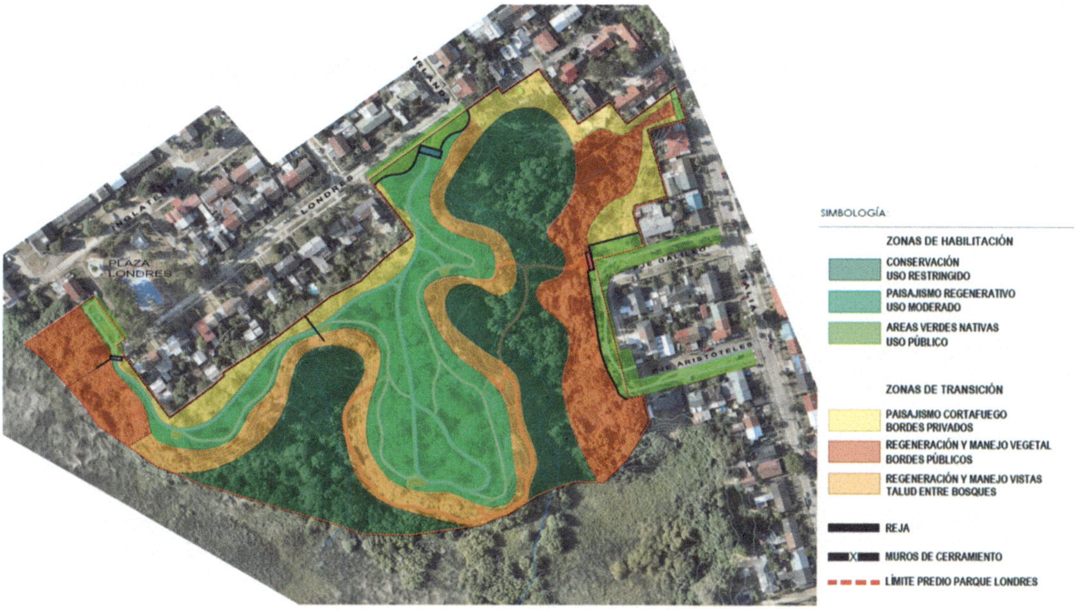

Fig. 21 Landscape zoning for Bosque Londres (colored green areas are enablement zones, colored yellow, orange and red areas are transition zones) (Source: Own elaboration, 2023)

Fig. 22 Photographs of participatory activities in Bosque Londres (Source: photos by the authors and Valdivia Municipality, 2023)

areas in the master plan correspond to existing trails and gathering areas to safeguard local practices in the forest. The latter was reviewed and aligned with the landscape zoning areas (Figs. 22 and 23).

6 Enablement Scale: A Minimal Intervention for Forest Conservation (Phase 1)

The municipal assignment defined a first stage of enablement to build during their administration covering 1.6 hectares of Bosque Londres that corresponds to the high terrace area historically used by the community for strolls and nature interaction. The enablement considers a transition from the urban to the natural environment, regulating intensity of use, with a focus on nature contact learning and traditional uses of the area (Fig. 24).

Thus, the architectural proposal aims to enable forest use and protection with subtle interventions allowing for pathways and meeting areas, while the landscape proposal is approached under the concept of ecological restoration with an emphasis on rehabilitation to recover the functions of the degraded ecosystem

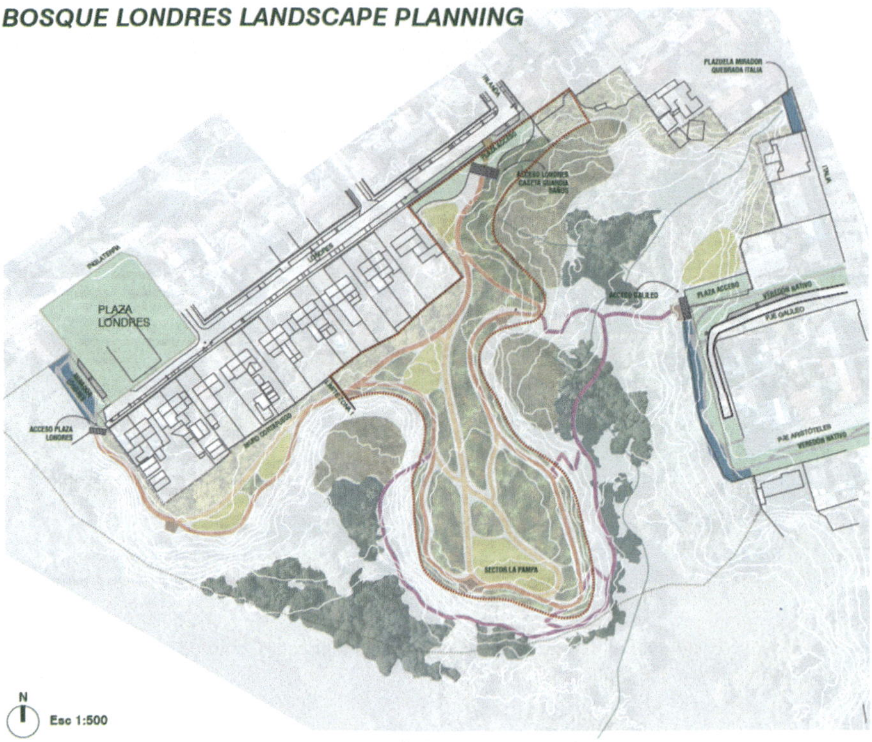

Fig. 23 Bosque Londres landscape planning—synthesis plan (Source: Own elaboration, 2023)

Fig. 24 Bosque Londres phase 1 site (Source: photos by the authors, 2022)

(MMA 2023). The following are the project enablement objectives:

(a) Integrate ecological restoration of the forest with human use.
(b) Enhance natural elements characteristic of the forest.
(c) Maintain the atmosphere of the place with minimal intervention.
(d) Provide safety and well-being for visitors (Fig. 25).

In this manner, the landscape architecture proposal is developed considering the following areas and components:

Fig. 25 Bosque Londres enablement plan, phase 1 (Source: Own elaboration, 2023)

Fig. 26 Access square and pavilion renders (Source: prepared by RIBA renders, 2023)

Fig. 27 Edge pathway, signage, and seating area with bench render (Source: prepared by RIBA renders, 2023)

6.1 Architecture Design Elements

(a) Pavilion and Access Square: The pavilion responds to the programmatic requirements presented by the local municipality that were validated by the community in the participatory process. It is located perpendicular to the access road, forming a portal, and set back from the sidewalk to create an access square as public entrance. The pavilion runs along the fence and its placement was carefully decided to distance it from large trees to avoid damaging roots (Fig. 26).

(b) Pathways: The proposal maintains existing pathways, making some adjustments to allow for gentle slopes in the access, create space for viewpoints, and maintain openness and visual control. The edge path consolidates the existing path with permeable red brick pavement that allows for universal accessibility. Interior paths remain rough trails with the addition of mulch in some areas (Fig. 27).

(c) Resting Spots: The proposal incorporates various resting spaces, such as an open-air classroom, two viewpoints, resting areas along the edge path, and benches strategically located in symbolic places in the forest. These places not only encourage contemplation but also play a vital role in visual control and the sense of security (Figs. 27 and 28).

(d) Grove Areas: Subtle delineations are designed in two grove areas—Olivillos and Ulmos—to highlight and visualize their value through simple architectural interventions with poles and rope, benches, walkways, and signage (Fig. 30).

(e) Signage: To highlight all the natural elements identified in the route, four different types of signage were designed, according to the local signage manual of the municipality (Fig. 31).

(f) Benches: The design challenge was to achieve harmony with the natural context, allow universal accessibility and comfort, using durable and anti-vandal construction systems. Two typologies—with and without backrest—were designed and placed in the resting areas (Fig. 32).

(g) Fences and Boundaries: Different elements were projected to safeguard the transition from urban to natural, from the fence that closes the enclosure for controlled access to temporal wooden barriers for ecological restoration areas.

Fig. 28 Outdoor classroom and forest viewpoint renders (Source: prepared by RIBA renders, 2023)

Fig. 29 Wetland viewpoint render (Source: prepared by RIBA renders, 2023)

Fig. 30 Ulmo Madre's wooden walkway and Olivillo's grove area render (Source: prepared by RIBA renders, 2023)

6.2 Landscape Design Elements

(a) Naturalistic and Restorative Landscaping: The overall landscaping proposal responds to the urban-natural transition, proposing a gradient that ranges from native flower gardens in the areas closest to the urban interface, to ecological restoration zones in the wilder forested area. A wide plant palette is proposed, seeking to maximize diversity with native species from the region. For this purpose, three general landscaping zoning areas have been defined (Fig. 33).

(b) Ecological Restoration Zones: These zones are located within the forest with the aim of restoring biodiversity functions of the degraded ecosystem. For this purpose, two large areas with restricted access were defined where regeneration is already occurring, and where orchids, Chilean bellflowers, and other protected, emblematic, or rare species have been identified. Additionally, shrub

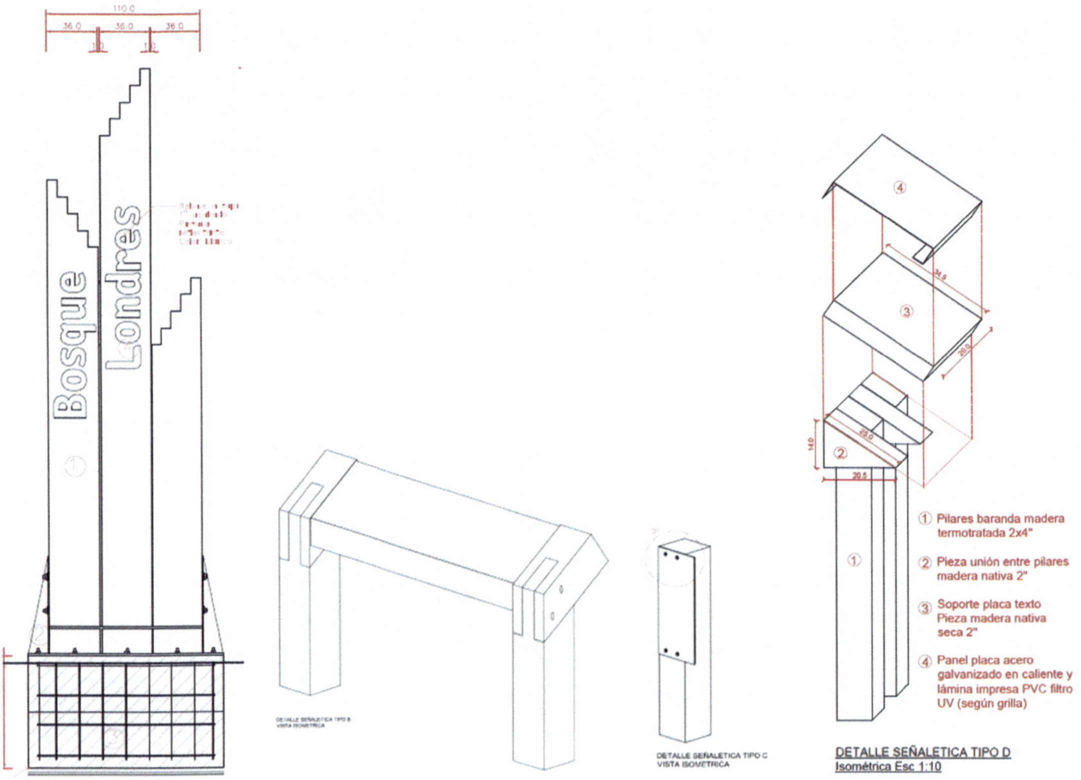

Fig. 31 Signage details type A, B, C, and D according to the local municipal signage manual (Source: Own elaboration, 2023)

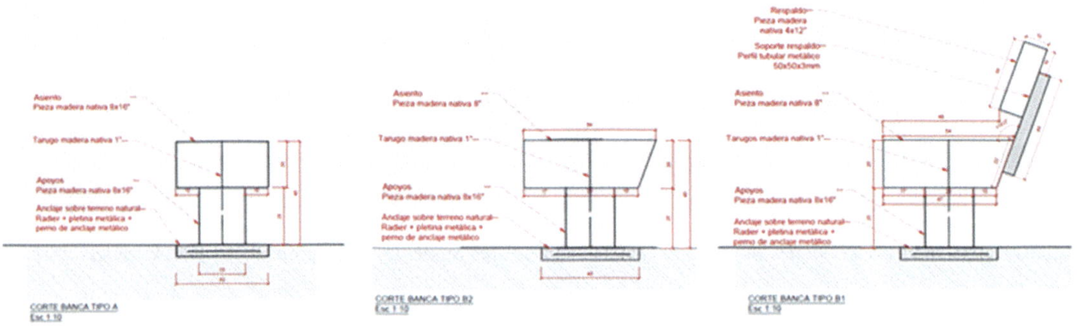

Fig. 32 Wooden benches details (Source: Own elaboration, 2023)

Fig. 33 Landscape zoning, in dark green ecological restoration areas, in light green lawn areas, in purple garden areas. (Source: Own elaboration, 2023)

masses are proposed in specific locations, which require blocking passage or framing places. An exclusive plant palette of the identified vegetation communities (Roble-Laurel and Roble-Coihue) is used to increase the diversity of existing species, prioritizing the insertion of species with low representation in the area. The restoration zones are classified into seven types: according to the density of plants required for the sector (intensive, medium density, or complementary); and whether multi-layer (with trees) or only low stratum (without trees) is specified (Fig. 34).

(c) Garden Areas: The gardens are in the access plaza, sidewalk, and surroundings of the open classroom and boundaries with adjacent lands. Only native species from southern Chile, highly resistant to the urban environment, are used, prioritizing species diversity that constitutes habitat and food, and plants with attractive flowering.

These are the sectors with the most aesthetic and spatial functions, and a naturalistic landscaping style is proposed with high density and a random dispersion of the different species (Fig. 35).

(d) Lawn Areas: Within the forest, specific areas are proposed where the existing lawn is maintained, resembling forest clearings, to keep these spaces open and sunny.

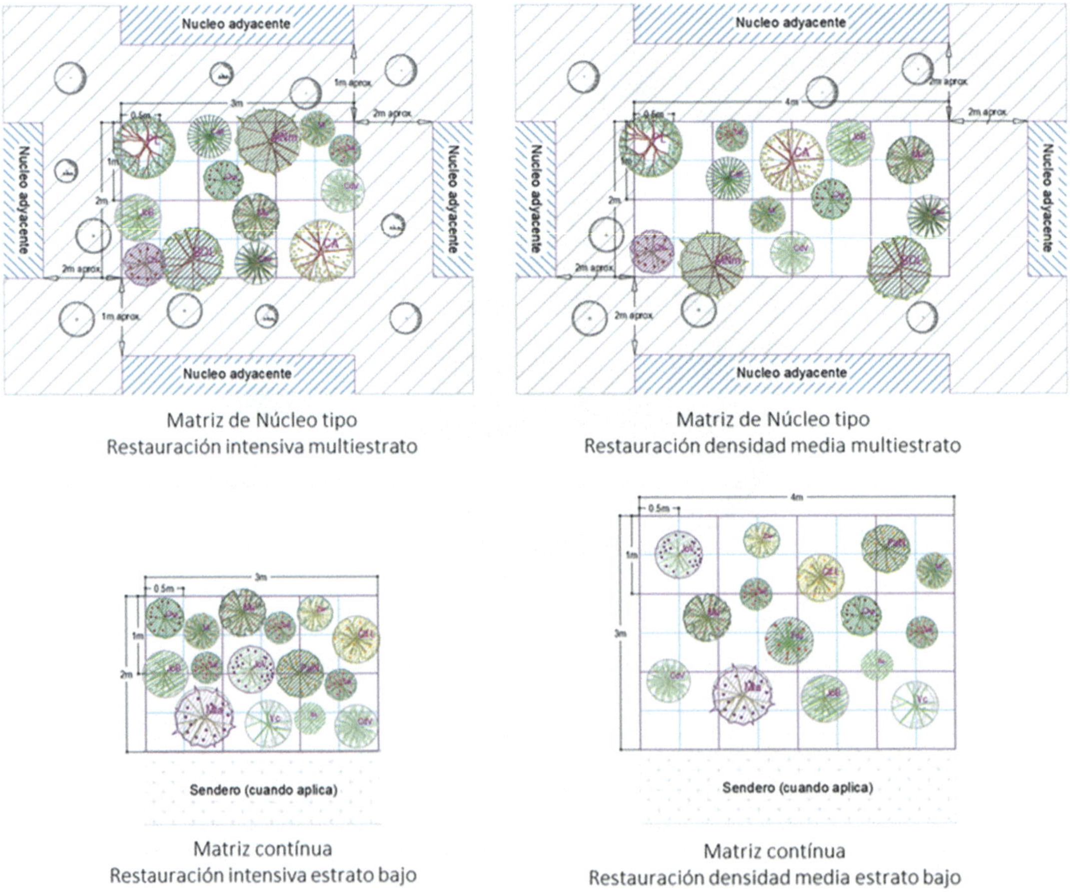

Fig. 34 Examples of plantation matrices for Ecological Restoration Zones, according to plant density and stratum types (Source: Own elaboration, 2023)

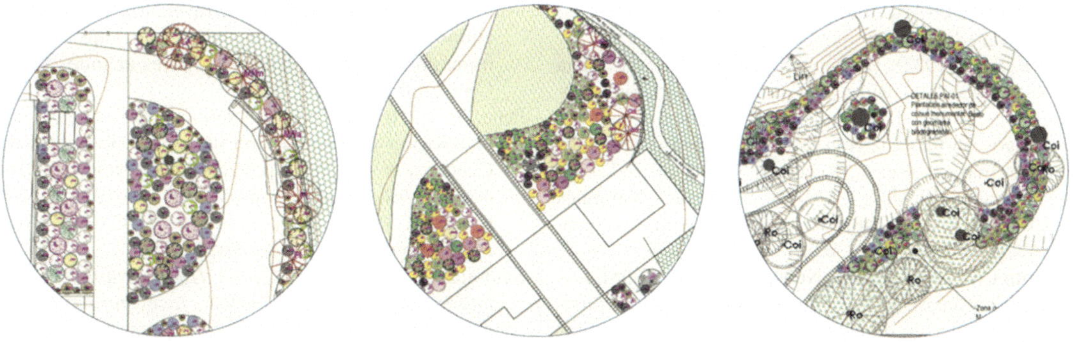

Fragmento planos de plantación
Núcleos Nativos Bandejón Bajo

Fragmento planos de plantación
Núcleos Nativos Arbustivo Plaza

Fragmento planos de plantación
Macizos Borde Camino

Fig. 35 Plantation plans examples for garden areas (Source: Own elaboration, 2023)

7 Findings

Despite the extensive literature on landscape architecture, ecological regeneration, and urban design, there is relatively little information on how to design with urban nature. This may be due, on one hand, to the absence of technical and legal guidelines to address these types of projects and/or, on the other hand, to the difficulty of integrating ecological criteria with those of architecture, which involve intervention for human use.

In this sense, Bosque Londres project seeks to clarify these issues and respond to the challenges of including natural systems in urban contexts in architectural projects for community use, in this case, through the study and design for the enablement of a relic forest on the edge of a wetland in the city of Valdivia.

The above is achieved by first examining the conceptual focus of the commission rethinking the design approach from an urban park to a natural forest reserve, which although it does not currently have legal support, it allows to prioritize its ecosystemic role with the imprint of re-naturalizing the concepts of spaces for public use. And later, by studying the forest space through a multiscale and participatory process where local knowledge along with technical studies are equally important.

Finally, the project aims to highlight the need to value these natural spaces that interface between urbanized and natural areas as public assets for the city that promote the learning of care and protection of its environment.

As a result, the design process has triggered a paradigm shift, evolving from a traditional focus on urban parks to one more aligned with natural conservation and the integration of existing natural systems within the city. Additionally, it has proposed design guidelines with the potential to be replicated in similar projects with the purpose of establishing an urban imaginary of a wetland city.

References

Gajardo R. (1994). *La vegetación natural de Chile, Clasificación y distribución geográfica*. Santiago, Chile: Editorial Universitaria.

Hauenstein E, M González, F Peña-Cortés & A Muñoz-Pedreros. (2002). Clasificación y caracterización de la flora y vegetación de los humedales de la costa de Toltén (IX Región, Chile). *Gayana Botánica*.

Ilustre Municipalidad de Valdivia. (2022a). *Plan de recuperación de áreas verdes, parques y áreas de valor natural*. PPT Official document.

Ilustre Municipalidad de Valdivia (2022b). PRC Valdivia Refundido. Online map. https://munivaldivia.maps.arcgis.com/apps/webappviewer/index.html?id=57f9244080d046bb9bacbe70f87e75d2. (Accessed March 1st, 2023).*Memoria Actualizada: Estudio Actualización Plan Regulador Comunal*. Working Report. https://www.google.com/url?sa=t&rct=j&q=&esrc=s&source=web&cd=&ved=2ahUKEwifxZ3_tPaEAxXxqpUCHUZSCBsQFnoECA4QAQ&url=https%3A%2F%2Fmunivaldivia.cl%2Fdoctos%2Fplan_regulador%2Fresumen_ejecutivo_estudio_prcv.doc&usg=AOvVaw1anIiB6g_

y1Xw5bVTHARZ5&opi=89978449. (Accessed February 1st, 2023).

Ladera Sur. (2022). Temporada de hongos en plena Región de Los Ríos: una diversidad de formas y colores. https://laderasur.com/articulo/temporada-de-hongos-en-plena-region-de-los-rios-una-diversidad-de-formas-y-colores/. (Accessed March 1st, 2023).

Organización Biósfera. (2011). Reservas Naturales Urbanas Valdivia. https://rnuvaldivia.wordpress.com. (Accessed March 1st, 2023).

Oyarzún C., Nahuelhual L., Núñez, D. (2004–2005). Los servicios ecosistémicos del bosque templado lluvioso: producción de agua y su valoración económica. *Revista Ambiente y Desarrollo* 20(3)–21(1) : 88–95. Santiago de Chile.

MMA Ministerio del Medio Ambiente. Definiciones marco para la Restauración Ecológica. MMA Website, https://restauracionecologica.mma.gob.cl/definiciones/. (Accessed December 1st, 2023).

Ministerio del Medio Ambiente, Inventario Nacional de Humedales. MMA Website, online map, https://arcgis.mma.gob.cl/portal/apps/webappviewer/index.html. (Accessed March 1st, 2023).

Peña, E. (2016). El bosque como regulador del ciclo hidrológico y la producción de agua. Universidad de Concepción website. http://udec.cl/panoramaweb2016/content/el-bosque-como-regulador-del-ciclo-hidrol%C3%B3gico-y-la-producci%C3%B3n-de-agua. (Accessed December 1st, 2023).

Sanhueza S. A. (2018). *Barrio Huachocopihue en el contexto de la Ciudad-Jardín: claves para entender el arraigo de sus habitantes*. Thesis. Universidad Austral de Chile. http://cybertesis.uach.cl/tesis/uach/2018/fcs226b/doc/fcs226b.pdf. (Accessed March 1st, 2023).

Walker, B. H., y Salt, D. (2006). *Resilience Thinking: Sustaining Ecosystems and People in a Changing World*. Washington, D. C.: Island Press.

Part III

Engaged Community for Sustainable and Water Sensitive Cities

Wetlands in Resistance: Critical Environmental Education and Community Participation in the Lower Paraná Delta

Luciano Iribarren, Katherine Guerrero Tamayo, Fernando Garelli, and Ana Dumrauf

Contents

Abstract

The present work is located in the Paraná Delta territory, a wetland macrosystem of strategic importance for flood regulation and drinking water supply for the Metropolitan Area of Buenos Aires (Argentina). In a scenario of growing environmental conflicts in Latin America due to the expansion of extractive activities, this territory is in dispute, threatened by real estate projects that intend to extract income, causing enormous environmental damage and dispossessing the land from the local population. The urbanization model in gated communities, which is pressing to be implemented, simultaneously produces the destruction of wetlands and the elimination of the islander way of life. The islander population, however, resists and maintains that their way of life, bearer of peasant and indigenous knowledge, guarantees life sustainability in the region. Interweaving the fields of Environmental Education and Political Ecology of the South, we present an analysis of how environmental knowledge was formed and constructed in the processes we have been carrying out with socio-territorial organizations that resist and r-exist in the wetlands of the Paraná Delta. We reference Latin American Environmental Thought, considering its approaches to environmental justice, political ontology and environmental rationality. Focusing on community participation, we share the results of the systematization of three experiences in which islanders,

L. Iribarren (✉) · K. Guerrero Tamayo · F. Garelli · A. Dumrauf
Grupo de Didáctica de las Ciencias, Universidad Nacional de la Plata/Instituto de Física de Líquidos y Sistemas Biológicos-CONICET,
La Plata, Provincia de Buenos Aires, Argentina
e-mail: rodiribarren@abc.gob.ar

C. Rojas Quezada (ed.), *Urban Wetlands in Latin America*, Sustainable Development Goals Series,
https://doi.org/10.1007/978-3-031-69590-2_12

environmentalists and academics participated.

Keywords

Resistance · Education · Community · Participation

1 Introduction

Due to growing environmental conflicts worldwide and especially in the Latin American context, Environmental Education (EE) has become a field of increasing priority. Since our group began in the 1980s, we have had a long journey from the Didactics of Natural Sciences to a field we call Natural Sciences, Environmental and Health Education (Cordero et al. 2011). We frame ourselves in different critical political-pedagogical currents, and from that focus, we intend to question the colonialist claim of the superiority of scientific knowledge over other forms of knowledge (Castro Gómez 2021) without denying its social importance but building a dialogue of knowledge(s) (Leff 2021). This position is expressed in the constitution of our group, which is intersectoral, interdisciplinary, intergenerational, interethnic and intergender, in which knowledge and experiences circulate and actions and decisions are jointly defined (Dumrauf et al. 2019). We try to contribute to the construction of decolonizing pedagogies nourished by the exchanges generated between those of us who weave from diverse spaces, recognizing different forms of knowledge as inclusive principles for life care.

In this chapter, we will address the characteristics of three critical EE experiences in the delta of Paraná River wetlands, focusing on community participation.

2 Historicizing the Disputed Territory

From an ecological point of view, the Paraná River delta is a wetlands macro-system, a source of valuable common goods, with a wide variety of beneficial ecological processes for humanity (Malvárez 1999). In his studies of the region's environmental history, Galafassi (2005) produced a historical narrative from the times of Chaná and Guaraní ethnic groups/nations to the end of the twentieth century and describes how the traditional peasant population of the region was formed. With lines of rupture and continuities with the ancestral indigenous population, they are popularly known in Argentina as "islanders", bearers of an "islander way of life". This population started to decrease after 1950, but it still has a significant representation in the region. Since the end of the twentieth century, we could speak of a new historical period in terms of changes in migratory flows, in which a reinvention and dispute over the "deltaic paradise" has occurred (Astelarra 2016).

In 2008, a real estate company tried to build a mega-gated neighbourhood named Colony Park,[1] forcing a territorial, environmental conflict with the local population. The dredgers illegally advanced over more than 300 ha of wetlands, violently evicting "junqueros" [sedge harvesters] families and fishermen living in the Anguilas stream. In this context, the Isla Esperanza Cooperative (IEC)[2] was created, a socio-territorial organization linked to indigenous and peasant movements (Cooperativa Isla Esperanza 2012). In 2012, in a context of intense social pressure and struggle, the federal court ruled in favour of the "isleros"[islanders], prompting the company to remedy the environmental damage produced. However, the remediation has not occurred yet, and the legal process is still open. At the end of 2016, after new episodes of violence from Colony Park, the IEC called local environmental organizations and wetlands researchers from academia for support. As a result, the Delta Wetlands Observatory (DWO)[3] was formed, an intersectoral organization which we integrate (Fig. 1).

According to Astelarra and Domínguez (2015), sedge harvesting families recognize themselves as "island makers", a modality of

[1] http://www.islacolonypark.com/la-isla.html

[2] https://www.facebook.com/CoopEsperanzaIsla/

[3] https://www.facebook.com/ObservatorioHumedalesDelta/

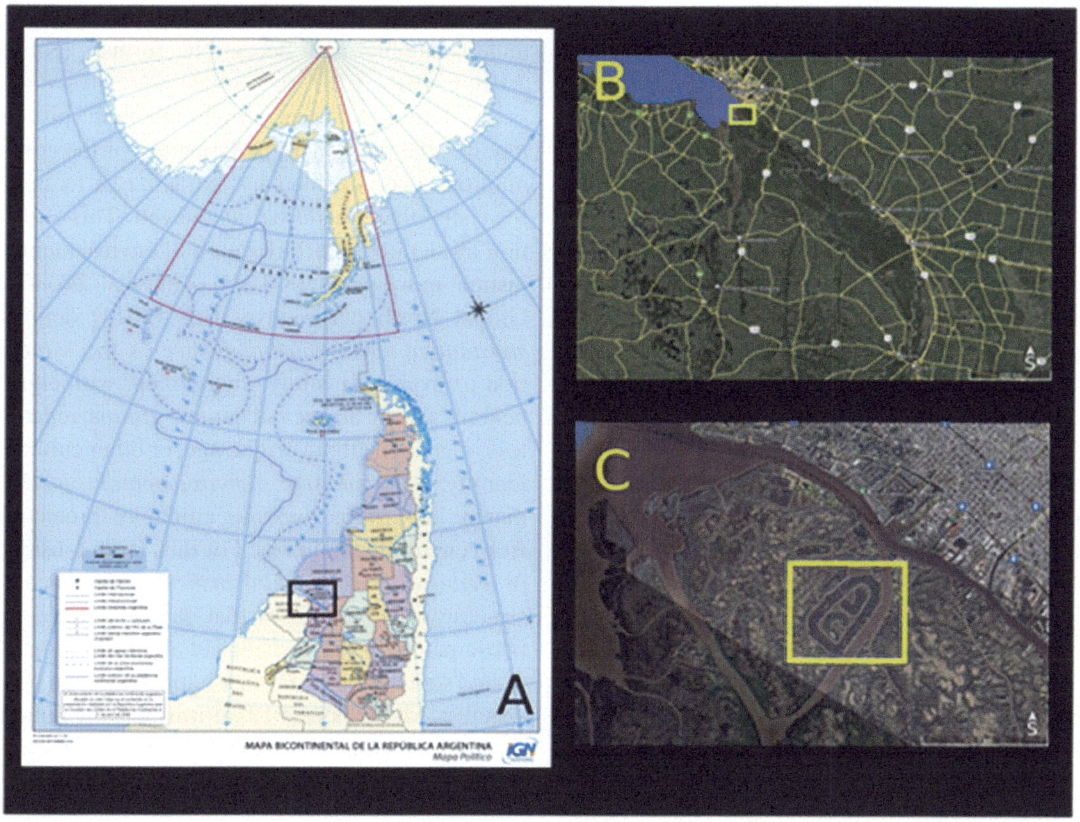

Fig. 1 Maps from the area in conflict, South orientation pointing up. (**a**) Argentina. (**b**) Paraná river delta. (**c**) Anguilas stream, Tigre, Buenos Aires. (Source: own elaboration from maps of the National Geographic Institute and Google Earth)

appropriation that does not respond to private property nor destruction as a way of new territory creation. According to this point of view, the land temporarily belongs to those who work it, inhabit it and wish to continue doing so. This territorialization modality entails deep ecosystem knowledge and can contribute to a more general process of social re-signification of nature with an emancipatory perspective. Islander way of life is related to environmental rationality, an alternative to instrumental rationality. The latter rationality has caused enormous environmental damage to the Paraná Delta and other wetlands in the world through urbanization (Rojas et al. 2019) and, in this specific case, real estate urban extractivism (Domínguez 2022).

3 Environmental Education from Critical Perspectives, Its Relationship with Participation and Human Rights

From the aforementioned territorial context, we intend to contribute to an Environmental Education (EE) understood as a political, epistemological and ontological practice that tries to build environmental rationality for life care (Leff 2014) and the decolonization of the Global South. This EE disputes centralized knowledge-power of control and regulation of the environment, intending to recompose views and words that speak of other possible worlds (Escobar 2018).

For this reason, based on the political-pedagogical frameworks of Latin American Popular Education, we frame ourselves in Latin American Environmental Thought (LAET—Leff 2012). Canciani et al. (2017) propose an Environmental Conflict Pedagogy (ECP) as an EE aimed at participation in environmental conflicts which intends to tackle the environment from a critical and conflictive view of reality. Linked to Political Ecology of the South[4] and LAET, this perspective develops pedagogical tools to understand and address the complex social interconnections in which environmental conflicts develop in our region.

From this viewpoint, addressing the relationship between EE and human rights is fundamental. Telías et al. (2014) sustain that any proposal of critical environmental education aims to form a participative citizenship capable of choosing and acting according to ethical considerations and collective and community interests. In other terms, it is political. Therefore, teaching about human rights and citizenship obligations regarding environmental issues and the differentiated responsibilities of the multiple social actors involved is necessary.

In early 2018, the United Nations Human Rights Council approved the Framework Principles on Human Rights and the Environment (Knox 2018). These principles express the fundamental interdependence between human rights and the environment globally, where many environmentalists are killed or suffer violence due to their activity. The report, among other issues, affirms that "states must ensure the full and effective participation of indigenous peoples and traditional communities in decision-making on the entire spectrum of matters that affect their lives" and "should assess the environmental and social impacts of proposed measures and ensure that all relevant information is provided to them in understandable and accessible forms" (Knox 2018, p. 19).

In Latin America, environmental justice requires the permanent participation of the communities overwhelmed by extractivism to fight for rights that should be guaranteed. This situation has led to the proliferation of environmental justice movements in Latin America. From a Global North perspective, Bell (2014) defines *substantive environmental justice* as the aspiration to live in a healthy environment with equal benefits among all socioeconomic groups, regardless of gender, ethnic origin, age, or other cultural identity. *Distributive environmental justice* means all groups must be equitably protected from environmental risks. For this, there must be *procedural environmental justice*, which means structures and processes that ensure fairness in participating in the decisions of all the aforementioned social groups. From a Latin American perspective, Merlinsky (2017) affirms that the notion of environmental justice implies the defence of living conditions and health to incorporate the discussion on access to resources (water, land, forest resources, basic infrastructure) in the framework of economic disputes, but also in terms of cultural and identity values.

Environmental justice movements generate instances of Environmental Education, producing new ideas and disputing this educational field. The same injustices that operate on subalternized groups regarding the environment are frequently reproduced at an epistemic level. In this way, the search for environmental justice entails the search for epistemic and cognitive justice. In this sense, it is necessary to interweave EE with critical and extended interculturality (Díaz and Rodríguez de Anca 2014). From a *critical intercultural* point of view, the focus on race must be extended as the asymmetry of power relations reproduced by the dominant culture is much broader and more complex. This intercultural perspective is also intersectional, opening more categories such as class, gender, and generation, among others, to be considered. Furthermore, *extended interculturality* proposes that interculturality should be primarily aimed at

[4]As a social practice, the Political Ecology of the South refers to a process and a product-always-in-process, of collective construction of knowledge, un-learning and inter-learning that are interwoven and developed from the dialogue of knowledge of community subjects who share the situation of literally "feeling firsthand" the effects of the systemic destructiveness of capitalism-colonialism-patriarchy (Machado Araoz 2017).

the dominant culture since it has already been practised by indigenous people and other subalternized identities, which have almost no option other than socializing in the dominant culture. This approach states that the primary goal of interculturality is the review and transformation of the dominant culture's practices. Tensions are expressed in environmental conflicts, where there are disputes over the *languages of valuation of nature* (Martínez Alier 2006) crossed by intercultural power relations, to the point that there are also *ontological conflicts* (Blaser 2013).

For this reason, Enrique Leff argues that Environmental Justice is inscribed in the field of political ecology as the space for controversies regarding the sense of fairness, as the space for the manifestation of socio-environmental conflicts derived from the injustice of the sovereignty of the ontological regime in which the legal rationality of modernity is inscribed; and as the challenge of constituting a non-hegemonic legal regime, allowing peaceful settlement of such conflicts and establishing rules of coexistence of diverse ways of understanding life and inhabiting the planet. Thus, environmental justice claims "the right to have rights", the legitimacy of the various codes of justice instituted through the social imaginaries of different cultures (Leff 2021).

Therefore, it is necessary to create ways of intercultural dialogue that express that search for environmental justice, where the sense of fairness is resolved in spaces that expand the current legal arena, which does not reflect the breadth of truly existing worlds (Álvarez & Coolsaet 2020).

Along this path, while theorizing, experimenting, and legislating based on new forms of law, such as the rights of nature, one of the most fertile ways to contribute to environmental justice is to work on community participation. In this sense, we seek the Freirean sense of participation for EE,

> Nor let it be said that this is a populist, or "democratistic" position. No, it is not democratistic, it is democratic. It is progressive. But it is the position of progressives and democrats who see the urgency of the presence of the popular classes in the debates on the destiny of the city. Their presence in the school is a chapter in that debate, and is a positive sign, and not something evil, something to be deterred. This is not the position of self-styled "democrats" for whom the presence of the people in facts and events, a people organizing, is a sign that democracy is not doing well (...) in its indispensable quest for a transcendence of the "knowledge of living experience" in order to arrive at a more critical, more precise knowledge, to which these persons have a right. This is a right of the popular classes that progressives have to recognize and fight for if they are to be consistent—the right to know better than they already know—alongside another right, that of sharing in some way in the production of the as-yet-nonexistent knowledge. (Freire 2021, p. 95)

The meaning of Paulo Freire's words is recreated in the conflicts in Latin American urban and peri-urban wetlands. In those places, producing environmental knowledge that guarantees life's sustainability requires a dialogue of knowledge(s). This dialogue should include different sciences, and academic disciplines interweaved with real participation of the peasant and indigenous population, bearers of *lived experience and ancestral knowledge*.

The Ramsar Convention proposals are actually in line with these prior statements. In its VII.8 Resolution "Guidelines for establishing and strengthening local communities' and indigenous people's participation in the management of wetlands" (Ramsar 1999), in the 4th item of this resolution, it is stated that "in many contexts, indigenous people and local communities are already involved in managing and using wetlands sustainably, and have long-standing rights, ancestral values, and traditional knowledge and institutions associated with their use of wetlands" (p. 1).

Paraná Delta is considered one of the Ramsar sites on the planet (Site No. 2255), and the three criteria mentioned are met: this wetlands system is inhabited; access to its common assets and the territory of wetlands is essential for the subsistence of the islander population; this population manifests through its words and public actions, a firm interest in participating in the management of the wetland.

4 Systematization of Experiences

As a methodological framework, we frame ourselves in the systematization of experiences perspective, linked to Participatory Action Research and Popular Education fields (Barragán Cordero and Torres Carrillo 2017). It is a participatory and critical form of knowledge production based on a narrative reconstruction of the process under study and an intersubjective critical interpretation of the experience. For the elaboration of this work, we resorted to previous systematizations of three experiences carried out between 2017 and 2018 (Iribarren et al. 2020, 2022, 2023), which we reconstructed from different records, mainly audio and video recordings, among other sources. We defined the links between EE and community participation that emerged throughout the process as our systematization axis for inquiry.

The DWO Education Commission, which we integrated with islanders and academics, planned the activities.

4.1 2017. Self-Training Workshop to Prepare an Environmental Remediation Plan

This workshop, held in July 2017, was part of an internal training process on wetlands (Fig. 2).

The IEC asked the DWO to assist in preparing a "Remediation Plan" to be presented to justice since a federal court had ordered Colony Park to remedy the environmental damage done in Anguilas stream. The IEC intended to avoid the recurring situation in which the companies that have caused the environmental damage later take care of the remediation with subsidiary companies that carry out *greenwashing*, not really repairing the damage. Thus, instead of facing environmental passives, they do not resolve them and profit from a fictitious process that does not benefit the local community and even harms it (Carrrere 2006).

The workshop was planned as a starting point for a dialogue of knowledge(s) about the territory between islanders, environmentalists, ecologists, lawyers, and sociologists. The subsequent systematization of this instance of popular education was an essential input for preparing the Remediation Plan (Iribarren et al. 2020).

The planning included a first part carried out by the IEC, while the DWO Education Commission coordinated the second. The IEC guided us through the territory focusing on the environmental damage produced. We navigated through the Anguilas stream and its surroundings to the beaches where the families collected sedges before the start of the conflict. The islanders narrated their experiences during the tour and showed the sedge harvesting process.

In the second part of the workshop, we worked in groups around *the notion of the river as a living entity.* During the previous tour, one of the islanders posed the idea of the river as a living entity, defying and denaturing the reification of nature. Recognizing its profound ontological and epistemological depth, we modified the original planning of the workshop. We proposed thinking about remediation based on the question, "*What would the river like us to do?*". Thus, new questions arose: What does "*the river is alive*" mean to the academy? How can we translate this idea into an environmental remediation plan? How can we express these ideas in a way that could be valued positively by justice? How can we get dissimilar knowledge(s), livings and actions to dialogue and build consensus and lines of action?

At the end of the workshop, we socialized and discussed group productions and agreed to use them as input for the Remediation Plan. A Remediation Commission was in charge of articulating this *dialogue of knowledge(s) and livings* (Merçon et al. 2014) in the first draft of the plan, which we then continued to rework in successive instances of participation (Fig. 3) until it was attached to the file of the legal case (Observatorio de Humedales Delta 2017).

Due to its content and construction method, we understand the Remediation Plan as a contribution to environmental and epistemic justice. It reflects the social productivity of environmental conflicts in a legal dimension and EE as a political-pedagogical field. The plan's construction was productive from the legal point of view

	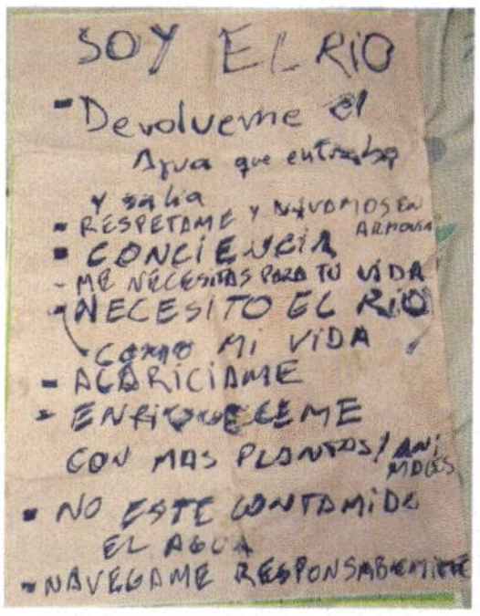	
Group work in Anguilas stream	One of the productions of the workshop	Participatory Remediation Plan presented to Federal Court

Fig. 2 Self-training workshop to prepare an environmental remediation plan for a wetlands territory.

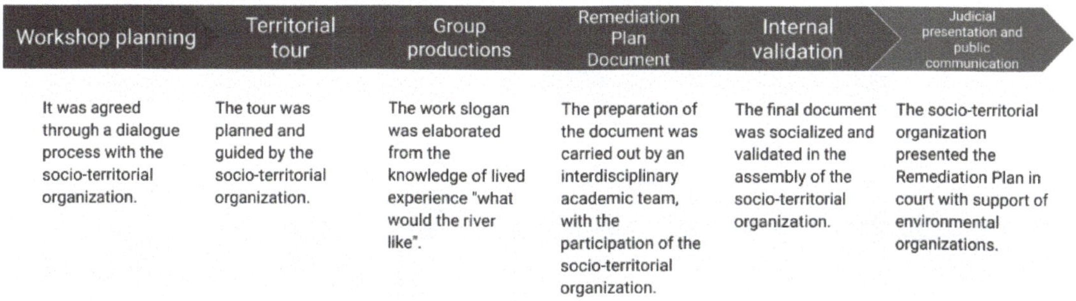

Fig. 3 Synthesis of the process of preparing a wetland remediation plan in a participatory manner

since the islanders' knowledge was articulated with scientific knowledge, which enjoys legitimacy in the legal arena. In this sense, it set a jurisprudence precedent and constituted a *local update of the law* (Merlinsky et al. 2018).

This process was an important precedent for the IEC and DWO, which would later guide teacher training actions. For us, it was essential to have first immersed in the environmental conflict with the actors in the territory. After this experience, it became clear to systematize environmental knowledge about Paraná Delta wetlands. It was necessary to continue the dialogue of knowledge(s) and deepen it in a broader environment. Furthermore, we concluded that future experiences in teacher training should include the greatest possible number of voices from the territory, academia, and environmental networks.

4.2 2017. "Feelthinking with the Wetland" Meeting

The following knowledge(s) dialogue instance was an activity called "Feelthinking[5] with the wetland". It consisted of a 2-day camp in Anguilas stream, with approximately 100 attendees, linked to local environmental organizations, that travelled from other islands of Paraná Delta and near urban continental areas to the territory in conflict (Fig. 4).

[5] To feelthink (*"sentipensar"*) is a category popularized by the Colombian sociologist Orlando Fals Borda (1925–2008). It is a notion that comes from riverside communities in the Colombian Pacific that joins rationality and emotions in the same process, not as separate instances.

The activities began with a general presentation and historicization of the environmental conflict given by the IEC. In the afternoon, a tour was carried out through the territory to assess environmental damage collectively. It was coordinated by a sedge collector member of the IEC and a wetland academic member of the DWO. The following activity consisted of a panel held by local environmentalists, the IEC and the DWO. The "Bonfire of Memory" activity began at sunset, where an oral reconstruction of islander life was carried out. That instance was in charge of the Islander Women Group. The closure of the bonfire was carried out by members of the "Punta Querandí" Movement in Defense of the Pachamama, protagonists of an environmental conflict a few kilometres from Anguilas stream.[6]

Day 2 began with an activity about the work of the sedge-collecting families. It was coordinated by sedge collectors ("junqueros") who spoke with a graduate in environmental sciences, a specialist in the fair trade of sedge. After lunch, members of the DWO coordinated an activity based on the Remediation Plan and at sunset, a collective assessment of the 2-day camp was made. Finally, a "mystical"[7] group closure called "The Heart of the Sedge" was performed.

In this 2-day event, various social organizations from the territory met and enriched the dialogues feelthinking the wetland. They have been protagonists since the return of democracy in the 1980s of the local environmental justice movement and other struggles for human rights (Astelarra 2021): "Asamblea Delta y Río de la

Fig. 4 Different moments of "Feelthinking with the wetland" (September 2017)

[6]This dispute was originated by the "Santa Catalina" closed neighborhood, which in addition to destroying wetlands damaged an important indigenous cultural heritage, now protected by an assembly of indigenous peoples and residents of the area.

[7]Term refers to activities that promote expressing feelings and self-perceived learning during the process. According to Gómez (2008), it can be understood as a communication, identity and expressive device.

Plata", "Liberpueblo" Civil Association, "Casa Puente" Cultural and Community Center and the "Punta Querandí" indigenous community. They shared their experiences of fighting for the right to a healthy environment, water, and territory.

The "Feelthinking with the wetland" also allowed us, as Merlinsky (2017) mentions, to identify other actors involved in local environmental conflicts (state institutions, security forces and businessmen) and their functional role in extractivism. First-person accounts of the violence suffered, and the emotions experienced enabled participants to recognize the particular importance of the social dimension of environmental harm (White 2013). It was made explicit during both days, at different times, that damages affected the wetlands as non-human nature and harmed the islander population and their ancestral way of life.

We reflected on the collective mobilizations against the most aggressive forms of extractivism, a point of convergence of the participants. A dialogue of knowledge(s) occurred, socializing and experiencing different forms of communality, inhabiting the territory, diverse conceptions and connections with "nature". In dissonance with contemporary societies where economic value prevails, other worldviews that consider the affective, aesthetic, sacred, and spiritual were shared. In ontological conflicts, a path opens to investigate which worlds are in dispute, how they narrate themselves and the conflict, and how our analysis can relate to those stories and intervene in the conflict (Blaser 2013). The presence of the Punta Querandí Indigenous Community, with their claims regarding their right to community land tenure, allowed participants to experience an intercultural EE. Lines of continuity were evident between socially and economically privileged sectors that, at the end of the nineteenth century, perpetrated an indigenous genocide in the misnamed "Desert Campaign" and current business-political-military-ecclesiastical associations that continue to overwhelm communities in Paraná Delta wetlands. The oral reconstruction of the conflict carried out by the Punta Querandí community clearly showed the continuity of racism associated with the coloniality of power in Latin America. The content of the conference marked a difference between *functionalists* and *relational* intercultural perspectives, evidencing the nonexistence of a friendly dialogue between dominant and dominated cultures (Díaz and Rodríguez de Anca 2014).

The meeting "Feelthinking with the wetland" constituted a new milestone, reflecting the vast participation of the political actors who intertwine and weave alliances in the struggles for urban wetlands: peasant and environmentalists collectives, indigenous communities, ecofeminists groups, cultural communitarian and human rights organizations (Fig. 5).

4.3 2018. Communitarian Teacher Training with the Islander Sustainable Secondary (ISS) Group

The following year, the socio-territorial movement proposed to articulate with local schools. From the DWO Education Commission, we made contact with the Islander Sustainable

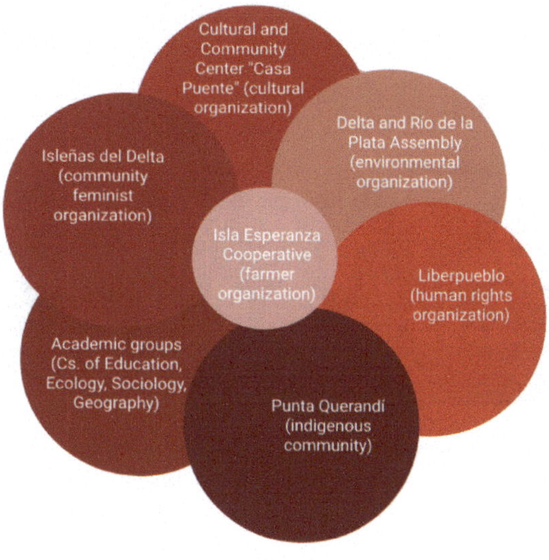

Fig. 5 Organizations participating in the meeting "Feelthinking with the wetland"

Secondary (ISS)[8] group linked to Secondary School No. 12 of the Sarmiento River (Tigre, Buenos Aires). We jointly designed a training device in EE aimed at teachers and the community in general. It consisted of four in-person meetings lasting four hours each and was disseminated as "Training Workshop on Environmental Education in the Delta of Buenos Aires". Participants were people from the islander community and nearby urban neighbourhoods, from the school (teachers, families) and the university (researchers). The first meeting focused on a collective mapping activity; the second delved into environmental conflicts identified in the mapping at different scales; the third was a field trip to the territory in conflict (Anguilas stream); and the fourth, a balance of the process and an invitation to develop EE projects (Figs. 6 and 7).

This formative process and its relationship with the Pedagogy of Environmental Conflict was systematized in previous articles (Iribarren et al. 2022; 2023).

After showing two brief environmental short videos during the first meeting, we proposed to analyse the gender perspective underlying each one working in small groups. One of them recalled the violence of islander men hired by Colony Park—also commanded by men—using the dredgers and backhoes that demolished the houses of islander families. Violent situations experienced by some participants and other women in the territory were shared and analysed. They narrated the recurrent mistreatment of women by the "*lancheros*", colloquial way of naming the precarious boatmen workers in charge of river public transport. Afterwards, during the collective mapping, some participating women promoted debates on the rights to territory, autonomy, food sovereignty, recognition of the rights of the various groups of women (islanders, indigenous, rural, urban, black), sexual and reproductive rights, self-care and self-knowledge.

Fig. 6 Communitarian teacher training workshops were conducted with the "Islander Sustainable Secondary" group from Secondary School No. 12 (Tigre, Buenos Aires)

[8]The SIS group is mainly the result of the struggle of island teachers and mothers against building deficiencies, lack of space and timetable problems of an island school, making many people unable to sustain schooling and migrating to urban areas.

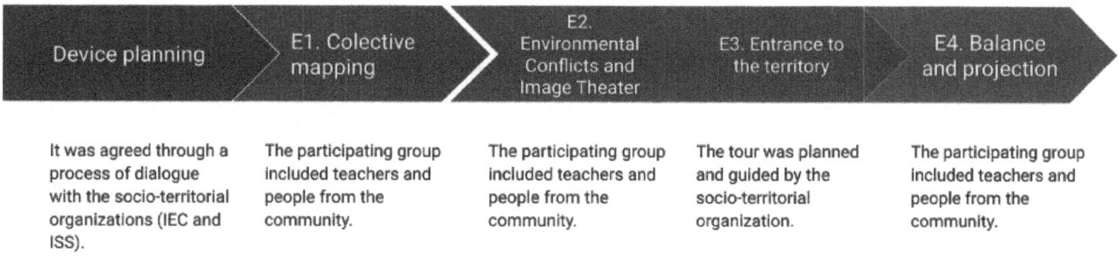

Fig. 7 Participatory process scheme carried out in the communitarian teaching training workshop

In the third meeting, during the field trip or "entry into the territory in conflict", a question arose. How could the IEC resist the menacing advance of Colony Park? Faced with this question, one of the members of the DWO made a historicization of the conflict. She added details about the public visibility of the conflict and the political-governmental responses; the legal arguments that allowed to paralyse the advance of Colony Park; the role of the IEC and other local social organizations, born from previous struggles and conflicts for the right to water and a healthy environment (Astelarra 2016).

5 Conclusions

The community participation approach we developed through these experiences is a contribution from Latin American critical Environmental Education to the modes of participation in the management of wetlands. To carry out instances of community participation requires flexibility and creativity in the specific moment and context in which the process is being carried out. There are no universal approaches or recipes to ensure success in community participation (Ramsar 1999).

We highlight the main points arising from the systematization: listening to the territory; constructing the conditions to engage in a dialogue of knowledge(s) including peasant, indigenous and ecofeminist knowledge; performing communitarian teacher training.

These critical points allowed, for example, the participatory construction of a wetland remediation plan based on scientific and islander knowledge. Listening to the territory included the population and the river itself as an entity, planning the remediation of urban wetlands from relational ontologies, and alternatives to the underlying dualist ontology in scientific disciplines (Blaser 2013).

The power of community participation framed in a *dialogue of knowledge(s)* was fundamental to promoting the construction of environmental knowledge and environmental justice in the senses referred to by Leff (2014, 2021). The "Feelthinking with the wetland" activity showed the richness of incorporating indigenous, ecofeminist and human rights organizations into the dialogue, enriching the process already started between islander peasants, environmentalists and academics.

To promote critical Environmental Education in Latin American urban wetland schools, the perspective of communitarian teacher training proved empowering and productive. This training in Environmental Conflict Pedagogy was enhanced by the participation of community actors involved in local conflicts resisting as supporters of life care and sustainability in the territory (Iribarren et al. 2023). As the experience analysed shows, participation can be included in all the stages of teacher training, from the initial planning of the actions to the implementation of workshops or other actions in diverse and intercultural workgroups.

From these experiences, we hope to have contributed to reflecting on Environmental Education's role in strengthening and expanding r-existences (Porto Gonçalves and Leff 2015) and processes of social reappropriation of the territory in urban wetlands of Latin America. We expect to collaborate in guaranteeing human rights to life, a healthy, dignified and diverse environment, and the river's right to be respected, loved and cared.

References

Álvarez, L., & Coolsaet, B. (2020). Decolonizing environmental justice studies: a Latin American perspective. *Capitalism nature socialism, 31*(2), 50–69.

Astelarra, S., y Domínguez, D. (2015). Los junqueros de las islas del Delta del Paraná: sujetos emergentes en un territorio amenazado. [The junqueros of the Paraná delta islands: emerging subjects in a threatened territory]. *Estudios Socioterritoriales, 17*, 0–0.

Astelarra, S. (2016). Disputas por la reinvención del "paraíso deltaico": de los lugares de la querencia a llegar a una isla y olvidarse de todo. El caso del conflicto "Colony Park" en la primera sección de islas del Delta del Paraná. [Disputes for the reinvention of the "deltaic paradise": from places of fondness to arriving on an island and forgetting everything. The case of the "Colony Park" conflict in the first section of islands of the Paraná Delta]. *In* G. Merlinsky (Comp.) *Cartografías del conflicto ambiental en Argentina 2.* Buenos Aires: CLACSO.

Astelarra, S. 2021. Devenires anfibios. Apuntes sobre lenguajes de valoración en torno al agua en el humedal deltaico y notas desde el ecotransfeminismo isleñe. [Amphibious becomings. Notes on valuation languages around water in the deltaic wetland and notes from island ecotransfeminism]. En Antuña, J.; Giordano, V.; Molinari, E. (compiladores) (2021) *Comunidad, territorio, futuro.* Buenos Aires: Teseopress.

Barragán Cordero, D., y Torres Carrillo, A. (2017). *La sistematización como investigación interpretativa crítica. [Systematization as critical interpretative research].* Bogotá: El Búho.

Bell, K. (2014). *Achieving environmental justice: A cross-national analysis.* Policy Press.

Blaser, M. (2013). Notes towards a political ontology of 'environmental'conflicts. *In* Green, L. *Contested ecologies: Dialogues in the South on nature and knowledge, 13–27.* HSRC Press.

Canciani, M. L., Telías, A y Sessano, P. (2017). *Problemas y desafíos de la educación ambiental. Un abordaje en 12 lecciones. [Problems and challenges of environmental education. An approach in 12 lessons].* Buenos Aires: Ediciones Novedades Educativas.

Carrere, R. (2006). *Greenwash. Critical analysis of FSC certification of industrial tree monocultures in Uruguay.* World Movement for Tropical Forests.

Castro-Gómez, S. (2021). *Zero-point hubris: Science, race, and enlightenment in eighteenth-century Latin America.* Rowman & Littlefield.

Cooperativa Isla Esperanza/Universidad Nacional de Luján. (2012) *Isla Esperanza. Trabajo, naturaleza y resistencia isleña. [Hope Island. Work, nature and islander resistance].* Buenos Aires: UNLU

Díaz, R., y Rodríguez de Anca, A. (2014). Activismo intercultural: Una mirada descolonizadora, crítica e interseccional. [Intercultural activism: A decolonizing, critical and intersectional look]. *In* Villa, A., y Martínez, M. (comp.), *Relaciones escolares y diferencias culturales: La educación en perspectiva intercultural* (171–197). Buenos Aires: Noveduc.

Domínguez, D. (2022). Mercado de islas y conservacionismo como despojo en el bajo delta del río Paraná: desterritorialización del dominio público y los bienes comunitarios de la naturaleza. [Island market and conservationism as disposession in the lower delta of the Paraná river: deterritorialization of public domain and assets community of nature]. *Revista Interdisciplinaria de Estudios Sociales,* (25), 9–46.

Cordero, S., Dumrauf, A. G., Mengascini, A., & Sanmartino, M. (2011). Entre la Didáctica de las Ciencias Naturales y la Educación Popular en Ciencias Naturales, Ambiente y Salud: relatos y reflexiones de un camino en construcción. [Between the Didactics of Natural Sciences and Popular Education in Natural Sciences, Environment and Health: stories and reflections of a path under construction]. *Praxis Educativa (Arg), 15*(15), 71–79.

Dumrauf, A., Cordero S., Cucalón Tirado, M.P, Guerrero Tamayo, K. y Garelli, F. (2019). Hacia nuevos territorios epistémicos: aportes desde un camino de construcción pedagógica descolonizadora en Educación en Ciencias Naturales, Ambiental y en Salud. [Towards new epistemic territories: contributions from a path of decolonizing pedagogical construction in Education in Natural, Environmental and Health Sciences]. *In* B. Monteiro, et al.: *Decolonialidade na Educação em Ciencias.* Sao Paulo: Livraría da Física.

Escobar, A. (2018). *Otro posible es posible: Caminando hacia las transiciones desde Abya Yala/Afro/LatinoAmérica. [Another possible is possible: Walking towards the transitions from Abya Yala/Afro/LatinoAmérica].* Bogotá: Ediciones Desde Abajo

Freire, P. (2021). *Pedagogy of hope: Reliving pedagogy of the oppressed.* Bloomsbury Publishing.

Galafassi, G. (2005). *La pampeanización del delta, sociología e historia del proceso de transformación productiva, social y ambiental del Bajo Delta del río Paraná. [The pampeanization of the delta, sociology and history of the process of productive, social and environmental transformation of the Lower Delta of the Paraná River].* Ediciones Extramuros.

Gómez, J. S. (2008). Rituales de lucha en el movimiento piquetero. Un abordaje a la "mística" de la lucha callejera como dispositivo comunicacional, identitario y expresivo. [Fighting rituals in the piquetero movement. An approach to the "mystique" as a communication, identity and expressive device]. IX Congreso Argentino de Antropología Social. Facultad de Humanidades y Ciencias Sociales - Universidad Nacional de Misiones, Posadas.

Iribarren, L., Guerrero Tamayo, K, Garelli, F., Dumrauf, A. (2020). Entornos de diálogo y construcción de saber ambiental con movimientos socioterritoriales [Dialogue environments and environmental knowledge building with socioterritorial movements]. *Temas de Educación*, Vol. 25, N°2. *In press*

Iribarren, L., Guerrero Tamayo, K, Garelli F., Dumrauf, A. (2022). Pedagogías del conflicto ambiental: aportes desde una experiencia participativa de formación docente en un territorio en disputa. [Pedagogies of the environmental conflict: contributions from a participatory experience of teacher training in a disputed territory] *Praxis educativa, Vol. 26, No 1.*

Iribarren, L., Tamayo, K. G., Garelli, F., & Dumrauf, A. (2023). ≦Qué nos cuenta Pachamama? Una experiencia de diálogo de saberes, vivires y formación docente descolonizadora en educación ambiental. [What Pachamama tells us? A dialogue of livings, knowledge(s) and decolonizing teacher training experience on environmental education]. *Tecné, Episteme y Didaxis: TED*, (53), 258–274.

Knox, J. (2018). Framework Principles on Human Rights and the Environment. Office of the High Commissioner for Human Rights.

Leff, E. (2012). Latin American environmental thinking: a heritage of knowledge for sustainability. *Environmental ethics,* 34(4), 431–450.

Leff, E. (2014). *La Apuesta por la Vida. Imaginación sociológica e imaginarios sociales en los territorios ambientales del Sur. [The Bet for Life. Sociological imagination and social imaginaries in the environmental territories of the south].* Mexico: Siglo XXI.

Leff, E. (2021). Racionalidad y Justicia Ambiental: La Elusiva Injusticia de la Vida. [Rationality and Environmental Justice: The Elusive Injustice of Life]. Historia Ambiental Latinoamericana y Caribeña (HALAC) *Revista de la Solcha, 11*(3), 19–38.

Malvárez, A. I. (1999). El Delta del Río Paraná como mosaico de humedales. [The Paraná River Delta as a mosaic of wetlands] *In* A. I. Malvárez, *Tópicos Sobre Humedales Subtropicales y Templados de Sudamérica* (pp. 5-53). MAB-ORCYT.

Machado, H. (2017) "América Latina" y la Ecología Política Del Sur. Luchas de re-existencia, revolución epistémica y migración civilizatoria. ["Latin America" and the Political Ecology of the South. Struggles for re-existence, epistemic revolution and civilizing migration]. *En* Alimonda, H.; Toro Pérez, C.; Martín, F. (coords.) *Ecología Política Latinoamericana: pensamiento crítico, diferencia latinoamericana y rearticulación epistémica.* Buenos Aires: CLACSO

Martinez Alier, J. (2006). Los conflictos ecológico-distributivos y los indicadores de sustentabilidad [Ecological-distributive conflicts and sustainability indicators]. *Polis, Revista de la Universidad Bolivariana, 5(13).*

Merlinsky, G. (2017). Los movimientos de justicia ambiental y la defensa de lo común en América Latina. Cinco tesis en elaboración. [Environmental justice movements and the defense of the commons in Latin America. Five theses in preparation]. *Ecología política latinoamericana: pensamiento crítico, diferencia latinoamericana y rearticulación epistémica, 2*, 241–264.

Merlinsky, G., Toledo López, V., Schmidt, M., Fernández Bouzo, S., Tobías, M., Langbehn, L., Pereira, P. y Capalbo, T. (2018). *Defender lo común: qué podemos aprender de los conflictos ambientales. [Defending the common: what we can learn from environmental conflicts].* Ciudad Autónoma de Buenos Aires: Universidad de Buenos Aires. Instituto de Investigaciones Gino Germani.

Merçon, J., Camou, A., Nuñez, C. y Escalona, M. (2014). ≦Diálogo de saberes? La investigación acción participativa va más allá de lo que sabemos. [Knowledge dialogue? Participatory action research goes beyond what we know]. *Decisio. Saberes para la acción en Educación de Adultos, 38, 29–33.*

Observatorio de Humedales Delta [Delta Wetlands Observatory]. (2017). Propuesta de remediación ambiental de los arroyos Anguilas y Paloma por el impacto ambiental negativo producido por los emprendimientos Colony Park y Parque de la Isla [Proposal for environmental remediation of the Anguilas and Paloma streams due to the negative environmental impact produced by the Colony Park and Parque de la Isla projects]. (Available in http://farn.org.ar/wp-content/plugins/download-attachments/includes/download.php?id=23015)

Porto-Gonçalves, C. W., & Leff, E. (2015). Political ecology in Latin America: the social re-appropriation of nature, the reinvention of territories and the construction of an environmental rationality. *Desenvolvimento e meio ambiente,* 35(1), 65–88.

Ramsar (1999). Resolution VII.8, Guidelines for establishing and strengthening local communities' and indigenous people's participation in the management of wetlands. Resolutions of the 7th Meeting of the Conference of the Contracting Parties to the Convention on Wetlands.

Rojas, C., Munizaga, J., Rojas, O., Martínez, C., & Pino, J. (2019). Urban development versus wetland loss in a coastal Latin American city: Lessons for sustainable land use planning. *Land use policy,* 80, 47–56.

Telías, A.; Canciani, M. L.; Sessano, P. (2014). *La educación ambiental en la Argentina: actores, conflictos y políticas públicas. [Environmental education in Argentina: actors, conflicts and public policies].* San Fernando: La Bicicleta.

White, R. (2013). *Environmental harm: An eco-justice perspective*. Policy Press.

The Role of Citizenship in the Protection of Wetlands. Comparative Analysis of Three Urban Wetlands in Chile, Colombia, and Peru

Caroline Stamm, Carolina Rojas Quezada, Juanita Aldana-Domínguez, Paola Moschella, and Macarena Sepúlveda

Contents

Abstract

Latin America leads the loss of wetlands worldwide and urbanization is one of the main causes of this change. The cases of the Aconcagua wetlands in Chile, La Ciénaga de Mallorquín in Colombia, and Pantanos de Villa in Peru are examples of the loss of Latin American urban wetlands in the last 20 years. In response to this situation, various initiatives have emerged to protect these three wetlands, which have involved the adoption of forms of territorial governance associating the public sector, the private sector, and civil society. The aim of the chapter is to analyze the role of citizenship in the protection of these three wetlands, with the aim of contributing to the construction of more resilient cities. Using semi-structured interviews conducted with local actors, the results show that, despite the differences between the governance mecha-

C. Stamm (✉) · M. Sepúlveda
Instituto de Estudios Urbanos y Territoriales, Pontificia Universidad Católica, Santiago, Chile
e-mail: castamm@uc.cl; macarena.sepulveda@uc.cl

C. Rojas Quezada
Instituto de Estudios Urbanos y Territoriales, Centro de Desarrollo Urbano Sustentable CEDEUS, Instituto Milenio SECOS, Santiago, Chile
e-mail: carolina.rojas@uc.cl

J. Aldana-Domínguez
Departamento de Química y Biología, Universidad del Norte Colombia, Barranquilla, Colombia
e-mail: ajuanita@uninorte.edu.co

P. Moschella
Departamento de Humanidades, Geografía y Medio Ambiente, Pontificia Universidad Católica del Perú, Lima, Peru
e-mail: pmoschella@pucp.edu.pe

C. Rojas Quezada (ed.), *Urban Wetlands in Latin America*, Sustainable Development Goals Series,
https://doi.org/10.1007/978-3-031-69590-2_13

nisms in the three cases, coordination between stakeholders and citizen participation are perceived as deficient. The interviewees propose to establish working tables with representatives from all sectors and productive dialogue instances and highlight the lack of participation of the neighboring communities of these wetlands. These communities are subaltern groups, as they are also the most socioeconomically vulnerable, and occupy, for the most part, the territory informally. Their participation, as well as motivating their interest in participating, is key to ensuring the protection of these spaces.

Keywords

Wetland · Governance · Community · Participation

1 Introduction

Latin America leads the loss of wetlands worldwide (Darrah et al. 2019) and urbanization is one of the main causes (Rojas et al. 2019). The cases of the Aconcagua wetlands in Chile, La Ciénaga de Mallorquín in Colombia, and Pantanos de Villa in Peru are examples of the loss of surface area of Latin American urban wetlands in the last 20 years (Rojas et al. 2020). In response to this situation, various initiatives have emerged to protect these three wetlands, which have involved the adoption of forms of territorial governance associating the public sector, the private sector, and civil society. Governance is defined as the emergence of new forms of deliberation, participation, and decision-making about problems of collective interest (Bevir 2012), breaking with traditional hierarchical forms, in which the State was in charge of territorial management. In the case of wetland protection, these initiatives have sought to install new mechanisms of horizontal dialogue between the three sectors, through the installation of various practices of coordination and cooperation.

However, even though these new forms of governance can be seen as advances and solutions, it is necessary to review these governance mechanisms, in line with the studies of critical geography and political ecology (Swyngedouw 2005; Bustos et al. 2019). Bustos et al. (2019) interpret the new territorial governance mechanisms as a reaction to environmental territorial crises, with the aim of pacifying and controlling the neoliberalized territory, and invite us to consider, from a critical perspective, the context in which governance instruments are implemented and the power relations that sustain them. Within these relationships, we will focus here, specifically, on the place of citizenship in the governance processes of the three urban wetlands: Aconcagua in Chile, La Ciénaga de Mallorquín in Colombia, and Pantanos de Villa in Peru. From the works of Agrawal and Gibson (1999), there is a wide international literature on the role of communities in the conservation of natural resources and, among other ecosystems, of wetlands. Within these works, there is a specialized literature on the topic of Latin American wetlands, related to citizen movements defending wetlands and socio-environmental conflicts (Cardona 2007; Skewes et al. 2012), changes in governance modes and community empowerment (Ruiz et al. 2011; Pinel et al. 2018), and social exclusion in governance processes (Eufemia et al. 2020). The aim of this chapter is to analyze the role of citizenship in the protection of the three Latin American wetlands, in order to contribute to the reflection on the participation of communities in the construction of more resilient cities. Using semi-structured interviews with local actors, we will focus on the existing citizen participation mechanisms in the management of each wetland, as well as their evaluation by local actors.

2 Methodology

2.1 Description of the Case Studies

The wetland of Concón, in Chile, has an area of 700 ha, but the protection area covers only 28 ha, where the Ecological Park La Isla is located. In 2003, the Council for the Defense of the State sued the National Oil Company (ENAP), which has a refinery situated in the wetland area, they settled and ENAP committed to enhancing the place. This led to the creation of La Isla Park, which was given in commodatum to the municipality of Concón for 10 years. At the end of this period, it was renewed in the form of usufruct for 30 years. In 2010, the Metropolitan Plan of Valparaíso PLANVAL, which was approved between 2012 and 2014, classified the wetland as Green Area. In 2018, the Municipality opted for the declaration of Municipal National Reserve (RENAMU). However, the figure of Nature Sanctuary has also been explored without being finalized as an option by the Regional Government (GORE) and civil society actors. Finally, lately, it was declared an Urban Wetland within the list of the Ministry of Environment.

The management of the Ciénaga de Mallorquín, in Colombia, operates around an agreement between the municipality, the Autonomous Regional Corporation of the Atlantic (CRA), and the District of Barranquilla. It is part of the RAMSAR area Ciénaga Grande de Santa Marta, complex of the Estuary of the Magdalena River. In 2017, the idea of establishing a Protected Area began to be explored, although studies have shown that the best figure corresponds to an Integrated Management District (DMI). In Mallorquín, territorial planning depends on several determinants; among them Ronda Hídrica, POMCA, Mangrove Zoning, and RAMSAR. There is a zoning in which are established three zones: Conservation, Preservation, and Sustainable Use. Since 2020, the integral intervention of the Ciénaga de Mallorquín was included within the District Development Plan 2020–2023 as the most relevant strategy of the program "Biodiverciudades". Recently in August 2023, the Ecopark of the Ciénaga de Mallorquín was inaugurated.

The Pantanos de Villa wetland, in Peru, is a Protected Natural Area with an area of 263.27 ha. In 2006, it acquired the category of Wildlife Refuge. It has two types of management: on one hand, the National Government through SERNANP, and on the other, the Provincial Government or Metropolitan Municipality of Lima, which in turn has a Decentralized Municipal Public Organization (OPD) that is Prohvilla, which includes the districts of the Special Regulation Zone. The Municipality of Lima owns 80% of the Protected Natural Area and SERNANP the remaining 20%.

2.2 Methodology

The methodology used is qualitative. Initially, a review of 151 press articles from the three case studies in the period 2015–2020 was conducted to identify the key actors in the management of these three wetlands. Subsequently, 30 interviews (10 in each case study) were applied with key actors previously identified. The sample selection focused on illustrating the diversity of the actors involved in the wetland management (see Table 1). The interviews were conducted in 2020 and 2021, in person and remotely, due to the sanitary conditions associated with the pandemic caused by Covid-19. The interview guide included a specific section on management and citizen participation. The interviews were transcribed and analyzed through the Atlas.ti program.

3 Results

3.1 The Concón Wetland (Chile)

The main participation initiatives come from the municipality of Concón around the administration of the Ecological Park La Isla, where environmental education visits and tours are developed with schools and universities. However, since 2015, the most important space for citizen participation is the Wetland Technical Table, convened by the local authority, the Regional Ministerial Secretariat of the Environment, and the state company ENAP, where actors from different sectors participate for the management of the protected area; among them, local organizations, foundations, Regional

Table 1 Description of the interview sample by case

	Public sector	Private sector	Civil society	Academia	Residents	Total
Aconcagua Concón	3	2	4	1	0	10
Cienaga de Mallorquín	6	0	3	1	0	10
Pantanos de Villa	4	0	2	3	1	10

Government and Regional Ministerial Secretariats.

The company ENAP has an active role in the protection of the wetland: it develops relationships with Neighborhood Boards and NGOs, supports the Ñamku Rescue Program, renews the signage, receives students in practice and has organized 2 wetland seminars, one in 2015 and another in 2017 on Wetland Management and Governance. ENAP also finances local projects and provides equipment for certain activities. Although its role, the state company was opposed to the declaration of Nature Sanctuary, because from its perspective, there is already a conservation entity which is the Ecological Park La Isla. However, the interviewees from civil society question the tension in relation to ENAP between generating awareness actions about the importance of conservation, and the fact that these actions are part of a corporate social responsibility strategy to remedy the negative impacts of the refinery and industries.

The lack of coordination among institutions is seen as something negative. The Wetland Technical Table is also criticized for not generating concrete actions and a lack of fluidity in communication with social organizations that form part of it. Likewise, it is underlined by interviewees that some institutions have instilled the idea that the wetland is the Ecological Park La Isla, although the wetland covers a much larger surface. The process of legally protecting the whole wetland area, under the figure of Nature Sanctuary, is slow, as the Regional Government in its 2018 report mentions that the best protection figure for the wetland is Nature Sanctuary, but it has no been successfull yet. Despite these initiatives, it is mentioned that there is a lack of a more permanent presence of public institutions, mainly around their supervisory function, as for example, the Maritime Governorate regarding the development of water sports such as kayaking in the wetland. The perception that sanctions and greater surveillance (fines, company closures) are lacking, as well as an authority with more power, was reiterated.

Regarding civil society and citizenship, concrete cleaning actions are carried out, especially in La Boca Beach. Many of the local actors also form part of foundations and research entities, such as Sidemar and Ñamku, which maintain coordination actions with international foundations, such as FIAA "International Foundation for Animal Aid". Currently, the various organizations present in the area seek to generate an Environmental Coordination Organization and rare pushing the declaration of Nature Sanctuary. The interviewees mention that citizen participation emanates from small groups, environmental groups that are already formed. The mechanism of mobilization is described as the most recurrent citizen action, which is criticized.

Other actors are also participating to the wetland protection: NGOs such as FIAA, TERRAM, the student NGO Vive Consciente, and the exchange work between U. Católica de Valparaíso and the United States within the framework of the United Nations Future Wetlands Program, and also the Observatorio de la Costa. It is identified as positive that these actors seek to work with citizens and teach that it is possible to enjoy natural environments without directly intervening.

Regarding the mechanisms that should be implemented to include citizens in the wetland governance, the interviewees propose that the participation mechanisms should be bottom-up processes, otherwise it is more difficult to maintain them over time and to ensure a real protection of the area. In relation to the proposed mechanisms, different options are mentionned: citizen oversight, a Consultative Council, that is a group of people who report monthly and provide resources, and also the figure of a foundation, which would be in charge of a management plan. Other mechanisms are related to the key issue of environmental education: training senior monitors, carrying out pedagogical work and organizing wetland talks open to neighbors to promote the insertion of the topic in local inhabitants.

3.2 La Ciénaga de Mallorquín (Colombia)

There are different institutions oriented to the protection of the Ciénaga. The Autonomous Regional Corporation of the Atlantic (CRA) is in charge of the care of flora and fauna, blue crab, surface and underground water management, as well as implementing the National Policy for Integrated Water Resource Management, which is a planning, administration, regulation, control, and monitoring instrument. For its part, Barranquilla Verde, the urban environmental authority of the District Mayor's Office of Barranquilla, carries out actions in favor of the defense of the Ciénaga, and initiates the process of declaration as a Protected Area together with the University of the Atlantic. Likewise, it is in charge of carrying out flagrancy operations regarding the cutting of mangroves, illegal fishing, and fillings. The Secretary of Urban Control and Public Space of the Mayor's Office of Barranquilla ensures that there is no construction and illegal occupation of the site. The Colombian Navy carries out actions to prevent maritime pollution, ordering, uses, and permits of public marine goods. The Basin Council is pending what happens in the Ciénaga (fish mortality, monitoring, water sedimentation, planning instruments). Among the relevant social organizations, the Batis Foundation is mentioned, and among the companies, Tecnoglass and Argos. Tecnoglass gave talks and reforestation of 2000 mangroves, and also the company Argos develops initiatives together with Fundambiente.

Regarding participation mechanisms, actions have been oriented toward institutional articulation, rather than the inclusion of citizenship. Since its formation in 2020, the Ecoblock is aimed at uniting inter-institutional efforts and executing in a coordinated manner the actions agreed upon in the Ecoblock Action Plan, which ensure the protection, conservation, and restoration of the Cíenaga. The members of the Ecoblock are environmental authorities, police, prosecutor's office, comptroller's office, private companies, and NGOs, among others. However, this space is described by the interviewees as an "inter-institutional elite". Previously, there were some initiatives to highlight: the realization of a Basin Management Plan (POMCA of Mallorquín) in 2007 and adjusted in 2017. The Barranquilla "biodiversity" development plan was also created in 2020 and the Pact for Mallorquín in 2007.

Two positive elements are mentioned regarding institutional participation mechanisms: first, the creation of the Basin Council because it allows involving citizens in all the instruments that the Ciénaga has, and in turn, restricts actions, identifies programs and projects that the CRA must include in its action plan; second, the Ecoblock that allows uniting the action of entities that were making efforts separately, particularly the joint work between Barranquilla Verde and CRA. In this relationship, the application of sanctions and operations against dump truck drivers filling the area with rubble, the recovery of public space after removing people who were filling or were located in the Las Flores area, have been positively evaluated.

According to the interviewees, there is a sector of the neighboring population that has active participation because they depend economically on the Ciénaga or because they want to avoid floods and keep clean the environment of their houses. Therefore, the authority mentions that the actions promoted by the local community have greater acceptance, since the sense of belonging would contribute to greater participation and action in order to protect the area. The Ciénaga is a place of supply and subsistence for the neighboring populations. Depending on the season, fishing is one of the most common activities, as well as the extraction of shrimp and snails, which is carried out in La Playa area, Bocas de

Ceniza, or on the other side of the Magdalena River. More than 80 people dedicated to the extraction of snails or “snailers”, in addition to 250 fishermen, are present in the area. The residents of the edge of the Ciénaga, particularly from La Playa sector, are fishermen who have been practicing the activity since they were 15 years old, therefore, there is a sense of belonging on the part of these inhabitants who depend on access to the Ciénaga for subsistence. The opposite happens among the residents of Las Flores sector where Venezuelan migrants and people displaced by violence in Colombia are located, who are associated with crime and drug addiction. There is a constant tension between residents who think they have possession of the area, versus authorities who see the space as an area of public interest. From the perspective of the interviewed authorities, it is mentioned that the illegal inhabitants of the area seek to achieve economic compensations and that it is a community quite difficult to include in the participation processes for the protection of the Ciénaga. From the perspective of the residents, it was mentioned that the low participation is due to fear of threats and violence, some neighbors, if there is support from the authorities, would be in favor of expelling people from the area, which has ended in conflicts among the same inhabitants. Civil society organizations organize workshops, where they try to involve all social actors (fishermen, academia, companies), including work focused on women.

Actions by other actors were also mentioned: the University of Atlantic (support process for Declaration of Protected Area (2018), monitoring and re-vegetation of mangrove), the mayor’s office of Barranquilla, The Coastal Concession (mangrove compensation), ARGOS and Tecnoglass (sponsor the planting, carry out waste collection campaigns), the Simón Bolívar University (sponsor the planting, waste campaigns), FUNDAMBIENTE (mangrove reforestation, education in Barranquilla neighborhoods), the Association of Technical Units of Environmental Promoters of the Atlántico Department, the BATIS Foundation, and the municipality of Puerto Colombia (generation of data on the swamp; bioclick; fauna monitoring) related to INVEMAR. Many groups and students are organized in relation to the Ciénaga, but there is a lack of consistency. Regarding mechanisms to implement, interviewees mainly comment on the need for more environmental education and a greater presence of the environmental authority in the area.

3.3 The Pantanos de Villa Wetland (Peru)

The institutional work of Prohvilla is channeled through a Board of Directors formed by the municipalities of Chorrillos, Santiago de Surco, San Juan de Miraflores, and Villa Salvador. This group meets every month to discuss plans and directives around Pantanos de Villa. The Municipality of Lima generates work tables and ecosystem discussions. Within the framework of the Educa municipal program, promoted by the Ministry of the Environment, workshops and discussions are taking place to train Environmental Promoters. Recently, the Ecological Serenos Program has been implemented, which seeks to strengthen the surveillance system among municipalities, prevention of crime or threats to Pantanos. It is mentioned that previously the efforts were only oriented to tourism, there was no interest in working directly with the neighbors, which probably changed with the pandemic context, turning to the local visitor.

The spaces for participation are questioned both from the authorities themselves and from the citizens. On one hand, the lack interest of the neighbors to participate is questioned. Other actors mentioned that in the process of elaborating the Master Plan, key actors (like Prohvilla) and other urban neighbors were often not invited. There is a perceived lack of dialogue between government units and disjointed work, as well as a lack of mechanisms for articulation and encounter between citizens and government. The processes of listening to the local population are often seen as a political strategy. However, positive elements of these instances are also recognized, such as the creation of the Training

Program; the Internship Program in Public Management and Wetland Conservation, aimed at last-cycle university students.

An attribute highlighted as positive by the interviewees is the work with children and young people, since the school stage allows the formation of citizens. Along the same lines, the positive aspect is the creation of the Guardian Children group, which gives voice to girls and boys, who are the main visitors to the site, in the protection of the wetland, through active participation in the PROHVILLA Board of Directors. This is a group of 40 children between 8 and 12 years old, who meet to propose alternatives for the protection of Pantanos de Villa. In 2022, the Guardian Children Group won the Ramsar Convention Prize for Innovation in wetlands.

From the authorities, they do not recognize conservation initiatives led by the local population, they only identify the initiatives from Pantanos de Villa in the education area. However, interviewees highlight a good relationship of the neighbors with Prohvilla to keep the Puquío canal or source clean and prevent floods. There are however conflicts with neighbors over land boundaries. Owners with legal lands feel their rights violated by not being able to install commerce or sell their houses for being within the protected area, especially on one side of the Marvilla Lagoon. Some collective actors and inhabitants throw debris on the lagoons and solid waste into the canals. There are also conflicts with other neighbors who have economic interests and develop landfill activities for land use change in the wetland. The above results in tensions between residents who have lived for many years in the area—prior to the declaration of Protected Zone—and the authorities. In addition, irregular settlements are estimated at 200–300 thousand people, who contribute to the pollution of the area after the throwing of garbage and the lack of sanitation, especially in the southern area of Pantanos.

Although the forms of participation promoted were not discussed in detail, other actors are developing initiatives including the Peruvian University of Applied Sciences (UPC), the Baptist University, the Italian NGO Terra Nuova and the Private University of the North. The positive aspects of these instances are mainly internships, as they train students on management of the wetland. For example, thanks to the collaboration between the local government and Terra Nuova, students from the University of the Amazon did internships at Pantanos de Villa. Regarding mechanisms driven by companies, the companies Backus, Romero, and Country, which carry out corporate volunteering, are mentioned. The Esmeralda CORP company, which is within the Special Regulation Zone, has carried out social responsibility actions by cleaning external channels of the wetland, and the Calera company has installed surveillance cameras. Finally, regarding mechanisms of inclusion and participation of citizens, it is mentioned that Prohvilla is the entity with the most contact among the local population, mainly to organize cleanings.

From the perspective of the interviewees, the inclusion mechanisms that should be implemented are: good solid waste management, appropriate use of water resources, socializations with high school students, and instructing children from the age of 5 about the responsibility of care and littering. More concretely, environmental audiences and the right to petition, which is a mechanism of citizen participation par excellence, are proposed.

4 Discussion and Conclusions

Regarding participation mechanisms, the case of Peru and Colombia stands out with the greatest number of actions and initiatives from the authorities for participation and management of the area, while Chile presents fewer initiatives from the authorities. Both in Peru and Colombia there is a large number of institutions trying to coordinate joint actions for the management of the wetland, and therefore, the main management problems have to do with the competence of each entity and the political interests that arise from it. In Chile, there are fewer institutions involved, while the state company ENAP plays a leading role in the management and conservation of the wetland, particularly in the Ecological Park La Isla in Concón.

From the citizens, in all three cases, cleaning actions, workshops and roundtables are common as ways to protect and conserve the environment. In the case of Peru and Colombia, stakeholders perceive a change in management, previously very focused on forms of control and surveillance, while now they are more focused on citizen participation and the inclusion of neighboring residents. In the case of Chile, the opposite happens, while the institutionality prioritizes participation (not always effective) in the area, citizens advocate for greater surveillance and control of the sector.

Despite the differences revealed in the three cases, with different governance mechanisms that include organized citizens, the issue of coordination of actors (to which problems of competences and political interests are added) and citizen participation in the protection of the three wetlands are deficient. The interviewees propose to establish working tables with representatives from all sectors, to install productive dialogue instances that can materialize proposals, but they also highlight the lack of participation of the neighboring communities of these wetlands. These communities are subalternized groups as they are also the most socioeconomically vulnerable and occupy, mostly, the territory informally. Their participation, as well as motivating their interest in participating, is key to ensuring the protection of these spaces in the short, medium, and long term.

However, it should not be overlooked that the relationship with neighboring communities can be complex. Some people think that wetlands should be used for sports and recreational and even commercial purposes, and on the other hand, conservation, restoration, and protection of the ecosystem require limitation and often elimination of these activities as it has been the case in Peru, where inhabitants of the Las Flores sector were expelled.

Regarding uses, there is a great difference between the three cases, since Pantanos de Villa and the Cienaga de Mallorquín are environments of supply and subsistence for the population adjacent, providing them fish, shrimp, and snails. Concón is predominantly used for recreational, scientific, and tourist purposes. These conflicts can make the inclusion of the most vulnerable citizens in wetland management complex (Eufemia et al. 2020), but they still pose the challenge of their inclusion.

Finally, the experience of working with children in the Peruvian case, and with women in the Colombian case, reveals the need to give voice to emerging groups in citizen participation in Latin America in view of sustainable development.

References

Agrawal, A., & Gibson, C. C. (1999). Enchantment and Disenchantment: The Role of Community in Natural Resource Conservation. *World Development* , 27(4), 629–649. https://doi.org/10.1016/S0305-750X(98)00161-2

Bevir, M. (2012) *Governance. A Very Short Introduction* . Oxford, Oxford University Press.

Bustos Gallardo, B., Lukas, M., Stamm, C., Torre, A. (2019). Neoliberalism and territorial governance: proposals and reflections based on the case of Chile. *Revista de geografía Norte Grande* , (73), 161–183.

Cardona, N. S. (2007). Construction of urban habitat in the public sphere: the case of the Córdoba wetland conflict. *Bitácora Urbano Territorial* , 1(11), 8–19. https://revistas.unal.edu.co/index.php/bitacora/article/view/18627

Darrah, S. E., Shennan-Farpón, Y., Loh, J., Davidson, N. C., Finlayson, C. M., Gardner, R. C., & Walpole, M. J. (2019). Improvements to the Wetland Extent Trends (WET) index as a tool for monitoring natural and human-made wetlands. *Ecological Indicators* , 99, 294–298. https://doi.org/10.1016/j.ecolind.2018.12.032

Eufemia, L., Bonatti, M., Sieber, S., Schröter, B., & Lana, M. A. (2020). Mechanisms of Weak Governance in Grasslands and Wetlands of South America. *Sustainability* , 12(17), 7214. https://doi.org/10.3390/su12177214

Pinel, S. L., López Rodriguez, F., Morocho Cuenca, R., Astudillo Aguillar, D., & Merriman, D. (2018). Scaling down or scaling up? Local actor decisions and the feasibility of decentralized environmental governance: a case of Páramo wetlands in Southern Ecuador. *Scottish Geographical Journal* , 134(1–2), 45–70. https://doi.org/10.1080/14702541.2018.1439522

Rojas, C., Aldana-Domínguez, J., Munizaga, J., Moschella, P., Martínez, C., & Stamm, C. (2020). Urban Wetland trends in three Latin American Cities during the latest decades (2002): Concón (Chile),

Barranquilla (Colombia) and Lima (Perú). *Wetlands Science & Practice* , Vol N°37, N°4, 283–293.

Rojas, C., Munizaga, J., Rojas, O., Martínez, C., & Pino, J. (2019). Urban development versus wetland loss in a coastal Latin American city: Lessons for sustainable land use planning. *Land Use Policy* , 80, 47–56. https://doi.org/10.1016/j.landusepol.2018.09.036

Ruíz, A. G., Hes, E., & Schwartz, K. (2011). Shifting Governance Modes in Wetland Management: A Case Study of Two Wetlands in Bogotá, Colombia: *Environment and Planning C: Government and Policy* . https://doi.org/10.1068/c10144

Skewes, J. C., Rehbein, R., & Mancilla, C. (2012). Citizenship and environmental sustainability in the city: the recovery of the Angachilla wetland and the local organization in Villa Claro de Luna, Valdivia, Chile. *EURE* (Santiago), 38(113), 127–145. https://doi.org/10.4067/S0250-71612012000100006

Swyngedouw, E. (2005). Governance Innovation and the Citizen: The Janus Face of Governance-beyond-the-State, *Urban Studies* , 42 (11), 192–200

Observing the Wetlands of the Huasco River, Chile. Co-creation of a Space for the Dissemination of the Environmental and Cultural Values of the Landscape

Ximena Arizaga and Osvaldo Moreno

Contents

Abstract

The project presented seeks to describe the experience of an observatory project in the Huasco wetland, which in its essence is more than a built space, and, is based on the fact of observing together. The act of observing implies in this case representing the information in order to transmit it and correct it, together, with the inhabitants of Huasco. In this way, the Huasco Wetland observatory is a collective look at the wetland and is proposed, also, as an opportunity to rethink the city and its relationship with the territory. Wetlands are fragile spaces of enormous environmental wealth that provide ecosystem services to the inhabitants. Understanding them, studying them, and reconstructing the relationship of the inhabitants with these spaces is the first step toward their active protection.

Keywords

Urban wetlands · Co-construction · Community project

X. Arizaga (✉) · O. Moreno
Escuela de Arquitectura, Pontificia Universidad Católica de Chile, Santiago, Chile
e-mail: axarizaga@uc.cl; omorenof@uc.cl

C. Rojas Quezada (ed.), *Urban Wetlands in Latin America*, Sustainable Development Goals Series,
https://doi.org/10.1007/978-3-031-69590-2_14

1 Introduction

The Huasco River Estuary Wetland is the most important coastal wetland in the Atacama Region. It is located north of the city of Huasco, capital of the province of the same name, which has 11,374 inhabitants (INE projections) and receives 27,119 tourists per year (EMAT 2019) and an average floating population of 695/day (Fondo Común Municipal 2018). This means that, in months like February, the population present in the commune is 12,277 people (residents, tourists, and floating equivalent days) which means it increases by 8%. This population accesses the wetland without specific preparation that allows them to get involved with its beauty and the care of the endemic species that make up this fragile ecosystem. The purpose of the project[1] was to implement a space of mediation between the ecology of the wetland, the local community, and visitors, by activating an area for the dissemination of its environmental and cultural values in situ, providing scientific and didactic information that would expand the experience of visit, thus contributing to its visibility and understanding.

The Huasco River Wetland Observatory is proposed as a meeting place to explore and appreciate the multiple environmental and cultural values of this surprising landscape. It is made up of a central space called "environmental classroom", characterized by a permeable gravel floor that favors the development of educational and recreational activities. Around this space, vertical elements as "rushes forests" are deployed, which on an increased scale represent the qualities of the marsh fibers that inhabit the edges of the coastal wetlands. As with these species, these forests cast dynamic shadows and harbor life in them. The various "interpretive stations" present in the Observatory expose—through a sequence of horizontal panels—content related to hydrology, geomorphology, ecology, and the history of the wetland.

Through this project of an Observatory the expectation is to re-construct the relationship between the landscape of the Huasco wetland and its inhabitants, realizing one of the main characteristics of the landscape, which is to exist through this interaction (Berque 1994); and, revealing at the same time the possibility of the project to recompose this cleaved relationship.

By considering that the landscape "does not reside only in the object, nor only in the subject, but in the complex interaction of the two terms" (Berque 1994: 5) the observatory operates as a kind of suture of this relationship. Therefore, it is not just about building an observatory, but through the construction of this observatory to study the landscape in different scales and temporalities making possible the appearance of a narrative of the Huasco wetland. The project operates as a kind of catalyst that brings out the different layers of this narrative through the inhabitants themselves: the observers. The different activities carried out, challenging instrumentalized participation, aim to make the architectural project of the observatory a "logical procedure through which all the particularities of the phenomenon itself are verified, without it being interrupted by experimentation" (Littré Dictionary, 1958 [1873], meaning of "observation").

Below the specific characteristics of the Huasco River wetland and its ecosystem values are described; secondly, the activities carried out and the results pursued and obtained from each one are explained. Finally, the Huasco River Wetland Observatory public space project is described.

2 Huasco River Wetland: The Largest Coastal Wetland in the Atacama Desert

Coastal wetlands are the home for a great diversity of resident and visiting bird species, being important areas for feeding and nesting (Bildstein et al. 1991) and one of the most productive systems on Earth; as well as one of the most fragile

[1] Proyecto "Observatorio de Humedales Urbanos Costeros de Atacama: divulgación de sus valores ambientales y culturales" financiado por el Concurso Nacional Ciencia Pública: Para el Desarrollo de Espacios Públicos Regionales de Divulgación y Socialización del Conocimiento 2021 del Ministerio de Ciencia Tecnología Conocimiento e Innovación (CP21- E014).

in the face of natural events, anthropogenic causes, changes in land use, among others.

Continental and coastal wetlands on a global scale cover more than 12,1 million km^2 (Ramsar 2018). In the South American continent, the coastal wetland system that extends between the Aconcagua valley in Chile in the south and the river Tumbes in the north, on the coast of Peru, constitutes an eco-region of 3800 km in length where the climate is Mediterranean and semi-arid in the central north of Chile, coastal desert and tropical—semi-arid in the extreme north of Peru (Tabilo et al. 2001). In addition to presenting valuable ecosystems that have supported the life of human communities for centuries, these wetlands bathed by the Humboldt Current are the habitat for migratory species of birds that travel from the distant coasts of Canada in the north and Patagonia in the extreme south. In total, there are between 90 and 100 coastal wetlands—depending on the cadastres—in the desert area of Peru and Chile, ranging from the Tumbes Mangroves (3°25′30″ S–80°16′16″ W) to the Illapel River (31°37′14″ S–71°33′21″ W). This implies an extension of the coast that houses various types of ecosystems located between the tropical semiarid zone of northern Peru to the semi-arid zone of Norte Chico in Chile: estuaries, lagoons, marshes, salt flats, wetlands, plains, rivers, deltas, etc. Of the total, 8 are Ramsar Sites: Tumbes Mangroves, San Pedro de Vice Mangroves, Virrilá Estuary, Villa Swamps, Paracas, Mejía Lagoons (Peru); Coastal Wetlands of Tongoy Bay and Limarí River Wetland (Chile) (Fig. 1).

The coastal morphology, characterized by open bays, sand and rock coasts, river mouths, cliffs, and ravines, has been modeled by various factors such as plate tectonics, water and wind erosion; and, in recent decades, by the effects of Climate Change. In this context, coastal wetlands are located on coastal plains, close to the mouths of streams and rivers, or inserted in valleys, close to surface, underground or underwater water sources. Factors such as high radiation and the evaporation process have influenced the temporary condition of the water mirrors (MMA 2023); as well as anthropogenic factors such as the construction of upstream reservoirs and water extraction from the wetlands themselves.

Regarding coastal wetlands, the majority correspond to lagoons (fresh and salt water), followed by marine, estuarine, and marsh wetlands. This great variety of aqueous bodies is manifested on the coasts of Peru in the presence of lagoons, mangroves, and swamps to the north; and descends in its geographical distribution in a north-south direction. The coastal wetlands of Chile are mainly of the estuarine type, mouths of streams and rivers, followed by riparian, lacustrine, marine, and marsh wetlands. The northern regions of Chile are characterized by an extremely arid climate. The water supply comes from the high Andes plateau with small and sporadic runoff throughout the year and significant floods in the summer, when the high-pressure system in Bolivia weakens. Regarding runoff, the area is characterized by the basins of the Lluta, Loa, Copiapó, Huasco, Elqui, and Limarí Rivers basins being the main flows that, either due to rainfall or snow, flow into the ocean. This contribution influences the dynamics of the estuary, being open or closed according to the existing flow and the water extraction processes in each basin. As for the tide, it has a micro-tidal regime that causes the formation of sand bars at the mouths of the rivers, and in turn influences as a saline wedge in the low-slope mouths (Cienfuegos et al. 2012) (Fig. 2).

In this context, the coastal wetlands south from Copiapó are located in semi-arid and desert areas, contrasting with the rest of the environment that surrounds them. Wetlands are located within one of the 25 “hotspots” areas with the greatest biological diversity in the world (MMA 2011). In the Atacama region there are 5 coastal wetlands: the emerging Chañaral wetland (39.24 ha) induced by the action of the inhabitants (Breciani et al. 2023), the wetland and Nature Sanctuary of the Estuary of the Copiapó River (174 ha), the Laguna Totoral coastal wetland and Nature Sanctuary (16.37 ha), the Carrizal Bajo Wetland and Nature Sanctuary (46.8 ha,) and the Huasco River Mouth Wetland (258.9 ha), which is the largest of the five and has also been nominated by the municipality to be a Nature Sanctuary.

The Huasco Valley is mostly an agricultural valley, with a predominance of olive trees for oil

Fig. 1 Huasco Wetland, located on the coast of the Atacama region, Chile. (Source: Authors' photography, 2022)

production, grape crops, and also horticultural crops. In the upper basin of the Huasco River, there are large-scale mining projects that extract gold and copper in addition to medium and small mining operations. To the south of the city of Huasco is the mining industry and port of the Compañía Minera del Pacífico and the Guacolda coal-fired thermoelectric plant (Fig. 2). Inside the valley, 20 km far from Vallenar, on the road that connects this city with the town of Alto del Carmen, is the Santa Juana Reservoir, the largest in the Atacama Region that regulates the flow of the Huasco River and therefore the comportment of the wetland. Despite the highly anthropized context in which the wetland is located—according to the baseline developed within the framework of the Nature Sanctuary application (MMA and CEA 2017)—it reveals a very dynamic and highly variable habitat, explained by the water quality which is suitable for irrigation and for recreation with direct contact.

The dynamics of the mouth of the Huasco River is influenced in part by the presence of a sand bar in the coastal zone that conditions the hydrological functioning of the wetland, with periods, particularly in summer, in which the exit to the sea is obstructed by the dune. This same dune separates the wetland from the beach, protecting it in a certain way from the intrusion of vacationers, which, added to the lack of clearly delimited paths, slows down visits and preserves the habitat of the bird species present in significant numbers (Fig. 2).

The data records reviewed in the context of the application for the category of Nature Sanctuary indicates for the Huasco River Wetland a total richness of 168 species of which 134 species (79.8%) correspond to birds, 16 species (9.5%) to micromammals, 12 species (7.1%) to reptiles, 4 species (2.4%) to macromammals (Guanaco, Zorro culpeo, Zorro chilla and Puma not confirmed in the field campaigns carried out in that context), and 2 species (1.2%) correspond to amphibians. Of these species, 150 are of native origin (91.5%), 14 endemic (8.5%), and four introduced (2.4%), the latter corresponding to micromammals. Regarding vegetation, the study in question identified 30 vegetation units, with

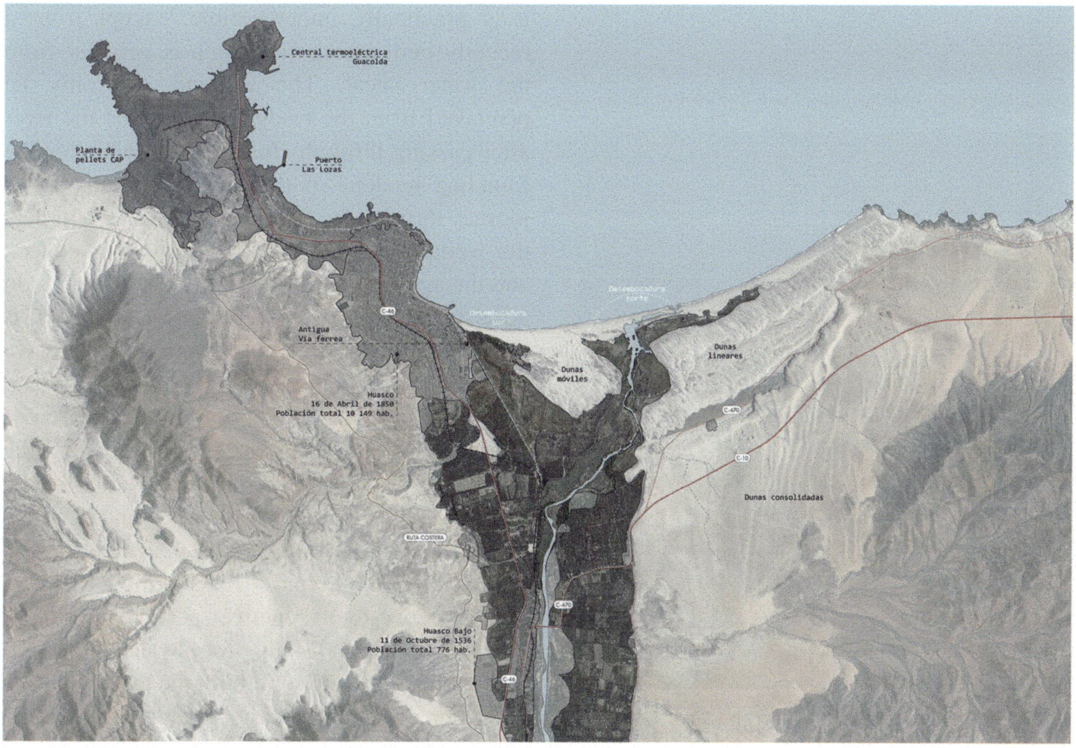

Fig. 2 Relation between Huasco city and its wetland. (Fuente: Constanza Garcés. Ejercicio del Taller Atacama, para optar al Título de Magíster en Arquitectura del Paisaje, Profesores Osvaldo Moreno y Ximena Arizaga)

few species that dominate in the area. Dominance corresponds to perennial herbs followed by shrubs as the most common; and the form with fewer representatives corresponds to succulents. In the field, water mirrors can be observed in which the combination of herbaceous plants, the presence Vatro (*Typha angustifolia*) that can reach up to 3 meters in height and the colonies of ducks and taguas (*Fulica armillata*), that gives its characteristic scenic beauty to the wetland.

In total, the area studied in the application for Nature Sanctuary of the Huasco River Estuary Coastal Wetland contains an area of 752.99 ha, of which 83.4% (627.84 ha) correspond to plant formations, mainly scrub, 4.5% corresponds to bodies of water (34.12 ha) and the remaining fraction corresponds to 10.5% agricultural crops (79.3 ha) and 1.6% urban areas (11.73 ha). Of this total, 461.9 ha are proposed as a Nature Sanctuary.

Adjacent to the wetland is the city of Huasco, which to date, except for some recently built social housing buildings, is totally unaware of the existence of this enormous space of nature that equals its surface. The main characteristic of the city of Huasco is to be an industrial-mining hub, which has marked its economic, social, and also urban development. The urban layout was initially organized around the railway and the port (Figs. 3 and 4). The train connects, tuntil today, the upstream mine with the Guacolda peninsula, located south of the city and where the Iron Pellet Industrial Plant of the Compañía Minera del Pacífico and the industrial port associated with the company are located. Also adjacent is the Guacolda thermoelectric plant that receives coal for its operation by sea. For its part, the functionality of the fishing terminal, from which agricultural products from the valley also left in the past, is today diminished by the presence of the industry.

In this configuration, the seafront, the city, and its inhabitants look toward the South, where the sources of income are located, turning their backs to the wetland located in the North. From the city

Fig. 3 Disused railway track. (Source: photography of the authors, 2022)

Fig. 4 Layout of the rail road entering the city from the south. (Source: photography of the authors, 2022)

it is practically imperceptible, except from a recently built viewpoint which is very far from the water sheet. The wetland can only be perceived from the bridge that crosses the river after passing through Huasco Bajo, which is the founding nucleus of the city. The train, on its journey through the fertile Huasco valley, borders the wetland, but does not transport passengers, and the agricultural properties that surround it do not offer access to the river. The wetland itself is in a certain way protected by these successive landscape units that separate it from the city and its inhabitants: the agricultural land, an area of failed forestry exploitation, the railway, and the dune. This has benefited the conservation of the wetland and its main body of water, and, has also distanced the inhabitants from their relationship with this natural space of enormous wealth, whose ecosystem benefits, particularly those associated with well-being for physical and mental health could help to rethink the territory.

In this urban context, in which environmental conflicts abound as a result of the existence of mining operations that support the local economy and employment; the wetland is not singled out in current urban planning; however, the municipality foresees the inclusion of a green area that protects it from its edges.

This situation of neglect is not consistent with the interest that the inhabitants show toward the wetland as expressed in the following section. The wetland, in short, is the space where everything that happens throughout a valley that is characterized by mining, olive oil production, and the cultivation of table grapes and pisco grapes since colonial times crystallizes; and where indigenous peoples lived before, whose descendants are still active as relevant actors in the territory.

3 Observing the Wetland with Its Inhabitants

The project states that the Observatory is not only the equipment enabled in the space built for the observing purpose, but also the co-design process carried out with various actors in the territory. In this sense, all the actions: workshops, meetings, tours, and reviews constitute the action of observing and therefore the observatory as such is expressed in the figure below.

The first activity of the Observatory consisted of collecting the existing information on the Huasco Wetland, which was, in particular, in the application file for Nature Sanctuary (MMA-CEA 2017). Nature Sanctuaries and Scientific Research correspond to areas protected by article 31 of Law 17.288 (1970) of National Monuments. The law defines them as "those terrestrial or marine sites that offer special possibilities for geological, paleontological, zoological, botany or ecology, or that have natural formations, which conservation is of interest to science or to the State." The Nature Sanctuaries are under the custody of the Ministry of the Environment and it will not be possible, without the prior authorization of the Environmental Evaluation Service, to begin construction or excavation work in them, nor to develop activities such as fishing, hunting, rural exploitation or any other activity that could alter its natural state. In accordance with the provisions of the law, the declaration of a Nature Sanctuary must always have a prior report from the National Monuments Council. There are 31 wetlands that have been declared Nature Sanctuaries; and, on April 10 of the year 2023, 100 have been declared Urban Wetlands under the Law N° 21.202 (2020). This law "On Urban Wetlands" modifies various legal bodies with the aim of protecting urban wetlands. The application file to declare Huasco Wetland Nature Sanctuary mandated by the Regional Ministerial Secretariat of the Environment of Atacama, responds to a desire for protection that has not yet been fulfilled.

The information collected allowed the construction of a series of interpretive panels that were exhibited in the first workshop and permitted discussion with participating neighbors and public services (Fig. 5). In this workshop, the relevance of the wetland on a continental and regional scale was explained, as part of a wetland system that extends along the entire Pacific coast and is part of a set of wetlands and protected areas of the Atacama region. Special emphasis was placed on graphically expressing the hydrographic basin of the Huasco Valley, which is born from the Tránsito and Carmen rivers, in the high peaks of the Andes and collects the waters of eight aquifers before reaching the sea. In the conversation, neighbors highlighted the concern about water vulnerability, which is recurrent in all territories. In this case it's expressed in great detail, highlighting the knowledge of the territory that the inhabitants of the Atacama region show which has been confirmed in other projects carried out in the region (Arizaga et al. 2021). This concern is recurring throughout the country and also worldwide: 58% of people express this concern worldwide (WWF 2023).

The valley was also divided into four large macrozones, which make up the landscapes of the Huasco Valley (Fig. 6). This division of the territory into landscape units that are built by the cutting of the territory based on its morphological characteristics and vegetation formations. This allows an understanding of the attributes of the landscape, recognizing those underlying values that, although not explicitly recognized by the inhabitants, tend to correspond to the same qualities they express in the speech. This is how, in the workshop, the landscape of the olive trees characteristic of Huasco Bajo appears, the nostalgia of the canals lined with vegetation versus the current piped ones, and the concern about irrigation and the natural malfunction of the wetland since the lowering of water in winter is not enough to break the sand bar.

From this workshop arises the need to explain the hydraulic functioning of the wetland, which is reflected in new panels that will be incorporated into the Huasco Wetland Observatory (Fig. 7). In this way, the observatory explains what citizens want to understand, based on the interaction, instead of being understood as a unilateral transfer from science to the inhabitants of the territory.

Indeed, the Huasco River wetland is dynamic, its behavior in winter and summer periods is a result of the amount of water provided by the rivers and their tributaries; and, also, by the Santa Juana reservoir that is located upstream and was built between 1992 and 1995 to regulate irrigation in

Fig. 5 Exhibition of the panels in workshop 1. (Photography: Huasco Wetland Observatory Project, Dissemination of the environmental and cultural values of the coastal wetlands of Atacama)

the valley. The current shape of the wetland is the result of coastal dynamics that are affected by tsunamis as we learned from the neighbors of Huasco, with the 1922 tsunami being the one that opened the estuary as we know it today; and also due to floods and landslides like the one in 2015. That is why the Huasco wetland today is not the same as the one that the oldest inhabitants of Huasco knew.

Thus, a content was woven that became much richer with the contributions of the participants, who requested different characteristics for the second workshop, and, that is why an international seminar[2] was held in Huasco. The seminar counted with the participation of experts and professionals from the Municipality of Huasco, from the National Forestry Corporation of Atacama, the Atacama Regional Ministerial Secretariat of the Environment, and national and international experts. In this seminar, knowledge and experiences were shared around: (i) "Geographic Information Technologies for the analysis, evaluation, and management of coastal cultural wetlands" explaining how new geographic information technologies allow us a better understand of the territory and support decision-making (Emilia Roman, UPM, Spain); (ii) "Conservation management and threat control in the Llanos de Challe National Park", close to the wetland (Gabriela López Acosta, CONAF, Chile); (iii) "Sustainable strategies for the development of architecture projects integrated into the landscape in coastal urban wetlands" accounting for the integration of environmental and

[2]Seminario: "Desafíos para la Rehabilitación y la Conservación de Humedales Urbanos Costeros en Contextos de Escasez Hídrica. Caso De Estudio: Huasco, Región De Atacama" Realizado el 11 de agosto 2022 y Financiado por la Vicerrectoría de investigación de la Pontificia Universidad Católica de Chile.

Fig. 6 The landscape of the Huasco Valley represented in four macrozones. (Photography: Huasco Wetland Observatory Project, Dissemination of the environmental and cultural values of the coastal wetlands of Atacama)

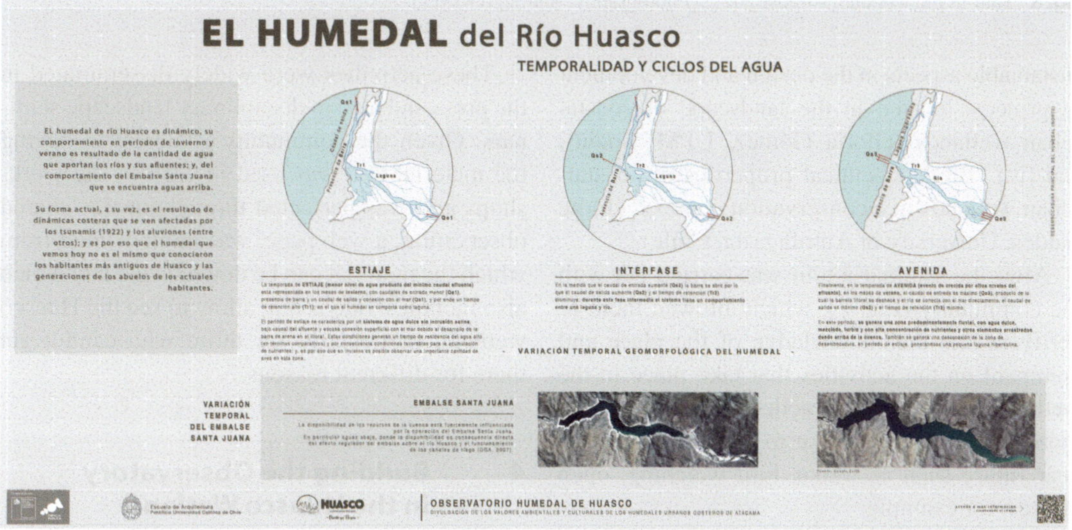

Fig. 7 Panel that explains the water cycles and the behavior of the wetland throughout the year. (Photography: Huasco Wetland Observatory Project, Dissemination of the environmental and cultural values of the coastal wetlands of Atacama)

Fig. 8 Tour in the wetland, August 2022. (Photography: Santiago Reyes)

sustainable aspects in the design and development of projects located in the landscape of coastal urban wetlands (Gloria Gómez, UPM, Spain); and (iv) "Biogeochemical properties of coastal-urban wetlands in a conservation context" (Jorge Valdés, University of Antofagasta, Chile).

After the seminar a tour was carried out with the community (Fig. 8), which allowed them to verify in situ their knowledge of the place and comment on the activities that take place in the wetland, such as the practice of running, and those that they would like to be developed in a near future such as walks, bird watching, open classes, for example.

The last formal workshop held was related to the validation of the interpretive panels and the verification of its legibility and educational properties with neighbors and public services. The design of the Observatory space was also exposed, that is: the formal built object located on the edge between the wetland and the city.

These activities were widely disseminated in the press and also in disciplinary landscape seminars. Given the community's interest in having the material that was used in each of the workshops and was perfected throughout this shared observation, a web page[3] was implemented from which the material can be downloaded and which also includes videos that allow to see the Huasco wetland, particularly for those who cannot get there for different reasons.

4 Building the Observatory in the Huasco Wetland

And finally, the Observatory as space was built, with the support of local companies and the active participation of the municipality and a

[3] https://guiastematicas.bibliotecas.uc.cl/observatoriohumedalescosterosdeatacama

Fig. 9 Painting activity organized by the Huasco Cultural Corporation. (Source: I. Municipality of Huasco)

neighbor who contributed with his machinery, thus facilitating the quick and efficient construction of the work. Thus, the desired objective was achieved and the observatory became a collective work that was inaugurated in January 2023 with the participation of the mayor, the municipal departments, particularly the environmental unit, business representatives, residents, the navy, firefighters, and even some tourists passing by (Fig. 9).

The Huasco River Wetland Observatory proposes a meeting place to explore and appreciate the multiple environmental and cultural values of this surprising landscape. It is made up of a central space called "environmental classroom", characterized by a permeable gravel floor that favors the development of educational and recreational activities. Around this space, vertical elements called "rush and reed forests" are deployed, which on an increased scale represent the qualities of the marsh fibers that inhabit the edges of the coastal wetlands (Fig. 10). And just as with these species, these forests cast dynamic shadows and harbor life in them. The various "interpretive stations" present in the Observatory expose—through a sequence of horizontal panels—content related to hydrology, geomorphology, ecology, and the history of the wetland.

This observatory transformed a micro-garbage dump into a public space: it is made up of a gravel oval that configures the plaza on which seven log supports are arranged that house fourteen folded and printed aluminum panels. The panels describe and explain the importance of the wetland at the mouth of the Huasco River, its role in the continental route of boreal migratory birds that travel to Chile from Alaska and also for southern migratory birds; the formation of the Huasco River basin, the functioning of the sand bar and its role as hydraulic infrastructure in the face of extreme weather events. It shows the different species that inhabit the wetland, the food chain, and the ecosystem services provided by the wetland; in addition to the cultural elements of the basin and a brief timeline of its history.

Through the experience of visiting this observatory, people—and especially children—will be able to understand the relevance of these systems, characterized not only by their great landscape beauty and the exceptional biodiversity they house. Also they are expected to be sensibilized by the important role they serve as ecological infrastructure, providing key ecosystem services for the sustainable development, resilience, and adaptability to climate change of territories and communities. At the same time, this wetland is understood as an expression of broader and more complex ecological system, related to regional hydrology and migratory movements of birds on a continental scale. From a socio-ecological perspective, wetlands are vital ecosystems for human survival, given that they constitute one of the most productive environments in the world: they are cradles of biological diversity, sources of water, and primary productivity on which countless plant and animal species depend for its survival. Despite the above, these systems are commonly exposed to severe threats and impacts to their components and functions that affect not only the ecology of the wetlands, but also the quality of life of the communities around.

Today the observatory is used for the development of art activities with older adults, activities with children, and community activities, it is a starting point to visit the estuary of the river and photograph birds; among other things (Fig. 11).

Fig. 10 Furniture and reed forest in Huasco Wetland Observatory. (Photography: Osvaldo Moreno)

Fig. 11 Huasco Wetland Observatory. (Photography: Osvaldo Moreno)

Fig. 12 Inauguration, January 16, 2023. (Source: Own elaboration)

5 Conclusions: To Observe by Observing

Wetlands constitute complex systems formed over thousands of years, whose structural components (geological, chemical, and biological) and their processes have managed to establish a level of functioning that allows them to get adapted to natural changes and fulfill a primary ecological function. This also makes them unique and fragile systems that currently face threats that put their existence at risk. Knowing them and understanding both their importance and their behavior is essential to guarantee active protection by citizens. In the case of the Huasco River Estuary wetland, the population is especially interested in learning in depth about the functioning of this ecological system and incorporating it into the functioning of the city.

In this sense, the wetland constitutes an opportunity where the city that looks toward the south, the Guacolda peninsula and therefore that which reminds it of its environmental conflicts, has the possibility of rebuilding its social fabric by turning its view toward the north and toward the wetland. As Sabatini (1997) points out, unlike economic development, which tends to be short-term, local communities visualize the costs and benefits of productive development and therefore tend to balance production with their quality of life. The wetland in this sense is precisely what collects the desires for quality of life and also a possibility of materializing it, highlighting the positive elements of the territory, both environmental and cultural.

In this way, the wetland has the potential to be an element of "*the convergence in the landscape of object and subject, of signifiers and meanings*" to fulfill one of the landscape "*greatest potentials for territorial planning, not only for defense and improvement*". It's the opportunity to involve the population, "*through the lived and sensitive relationship with the territory that the landscape experience entails, in the task of developing democratic territorial models*" as Mata (2006) points out.

If the wetland brings together so many actors it is precisely because it has the possibility as a landscape to "*establish a global, integrated, non-sectorized approach to the territory, which articulates the interrelationships between the elements and processes of an ecological, economic, social, cultural, urban nature. , aesthetics, etc.*" (Taroja 2006) (Figs. 9 and 12).

By forming the observatory, it is possible not only to observe the wetland, but also the relationship of the inhabitants with this element that they recognize as essential in the composition of their territory (Fig. 11).

References

Arizaga X., Moreno O., Palacios S. Gómez T. (2021) "El Paisaje como recurso estratégico: Diseño de Ruta turística en la Provincia de Chañaral, región de Atacama,

Chile". En "L'impact sociétal de l'aménagement touristique, Acceptabilité, prise en compte, recevabilité et long terme". Benamara K. , Bachimon P. El Boukhari M. Editores. (pp. 157–186). Ediciones Harmattan, Francia. ISBN: 978-2-343-23280-5

Berque, A. (1994); "Cinq propositions pour une théorie du paysage" en Berque A. , Conan M., Donnadieu P., Lassus B., Roger A.; Paris, Francia. Editions Champ Vallon.

Bildstein, K.L., Bancroft, G.T., Dugan, D.H., Gordon, R.M., Edwin, R.M., Noel, E., Payne, L.X. & Senner, S.E. 1991. Approaches to the conservation of coastal wetlands in the Western Hemisphere. Wilson Bulletin 103(2): 218–254.

Bresciani, L.; Arizaga X.; Moreno, O.; Valdés J.; Marambio-Alfaro, Y.; Bravo-Naranjo V; Rivera, M.; Vasquez, A.; Quevedo G; (2023), Littré, Émile. Dictionnaire de la langue française. Paris, L. Hachette, 1873–1874

Cienfuegos, R., Campino J.R., Gironás J., Almar R., y Villagrán M., (2012). Desemboca- duras y lagunas costeras en la zona central de Chile. p. 21–66. En: Fariña, M.J. y A. Camaño (eds.), Humedales costeros de Chile. Ediciones uc, Santiago, Chile.

EMAT. (2019). Encuesta Mensual de Alojamientos Turísticos. Instituto Nacional de Estadísticas, INE. https://www.ine.gob.cl/estadisticas/economia/comercio-servicios-y-turismo/actividad-del-turismo

Fondo Comun Municipal. (2018). Data SERNATUR. https://www.sinim.gov.cl/download/BASE_TECNICA_FCM_2022_Ver2_29122022_H1257_SINIM.xlsx

MMA - Centro de Ecología Aplicada. (2011). Diseño del inventario nacional de humedales y el seguimiento ambiental. Ministerio de Medio Ambiente. Santiago. Chile. pp. 164. https://bibliotecadigital.ciren.cl/items/6431ccde-0763-4cac-a481-36b48050546d

MMA, Ministerio de Medio Ambiente & CEA, Centro de Ecología Aplicada (2017) Caracterización del humedal costero estuario del río Huasco para su postulación como Santuario de la Naturaleza. Informe final.

Ministerio del Medio Ambiente (2023). Ecosistemas acuáticos y continentales. https://humedaleschile.mma.gob.cl/ecosistemas/humedales/

Mata R., (2006), "Un Concepto de Paisaje para la Gestión Sostenible del Territorio" en Mata, R. Y Domènech, M. (dirs), "El paisaje y la gestión del territorio. Incorporación de criterios paisajísticos en la ordenación del territorio y el urbanismo", Barcelona, España: ediciones Diputació de Barcelona.

Tabilo, E., Jorge, R. & Mondaca, V. 2001. Aves acuáticas en humedales costeros de la región de Coquimbo, Chile. Boletín Chileno de ornitología 8:13–17.

Tarroja R, (2006), "Transformaciones territoriales y valoración social del paisaje" en Mata, R. Y Domènech, M. (dirs), "El paisaje y la gestión del territorio. Incorporación de criterios paisajísticos en la ordenación del territorio y el urbanismo", Barcelona, España, Ediciones Diputació de Barcelona.

RAMSAR Convention Report (2018). 13th Meeting of the Conference of the Contracting Parties to the Ramsar Convention on Wetlands. COP13, Dubai.

Sabatini, F. (1997). Conflictos Ambientales Y Desarrollo Sustentable De Las Regiones Urbanas. Revista EURE - Revista De Estudios Urbano Regionales 23(68). https://doi.org/10.7764/1157

WWF, 2023, https://wwf.panda.org/es/?8113466/Preocupaciones-sobre-el-aumento-de-la-escasez-de-agua-segun-nueva-encuesta-global (vistado el 03/09/2023).

Mapocho Wetlands Bioroutes: Itineraries Proposal and Participatory Records

Sandra Iturriaga Del Campo

Contents

Abstract

This chapter presents a set of generative experiences that seek to value the Mapocho River and its wetland system in Santiago de Chile, based on the formulation of a set of itineraries and participatory records. These experiences were carried out within the framework of the research "Mapocho Downstream. Visual Atlas for the Revaluation of the Riverside Heritage and Landscape", carried out by Mapocho 42K Lab UC between the years 2021 and 2023, led by the architect Sandra Iturriaga and integrated by the landscape architects Raúl Brito, Yohanna Carvajal and Aníbal Retamal. The main objective was to revalue the natural and cultural heritage of the Mapocho River in its downstream section, characterized by a condition of ecological corridor and riverside wetland.

As part of the methodology addressed in the research, an exploratory and empirical experience was approached, which addressed the design, realization, and record of a set of itineraries and potential tours around the wetlands as generative experiences, which the research developed under the concept of *Mapocho River Wetlands Bioroutes.* These experiences were carried out under a participatory approach that links art and citizen science, in conjunction with local organizations and a collective of artists, with the aim of generating a tour experience that allows to know the value of the riverside wetlands from the interaction and perception of people with the environment and consider the value of these experiences for the potential implementation of tours in the wetlands.

What is presented below is the conceptualization and methodology addressed to carry out these participatory experiences, and mainly the record and its systematization as the main methodological tool for verifying the value of the wetlands from the perception of those who carried out the experiences of the Bioroutes.

S. Iturriaga Del Campo (✉)
Faculty of Architecture, Design and Urban Studies, Pontifical Catholic University of Chile, Santiago, Chile
e-mail: siturria@uc.cl

C. Rojas Quezada (ed.), *Urban Wetlands in Latin America*, Sustainable Development Goals Series,
https://doi.org/10.1007/978-3-031-69590-2_15

Keywords

Riverside wetlands · Participatory itineraries · Bioroutes

1 Introduction

In Chile, there is currently a growing interest in the protection of wetlands as areas of natural value, which has been largely driven by the promulgation and implementation of the "Urban Wetlands Protection Law 21.202"[1] (De Urresti, 2019). This law has allowed to put into action a set of regulations and ordinances in different cities and localities of the country and at the same time, has encouraged the interest of different communities to collaborate in this protection effort.[2] However, this has not necessarily had a correlation with a detailed survey of these ecosystems, with their potential access by the citizenry, or with the implementation of appropriate infrastructure for this to occur.

In this context, the research "Mapocho Downstream. Visual Atlas for the Revaluation of the Riverside Heritage and Landscape", carried out by the author, along with a multidisciplinary team of professionals under the umbrella of the Mapocho 42K Lab UC, addressed the opportunity to revalue the Mapocho River in Santiago, Chile in its downstream section—which crosses 55 km through 6 peri-urban localities—as a great riverside corridor and urban wetland.[3] This downstream section of the Mapocho River, characterized by an unchanneled course, formed by meanders, a seasonal water regime, and low slopes, contains a set of wetlands of great ecological and landscape value. However, this peri-urban stretch of the river and its wetland system remains largely invisible, being exposed to great fragility and threat.

In the research, the main proposed method of revaluation was twofold: on the one hand, the construction of an Atlas, which addressed the surveying and recording of a set of geographical, ecological, landscape, and cultural attributes and components, which allowed characterizing this riparian corridor and its wetland system from a set of edited cartographies. On the other hand, it considered the design, execution, and recording of a set of itineraries conceived as generative experiences around the wetlands with a participatory approach, which the research developed under the concept of Mapocho Wetlands Bioroutes, which allowed to know and record the perception and value of the wetlands from the direct interaction of people with these riparian ecosystems.

While there are many studies that address the ecosystem contributions and benefits of wetlands for people (AAVV Millennium, 2005; Groot et al., 2006), there is less exploratory research and qualified evidence that allows knowing the perception and interaction of people in the context of wetlands and how this perception can contribute to the design and implementation of itineraries in these contexts, favoring a sensitive approach and stimulating more conscious behaviors with the environment. In this way, this chapter aims to describe the generative experiences developed as a contribution to this approach, considering the objectives and the three successive stages that it implied:

The first stage considered the preliminary recognition of a set of places and wetlands that are located along the riparian corridor of the Mapocho, addressing aspects related to both its biodiversity

[1] Since the promulgation of Law 21,202 in January 2020, a set of 109 urban wetlands have been established under this protection figure throughout the country for their preservation as areas of natural value.

[2] This boom has had different effects depending on the context of the country and its field of application has been especially fruitful in wetland-cities like Valdivia and Concepción. There is also a significant set of local environmental organizations that have strengthened, with the aim of actively collaborating in the protection of these ecosystems, either from individual or collective initiatives, such as the Plurinational Network of Wetlands.

[3] During the development of the research, the first declaration of urban wetland for the Mapocho River was generated, driven by 3 per urban communes: Peñaflor (Nov 2021) Talagante and El Monte (Jan 2022). In January 2023—almost at the end of the research—the declaration of urban wetland for the entire Mapocho river is requested, endorsed by a set of 13 municipalities.

and its landscape, with the aim of having the base material for the design and realization of potential itineraries. In this early stage, a set of local organizations that were already working in the recovery of some wetlands were incorporated, adding a participatory approach into the proposal. The second stage considered the conceptualization, methodological design, and execution of a set of Bioroutes, as triggering experiences carried out under an approach of art and citizen science through an open call, for which a collective of artists was additionally integrated. The third stage considered the systematization of these experiences from the record made, producing material of great interest that sheds light into the perception and valuation of the wetlands by people. It was considered in a way complementary to the above, carried out with a temporal distance, in order to verify the sustained interest and connection generated by the experiences of the *Bioroutes* through a survey methodology carried out with the participants. The results of this phase of the research under this exploratory and empirical method of the Bioroutes led to the formulation of a set of routes of approach to the wetlands that were integrated into the final edition of the Mapocho Downstream Visual Atlas (Iturriaga, 2023), whose main orientation is to serve as support to the communities for the conservation and recovery of these ecosystems and for the potential implementation of these *Bioroutes*.

1.1 The Mapocho River Downstream: Ecological Corridor and Riparian Wetland

The case study focuses on the Mapocho River, the main tributary of the Maipo River basin in which the city of Santiago is located and on whose margins the city was established since its foundation.[4] From its birth in the mountain foothills, the Mapocho River runs a length of 97 km through the valley until it converges with the Maipo River, with a steep gradient and seasonal flow variations that define it as a torrential river. Along this route, it presents three well-differentiated sections: the upstream section, from its birth at the confluence of the Molina and San Francisco rivers until it appears in the valley, boxed in by a pre-mountainous sector; the central urban section runs from east to west, channeled and with highly urbanized edges; and the downstream section, formed by a channel parallel to the coastal mountain range in a north-south direction, which with a lesser slope and unchanneled banks, presents unique geographical and ecological characteristics within the basin.

This downstream riparian corridor runs 55 km until it converges with the Maipo River, crossing 6 peri-urban communes: Pudahuel, Maipú, Padre Hurtado, Peñaflor, Talagante, and El Monte.[5] Along this route, it presents a channel of variable widths that fluctuate between 70 and 700 m, with a morphology of low-depth meanders and gentle slopes, which together with its effluent course condition,[6] favors a rich system of wetlands along it. These characteristics enhance the presence of dense riparian vegetation and a rich biodiversity of species—especially avifauna—giving it an important environmental and ecological role, as well as an undeniable landscape value that interacts with the urban environments that coexist along it.

With the aim of identifying and characterizing the different areas of interest along the downstream section, a survey and record of the total corridor along its 55 km was carried out, based on field research and complementary sources. This resulted in the creation of a set of preliminary cartographies for a total of six sections,[7] in which the main environmental, ecological, and

[4]Although the Maipo River is the main water resource of the city of Santiago, the Mapocho River is the main urban river.

[5]This confluence is located southwest of the city of Santiago.

[6]Effluent condition implies that the gentle slopes and saturated soils generate a contribution of water to the channel, promoting the appearance of surface outcrops (Brito, 2019).

[7]These cartographies were developed at a scale of 1.15.000 and their edited versions are part of the Visual Atlas (Iturriaga, 2023).

landscape attributes that characterize them were identified, with the identification of three landscape units: wetlands and lagoons, tips and hills, and priority border areas. Within these landscape units, the riparian wetlands stand out as biodiversity hotspots to be conserved and restored.

These cartographies had a dual role: as an instrument for verifying the pre-existing conditions and attributes associated with each section, and at the same time, as a device that allows establishing the potential for revaluation of the different zones and areas of interest of each section. From this cartographic analysis, it is established that, to enhance the protection of the ecological and landscape values of the wetlands and lagoons along the corridor in the face of increasing urbanization, it is necessary to define access and approach routes favorable to these ecosystems. This led to consider the design and implementation of Bioroutes, conceived as potential routes that allow access and at the same time protect these ecosystems.

1.2 Mapocho Wetlands Bioroutes: Itineraries as a Proposal for Revaluation

The 'Mapocho Wetlands Bioroutes' emerge in the context of the research as an opportunity to carry out an exploratory qualitative method, which would allow recognizing the value of the riparian ecosystem and its wetland system from generative experiences carried out in interaction with the community. In this way, it was proposed that through the design and testing of sensitive itineraries as a method of verification, it is possible to derive qualified evidence that allows knowing the perception and interaction of people in the context of wetlands and how this evidence can contribute to the implementation of routes for the conservation of these systems, stimulating more conscious behaviors with the environment.

The concept of 'Bioroute'[8] was defined as an itinerary capable of generating a meaningful experience between people and their environment, through an approach route that allows recognizing its values and attributes and exploring its rich biodiversity, enhancing a sensitive approach to the environment by those who carry it out: "a space of interaction between biophysical and cultural systems, conceived as a meaningful experience of recognition, exploration, and perception, which allows integrating people with their environment and, at the same time, favoring the conservation of the ecosystem to which it refers, generating an experience of attachment" (Iturriaga, 2023: 78). Unlike the notion of interpretive routes or itineraries (Brunelli, 2019),[9] the main objective of the experience developed was not only focused on the dissemination or exploration of the environment as a way of recognizing the values associated with nature, but mainly, on verifying how these values are perceived by those who travel and interact with these environments, based on a precise methodology of recording the experience.

In this way, for the purpose of this research, the Bioroutes considered a dual objective: on the one hand, to design a triggering experience to recognize and value the attributes and richness of the Mapocho River wetland system—often unknown by the very communities of the environment—through an itinerary designed with a participatory approach integrating local organizations and collectives, and on the other hand, to apply an exploratory recording method as a fundamental part of the experience, which allows gathering valuable information about the perception and values that people recognize in these ecosystems and how this can contribute to the design and implementation of actions that promote their conservation.

For the design and planning of the developed itineraries and their corresponding record, focused on enhancing a sensitive approach to the environment, some contributions from recent

[8]The concept of 'Bioroute' was established as a detonating concept of the research, which allows to collect contributions from different disciplines, considering a sensitive and exploratory approach toward the environment.

[9]This concept is associated with *guided activities in nature* applied in the first national parks in the United States, as is extensively described by Brunelli (2019).

decades were reviewed through concepts such as 'soundscapes'. Similarly, the review of experiences that address and record people's perception of their environment, such as 'walking interviews' or commented walks (Evans and Jones, 2011),[10] served as a counterpoint to define the design of the experience to be carried out.

1.3 Collaborative Design of the Bioroutes: Narrative, Experience, and Record

The design of the Bioroutes was defined under a participatory approach that allowed the integration of *art and citizen science,* with the aim of generating a meaningful experience in relation to wetlands, both from a sensory approach and from informed knowledge of the value of these ecosystems. For this, a group of twelve local environmental organizations[11] involved in the conservation and protection of the Mapocho River were called upon, along with the participation of the collective of artists,[12] making them part of a methodology based on three pillars: the construction of a narrative, the design of the experience during the journey, and the recording of the experience. Along with this, the context of the Bioroutes to be carried out was planned, as a form of script verified on site.

1. *Construction of the narrative*: it considered developing a sensory approach to the experience of wetlands, promoting content as a 'script' for each of the senses, which could support the realization of the Bioroutes by the monitors of the different organizations. This script was designed from open questions, to stimulate a range of possibilities and applications both for the design of the activities and for the development of the activity. A maximum of 40 participants was also defined for each Bioroute, to generate a qualified interaction from the narrative in charge of the monitors of each one.
2. *Design of the experience*: it was proposed to sequence each Bioroute from a set of 5 sensory activities as 'activations' during the journey, whose main objective was to awaken sensitivity about the relationship with the environment, valuing its visual or scenic beauty, auditory or sound, olfactory and aromatic, gustatory, and tactile. With the help of the collective of artists, different activities and devices were designed as support for the sensory activations: a Viewer to look at the environment, a 'Hydrophone' for listening to water, a mini-herbarium for taste, 'frottage' and barefoot walking for touch, and 'capsules' to awaken aromas. For each Bioroute, a journey structure was defined as a script, defined from the sequence of these 5 'activations'. As part of the design process, the sequence of the route and its narrative, along with the devices of the activities, were previously tested in a pilot session with the set of organizations to adjust changes or variations prior to execution.
3. *Logbook Record*: a fundamental part of the methodology considered the recording of the experience. For this, a *Bioroute Logbook* was designed as a vital expression of the experience that would accompany the participants from the beginning of the journey and that could allow recording of the main explorations and discoveries with the senses. The

[10] This method is based on interviewing participants while walking, to capture their perceptions of the environment. Unlike this, the method defined and applied for the Bioroutes is not based on interviews, but on personal recording as an invaluable source of information from those who participate in the experience.

[11] The experience of the Bioroutes was carried out hand in hand with 12 environmental organizations already involved in recovering the Mapocho River and its wetlands, which were contacted to carry out this experience together: Talagante: Frente de Rio, Cabra de Cerro; El Monte: Verde Nativo; Peñaflor: Mapuko, Te quiero Verde, Recuperación Silvestre; Maipú: Parque Mapocho el Trebal; Naturaleza Maipu, Padre Hurtado: Padre Hurtado Comuna Verde, Isla de Maipo: Geonativo.

[12] The design of the experience had the collaboration of COLECTIVA UN LUGAR, composed of visual artists Melissa Ferreira, Consuelo Pedraza, and Antonia Telliery, with extensive experience in participatory mapping and sensory exploration devices. They also collaborated with a group of 20 volunteer Architecture UC students.

design of this Logbook constituted a field of testing and experimentation with the aim of activating the imagination of the participants, for which different methods were applied. This Logbook was designed in a format with five foldable sections, in which each section proposed a type of intentional record that accounted for the perception or observation associated with each of the five senses.

4. *Itinerary Proposal.*

The preliminary survey of the sites of interest that are part of the Mapocho riverside corridor downstream and its wetland system allowed to identify the itineraries and their sequence of route. Based on this survey and its characterization, a site of interest or priority wetland was defined for each district, which resulted in a total of 6 Bioroutes or guided itineraries along 6 highlighted points on the Mapocho River Downstream: in the districts of Maipú (M), Father Hurtado (PH), Peñaflor (PE), Talagante (T), El Monte (EM), and Isla de Maipo (IM). For each of them a route ranging between 1.5 and 2.5 km was established, to be completed in approximately 90–120 min. The script of each Bioroute was previously verified on site, since none of them have specific infrastructure to carry them out (Figs. 1 and 2):

B1_Bioroute El Trebal Oriente_MAIPU/Itinerary 1.88 km.

Monitors: Organization Parque Mapocho El Trebal, Nature in Maipu.

In the district of Maipu, next to the Mapocho River, there is a particular wetland generated by the presence of a network of streams that drain their waters toward the river among the vegetation of a dense willow grove. The Bioroute was defined starting from the upper terrace of the channel descending into the fluvial space, moving from an open and panoramic space to a contained one by the chiaroscuro of a mixed forest of native and foreign willows.

B2_Bioroute El Trebal Poniente_PADRE HURTADO/Itinerary 2.24 km.

Monitors: Organization Padre Hurtado Green District.

The itinerary runs along the west side of the Mapocho River, on the slopes of the Coastal Mountain Range, around a lagoon-wetland of almost 8 ha, generated from the extraction of aggregates. The route takes advantage of a geographical balcony condition over the river and develops amidst xerophytic vegetation of thorns and shrubs, which guide the route to an extensive bay that contains the lagoon.

B3_Bioroute Los Pozones_PEÑAFLOR/Itinerary 1.73 km.

Monitors: Organization Mapuko, Wild Recovery, Peñaflor I love you green.

The south bank of the Mapocho River in the district of Peñaflor is characterized by the presence of wetlands or 'pozones', as the chosen sector for the Bioroute is called. This place forms an ecosystem of great biodiversity, which can be accessed by some informal paths that allow approaching the channel, crossing a set of 'pozones' among a dense forest of Chilean willows, curly willows, and maitenes, alternated by clearings of vegetation.

B4_Bioroute Los Pidenes_TALAGANTE/Itinerary 1.75 km.

Monitors: River Front Organization.

The Bioroute develops along the southern bank of the Mapocho River in Talagante, where it is possible to traverse a set of wetlands formed by branches of the river. The route begins from the upper bank that offers panoramic views over the river, then descends to the intermediate terrace, crossing pools and springs formed by river outflows, accompanied by a landscape characterized by the dense presence of reeds, rushes, and Chilean willows.

B5_Bioroute Pedro de Valdivia_EL MONTE/Itinerary 2.53 km.

Monitors: Native Green Organization, Ecoclub El Monte.

The north bank of the Mapocho River in the district of El Monte is characterized by presenting a system of wetlands from old

Fig. 1 Bioroutes open Call with map of participating organizations © Mapocho 42K Lab UC

river branches, delimited by a river defense. The journey through the Bioroute allows crossing this series of pools formed by natural river outflows, from which it is possible to spot the rich variety of the sector. This Bioroute traverses the intermediate terrace, offering a unique experience combining natural landscapes with containment infrastructures.

B6_Bioroute Tajamar Confluence_ISLA DE MAIPO/Itinerary 2.24 km.

Monitors: Geonativa Organization.

This Bioroute is in the confluence area of the Mapocho and Maipo rivers, in a sector where the width of the channel, the wetlands, the river branches, and the riparian vegetation intensify. The route takes advantage of a breakwater that delimits the channel and allows a set of 'stops', in proximity to a zone of riparian wetlands and dense reed beds. From there, you reach the main channel, through an informal path accompanied by a large presence of tar on its edges.

Fig. 2 Bioroute Map Talagante, based on the activities carried out © Mapocho 42K Lab UC

1.3.1 Open Call

The Bioroutes were carried out in two differentiated instances in January, corresponding to the summer season. On each of these occasions, an open call was made, considering groups of a maximum of 40 people for each route and a total of 6 Bioroutes. The call was made through an online form published on the research team's digital platforms, through which one of the 6 possible Bioroutes options was chosen. The universe of registrants on the two days in which the activity was carried out was the maximum, with a total of 420 people, who were coordinated by groups for each Bioroute by a team of monitors.[13] Although people from nearby localities participated in the activity, through the registrations it was possible to know that a majority (65%) came from other districts of Greater Santiago and, for many, it was the first time they visited the Mapocho River in the sector.

[13]This coordination was carried out through WhatsApp groups, in which material was disseminated prior to the visit.

1.4 Bioroutes Logbook: Record of a Sensory Itinerary

The logbooks were the main recording method developed to verify people's experience during the tour of the Bioroutes. They thus represent a privileged support or 'impression' of the sensory experience and interaction of people with the wetland environment. Through this individually developed method, the recording of each sense was encouraged in a differentiated way:

The result of these logbooks developed by each participant and their subsequent systematization form an eloquent body of evidence of people's perception and interaction with the wetlands and a qualified expression of the valuation of these environments. For this, 25 logbooks were chosen for each of the routes, considering their

Fig. 3 Bioroutes carried out in the 6 districts ©Mapocho 42K Lab UC

highest degree of completeness to collect qualified information. The following are some results and observations derived from this experience of recording the Bioroutes (Fig. 3):

1. Visual Record: Intensive Focus on Wetlands.

For the sense of sight, an attentive observation of the environment was proposed, stimulated through the activity of the amplified viewer on site and along with it, through a recording method called 'Focus',[14] which consisted of drawing a random fragment of the place from a frame-cut in the Logbook as if it were a magnifying glass, observing and recording even the smallest of elements. The focus exercise required first selecting a particular place and moment on which the environment was 'focused' for the observation and

[14]This recording method was designed with the contribution of industrial designer Andro Yurac, who devised the 'Focus' exercise in his Degree Thesis as a field of experimentation for conscious attention (Yurac, 2018).

Fig. 4 Bioroutes Logbooks, Visual Record © Mapocho 42K Lab UC

intentional recording of details. This meant that no two 'portraits' of reality would be the same or one better than another, or repeated. All places present to the observer a potential and equivalence to be observed and recorded, and by focusing on this objective, it generated a favorable response in people to undertake the exercise, as there was no reality that warranted more interest than another. The records in most of the logbooks are made with care and attention, recognizing details, and drawing specific situations from the selected environment. The results obtained by this method in most of the logbooks denote an interest in observing nature and its peculiarities as if it were an amplified magnifying glass, and the drawings are a faithful expression of this. The variety of the drawings are also an eloquent expression of the richness observed in the environment from this record, visualizing a set of random elements that acquire greater imprint and value in these "focused portraits": fragments of leaves, trunks, flowers, branches, stems, alternate with the recording of stones, rocks, seeds, grains, moisture, and textures. The richness and expressiveness of the records is a value that is present in most of the logbooks and the observed reality is drawn with multiple nuances, even with the majority preferring the use of only one color. During the development of the experience, most people naturally assumed the proposed exercise and no resistance to drawing was perceived, with more interest in the attentive observation of what was drawn over the skill in drawing (Fig. 4).

2. Auditory Record: Attentive Listening to Wetlands.

The exercise invited attentive listening with blindfolded eyes, or through the 'hydrophone' device and then recording the main perceived sounds in writing. The auditory records are eloquent evidence of the favorable perception that people assign to the sound condition of the riverside environment and its wetland system. A persistent repertoire in the logbooks of all the Bioroutes is related to the presence of avifauna, with an attempt to recognize a great diversity of species through their sounds evident in the records. In some cases, it appears as a general description, mostly as a literal description of how they are perceived and in others, as an explicit recognition of a species through its sound: "Birds singing sharply and continuous" (PH5), "Sweet song of birds" (IM12), "Fluttering of birds in trees" (EM1); "Little bird - chirp chirp, cuy cuy" (T4). "Birds: *Keu keu -Ñie ñie -Tui tui -Uhuú uhuú"* (PE6), "The sound that birds make - *qui qui, tri tri"* (M10), "Birds - *priii, fiu- Prú prú- Tru tru*" (EM8), "The flapping of a bird's wings while flying - *clap clap clap"* (PE11), "There is a bird (thrush?) that whistles like 2 sounds, one gurgling and then whistles" (PH1), *"Sha sha sha* -Queltehue, *Chirruu chirruu* -Chercán, *Goa goa* -Great Heron" (T5), "Birds with a sound similar to the little sparrow"(EM3), *"Tuutú tuutú* - Northern Cuculí, *Sh sh sh* -Chercán" (PE7), "Queltehue - *Fpiu fpiu* - Woodpecker - *Titururu"* (PE16), " *Pi pi pi pi pi* - Pidén bird" (IM3), "Taguas - *Qua qua qua..."* "Thrush - *Chio chio"* "Queltehue - *Tuie... tuie... tuie..."* "Pidén - *Chuee... chuee"* (IM13). This is especially intensive in the logs of wetlands like Peñaflor and Isla de Maipo, where the environmental conditions—immersed among trees that mediate with neighboring areas—, present an advantage for the appreciation of this soundscape. Attention to birdlife is complemented to a lesser extent by the recording of insects and amphibians: "I heard crickets, they make *quer quer"* (IM23), "Bumblebee - *mmm mmm"* (T10), "Frog – *Wrenec"* (PE21). In districts where the wetlands are closer to the urban area—like Talagante or El Monte—, the record shows the presence of wild or farm animals that cohabit with the wetlands: "There was a small cat meowing, *meow, meow"* (T6), "I distinguished the sound of cows along with different birds" (EM11). Despite the risk this poses to biodiversity, it is not perceived as negative by people.

The repertoire of sounds related to water is also recorded in a recurring manner in the logs of all the riparian wetlands as a favorable aspect, evidencing a special attention to its sound condition in motion that is intensified by the proximity of the river and its rocky bed. This record is manifested through descriptions with nuances and contrasts that express the interest with which water is perceived: "Sound of the water flowing" (M1), "*Slisshshilhshlippphshih*, the water forcefully flowing" (PE4), "We heard water. The children hear the water." (PE10), *"Chiiiii* - Water" (T17), "Sound of how the current of the river collides with the small plants" (EM8), "Water: *trrrrr trrrr - glu glu glu trrr"* (IM5), "The sound of the water. It sounds like a snore (M5)," "River sound. Rock or stone - *pom pom*" (T10), "Sounds of how the water hits the stones" (EM3). Similarly, the sound of water stimulated through the device 'hydrophone' is perceived as an amplification of one's own body, with records of varied interpretations: "Listening to the water, the amplification was great (hydrophone). It was like being part of the river" (PH1), "The sound of the river through the listening tool (hydrophone) reminded me of the sounds of mantras, something like OM" (PH2), "Sound of moving water, similar to dance" (PE18), "The hydrophone makes me feel like I'm an insect or small animal" (EM11), "Listening with the device (hydrophone) relaxes me because the sounds of nature are very strong and you feel connected with it" (EM1), "Hydrophone: A unique, unexpected, beautiful, exacerbated sound of water" (IM25). Another aspect widely recorded in the logs is the interest in the sounds related to the trees and vegetation that are part of the environment. Their record is rich in nuances, particularly the movement they produce in contact with the wind as an essential component: "Tree: the sound is like passing under a tunnel" (PH6), "Trees clapping" (PE3), "The sound of the tree is like an increased dry snoring. The sound of the leaves is lighter" (M5), "Leaves with the wind: *Shhh"* (IM6), "The wind in the foliage of the trees" (PE9), "Trunks moving sound like sparks" (PE22), "Sound inside reeds - *clu clu clo clo"* (T10), "Bark: Like musical instrument" (EM7), "*cra cra cra cra,* tree creaking" (PH11), "Wind moving the willows" (PH12). It is significant to note that the sounds perceived in these wetland environments are often associated in many records with the well-being they produce in people, which is of great interest to value the connection with these ecosystems: "Sounds that give peace and tranquility, all different from each other" (PH4), "You can hear the wind, plus the sound of the branches, it's very relaxing" (M5), "Movement of the leaves and branches. Water. Tranquility" (EM7) "the first stop the sound of the river all wonderful!!, you can faintly hear the water, no one in the group was talking, we are all ears" (PE2), "water: the sound is relaxing, it feels very good" (PH6), "The experience relaxed me, so much so that I didn't want to take off the blindfold, I felt calm, at peace" (EM11). Other evidence in relation to the state of well-being produced by the 'sound landscape' of the wetlands, is related to affective memory and the association with the rural, which is perceived as something positive: "Listening to the queltehues, reminded me of my childhood in the district of El Bosque, when after a rain I heard them" (M8), "Memories of childhood trips, to the Maipo river with my parents and siblings, it was every Sunday" (M9), "It reminded me of my childhood, because I lived in the countryside of San Fernando" (EM11). Contrary to this, in some wetlands there is resentment of the relationship with urban areas and there is a negative perception of the mixture of sounds that invades: "Sound of nature in the middle of city noises, cars and planes" (PH5), "Airplane turbine NOT NATURAL!" (M6), "Sounds of aqueduct pipes" (PH8), "When closing my eyes, yes, it sounds nice and quiet, but pollution ruins it" (M2), "Native birds. Planes. Sharp and short sounds" (EM6). This is relevant if we consider the absence of buffer or transition zones in some of these contexts.

3. Scent Record: The Aroma of Memory

The exercise with the sense of *taste* involved the collection of different species along the route, recognizing the flavors of some plants present in the wetland environment and their record was carried out in the logbook as a mini-herbarium.

The record of smells and aromas also takes center stage in the logs of all the Bioroutes, evidencing the importance they have in people's perception and valuation. A diverse range of smells that are 'awakened' through the tour of riparian wetlands is recorded: "The whole forest has a very sweet aroma. Peri: Citrus, corn. Chamomile: Sweet. Hawthorn: Toasted flour. Willow: Citrus" (PH1), "River smell. Wet earth" (PH10), "Tar: like mint and eucalyptus. Campfire smoke" (PE1), "Tar: it seemed a fresh aroma. Palqui: unknown aroma, but intense" (PE2), "Humidity. Mud. Wood" (PH12), "Palqui: Rue, strong smell. Earth smell. Green forest. Smell of wild and free flowers" (PE10), "Wet earth. Citrus and aniseed smells. Sulfurous. Summer" (PE24), "Mud: Memory of natural springs or sources" (T12), "Mint aroma. Paico aroma. Field smell" (EM11), "Green fruit smells. Water with mud: mineral smell. Celery and cilantro smells could be experienced" (IM9), "River smells, it reminds me of the river" (IM15).

In many of the descriptions, the newly discovered aroma is attempted to be assimilated to something "familiar" or known, establishing a rich relationship of belonging inherent to this associativity: "Tar: fresh, Palqui: green beans" (PE3), "Smell of country kitchen fire"(PE7), "Tar: Serrano ham Twistos. Palqui: Smells like raw potatoes" (PE9), "Fennel: Smells like anise candy" (PE14), "Mojito. Mint candy. Caipirinha" (PE20), "Palqui: chocolate filling" (PE21), "Tar: Smell of turpentine" (PE16), "Tar. Baby and talcum smell" (PE22), "Hawthorn coffee, smell of toasted flour. 'Natural' smells from vegetation: Manzanillón, peri, mint. Sweet and acidic smells" (PH8), "Intense and pleasant smell of peppermint, it's exquisite. The same goes for the grass aromas" (M8), "Paico: remedy. Lawén. Willow: grass, green apple" (T2), "Herbs. What does it remind me of? Field, horses" (T17), "Tar: Menthol. Environment: Plants and earth, scout camp. Mud: Facial rejuvenation mask" (IM1), "Wetland plants: Celery salad with lemon, parsley, sweet half hour, lavender. Path plants: Dish soap" (IM6), "The tar reminds me of glue" (IM25).

To a greater extent than the auditory record, the record of aromas and smells is recurrent in recognizing an emotional memory and many of the descriptions in the logbooks refer to situations from childhood or affective memories: "I smell you green. Green of my childhood, aroma of the joy and beauty of the simple, of the source. I smell you green and fly to the past, to the joy of contact, to the peace of silence, of balance. I smell you green and I hear laughter, hours of games with friends, cousins, and pets. I smell you green and I feel like a child again, happy, relaxed in peace. I smell you green and I reconnect with myself" (PE4), "Smell of tar and rosemary: Childhood memory, by the Mapocho (Maipo) river in Melipilla" (IM7), "Childhood, river trip, *No worries*, friends. Home, family, tea, night, mommy!" (T5), "Paico. It's like a candy, it reminds me of my childhood, when I was a girl in the countryside" (T4), "Brea: I remember the countryside, childhood, family, having tea" (IM11), "I smelled mint, it reminds me of the countryside, my childhood" (EM11), "Brea: I remember childhood, blowing its flowers and making them fly" (IM13), "Paico. Lemon verbena. It evokes memories of my childhood. I remember my grandmother Meche and her house in Quintero" (T7), "Rosemary. Acacia. Hawthorn. Brea. I remember childhood" (IM19). "The smell of mint reminds me of my grandfather who liked to garden. The smell of Paico reminded me when I got a stomach ache when I was little" (M10).

The olfactory record in the logs is also a good indicator of the states of well-being that people perceive during the tour, associating these aromas with disconnection, tranquility, and calm: "Relaxing smells, they bring calm" (PH10), "Grass. Calm. Disconnection" (M6), "Aromas and smells that envelop you" (PE13), "Relaxing essence, lavender, parsley, rosemary. Coffee with earth. Willow leaf" (EM7), "It smells sweet, a perfume, made with tranquility and special treatment" (T6). There are few descriptions that refer to smells in negative terms, even when there are, the perception is nuanced with other smells: "Stagnant water: Green drain. Stone path: Dust. Brea: Candy" (IM20), "Mud: it smells like lucuma. Water: decomposing flowers. Purple flowers remind me of something bitter" (IM5), "Palqui: Food, strong smell, not pleasant. Celery: Salad" (PE15), "Smell of mud in the wetland, not unpleasant. Rosemary. Brea. Acacia" (IM25) (Fig. 5).

4. Taste Record: Collecting Ancestral Knowledge

The sense of taste was an experience carried out during the tour in which they were encouraged to try and collect different varieties of plants or herbs and their record was reflected in the logs in the form of a personal "mini-herbarium". This collection shows people's interest in getting to know the plants and herbs which is reflected in the indication of the names that account for what was collected. These "mini-herbariums" detail a significant variety of herbs that coexist with these environments: "Mint. Chamomile. Lavender. Coffee" (PH5), "Peri. Chamomile. Cow's tongue. Oats. Thistle" (PH6), "We crossed a field of chamomile" (PH9), "Rue. Chamomile. Rosemary. Mint" (PH10), "Mint. Chamomile. Herb water. Spearmint. Lavender" (PH12), "Mint, only missing sweetener, but tasty. Hawthorn: toasted flour. Chamomile, reminded me of infusion tea. Peri, a bit bitter" (M4), "Toasted hawthorn seed. Peri. Chamomile: mixed with cold and fresh. Mint" (M6), "Fennel. Plantain. Sage. Lemon verbena" (PE3), "Fennel. Sage. I smell you green. Plantain. Mallow. Lemon verbena. Palqui (PE5)," "Palqui: Complete seasoning. Brea: Sweet mint. Paico: Herb water. Celery: Salad" (PE6), "Mallow. Mint. Medicinal plants. Paico. Fennel" (PE8), "Pennyroyal. Paico. Fennel. Mallow: Lemongrass" (PE12), Paico: Familiar, mint. "Plantain. Sage. Fennel" (PE18), "Mint. Chamomile. Lemon balm" (EM2), "Mint. Large chamomile. Landa. Peri. Rosemary" (EM3), "Mint. Anise. Mushrooms. Basil". (EM6), "Mint. Hawthorn. Peri. Large chamomile. Coffee, earth. Rosemary. Lavender. Basil, spearmint" (EM7), "Culén, culén punch" (IM3), "Rosemary - anti-asthmatic, antibacterial, antiviral. Culén - indigestion, diarrhea, lack of appetite. Brea: menthol candy" (IM4), "Rosemary: sedative, antiparasitic. Culén: healing, laxative, intestinal pain, indigestion" (IM14).

Fig. 5 Mapocho Wetlands Bioroutes, Activities with the senses ©Mapocho42K Lab UC

Along with this, some of the plants collected in the herbarium of the logbook and their descriptions, show a familiarity and positive valuation of their medicinal uses, as well as with the rurality to which they refer: "Mint: mom, home remedy. Coffee: city, warm. Paico: open field" (M11), "Cedrón: It reminds me of grandma Meche's house. Mint: It reminds me of medicinal teas" (T11), "Mint: Infusions when one was sick and my mom and grandma made me herbal tea to get better, this when I was little" (T8), "Plantain. Paico. Childhood, Cold, Flower tea, expectorant" (T9). In a significant proportion of these herbariums (40%), the collection is not accompanied by a description that identifies what is collected with the names of the plants or herbs. This could be partly because there is a closeness with these knowledge and ancestral wisdom, which have been transmitted largely by generations of grandparents or parents and that do not require to be recorded.

2 Touch Record: The Conscious Footprint

The activity linked to *touch* involved barefoot walks around the place recognizing textures, temperature and in all the Bioroutes the monitors stimulated direct contact with the water in the vicinity or crossings of the river, which was especially well received by the participants. In all the districts, the context in which the experience was developed does not have suitable infrastructure to approach the river; therefore, these situations were guided by the monitors to allow good management of the people. The activity also involved making a 'frottage' record in the logbook, which encouraged to experiment different textures of the place, promoting a playful and experimental exercise. Like the exercise of the 'focus' record, the touch instance involved choosing the situation that one wanted to represent and generated, therefore, an initial movement of exploration in the place for the selection of the observed object. Although this movement of exploration in the place is not reflected in the logbooks, it was well received by the participants of the experience, motivating to search for a specific place that warranted its selection. People freely exercised this option to portray the reality through touch, sometimes literally recognizing an object from the environment -leaf, stem, trunk- or generating various textures on paper as trial and error.

2.1 Bioroutes Wetlands of the Mapocho as a Memorable Experience

As a complement to the experience of the Bioroutes and with the objective of verifying the impact and interest generated in people about the Mapocho River and its wetland system, a survey was conducted 20 months after these itineraries.[15] For this, a group of participants was contacted through the same groups generated for the coordination of the Bioroutes, which surprisingly were still active.[16] In this group of respondents, all districts were represented: Maipú (18.6%), Padre Hurtado (13.9%), Peñaflor (18.6%), Talagante (20.9%), El Monte (14%), and Isla de Maipo (14%), with which a representative sample could be generated to gather qualified information about the experience.[17]

The Survey included a section related to the evaluation of the Bioroutes experience, in which it was asked to highlight "3 main aspects that you value from your experience" from a set of 9 alternatives to choose from: Novelty, Learning, Beauty, Recreation, Well-being, Dynamics, Identity, and Involvement. The results show that the main factor chosen is "Learning: Learning about the biodiversity and attributes of the river and its wetlands" (82%), which reflects the importance that people assign to better understanding these ecosystems directly and at the same time, that the 'learning' proposed by the Bioroutes is valued for these purposes. In second place, a large percentage of participants chose "Novelty: Visiting a natural place that I did not know" (53%), which demonstrates a genuine interest in better understanding the places with which they coexist and not just the usual public spaces. In third place "Well-being: Connecting with Nature as a contribution to my well-being" (44%), appears as relevant, as well as "Involvement: Being part of an action that protects the river and its wetlands" (44%). This suggests that there is a weighted value considered by people between these factors; what nature provides for my well-being and at the same time, actions that can be carried out for the well-being of nature itself. These are followed by the factor of "Beauty: appreciating the scenic beauty of the riparian wetlands" (40%) and the "Dynamics: Participating in an experience to perceive and value the natural environment" (33%), not as minor aspects of the experience. The least relevant factors were: "Identity: Feeling part of a place to which I belong as an inhabitant" (29%) and "Recreation: Pleasant moment with family, partner or friends" (19%). This suggests that the value assigned to the environment of rivers and wetlands is not necessarily

[15]The development of this survey was carried out with the collaboration of anthropologist Jens Benhor.

[16]A good indicator of people's interest in relation to the experience of the Bioroutes is that after 20 months, 67% of the people kept in touch with the WhatsApp groups generated to coordinate the activity. Of that universe, 40% answered the survey through an online system sent by this same platform, of which 53% were women and 42% men (−5%), in predominant age ranges of 25–34 years (31%) and 35–44 years (27%).

[17]In this article, only some selected questions with their results are included, given that the survey carried out considers a greater extension and its systematization is still the subject of ongoing research.

associated with a particular identity and, that the idea of a recreational place does not prevail, which is decisive when implementing infrastructure that goes beyond the recreational.

This valuation is also addressed by the question: "Regarding the activity carried out, select 2 aspects that you remember as important in your experience of valuing the Mapocho River through the senses", contemplating 5 alternatives. The answers show a marked emphasis on the option: "Listening attentively to the sounds of the environment allowed me to value its biodiversity" (96%), coinciding with the valuation that people recorded in the logs through direct experience. This was seconded by: "Contemplating and drawing a fragment of the place made me perceive better what we usually do not see" (47%), which denotes interest in a careful observation of the environment from the 'focus' exercise, different from a photographic image. In third place is the option: "Some smells of the river connected me with moments and personal experiences" (38%), which appeals to the aspect of memory in the records made in the logbook. The experience of touch is chosen in fourth place: "The textures of the place when rubbing them with a paper connected me with a playful sense" (22%), although this is later reinforced with the valuation of walking barefoot in contact with the water that many describe as a memorable experience.[18] Finally, "feeling the taste and collecting some plants allowed me to learn about their value" (18%) appears to a lesser extent, although the record in the logs is eloquent.[19]

To delve deeper into the previous point, participants were asked to freely describe a memorable situation: "In relation to your own experience of the Bioroute through the senses, mention an experience or learning that you remember as valuable". A considerable part of the responses (40%) allude to the memory of sound as a very valuable component of the Bioroute experience: "the sound of water, birds, it was ecstasy", "the exercise of listening to the river's waters sing"; "listening through a hydrophone to the texture of the flowing water", "listening to the sound of birds and discovering the biodiversity found in these wetlands", "I remember being amazed by the sounds that were enveloping, as well as being very dynamic and entertaining", "listening to the water as it flowed and produced a special moment". Just like in the logs, the value of sound in the well-being it produces is recognized: "the best thing was to experience the sounds of the river and its trees, to listen beyond, everyday life often does not allow us to hear these beautiful sounds", "listening to the wind rocking the trees next to the water with my eyes closed. Feeling of mental rest, pleasure, tranquility", "the gradual loss of city noise as you enter the wetland ecosystem", "The connection nature-being as a whole". Along with this, the beauty offered by these places and the need for their necessary care is recognized: "knowing such beautiful places in the territory I inhabit made me appreciate the place a lot", "a part of the place was very beautiful", "the walk and the view was very pleasant" "Connect and value better what surrounds us", "observe its fragility and degree of intervention", "find and enjoy an environment lost to me, blocked by the cement of the roads". On the other hand, the value of relating to nature through direct contact with it is highlighted: "I greatly valued the moment of taking off our shoes and walking around the place", "crossing a small stream with aquatic plants and gambusias[20] barefoot", "touching the water barefoot crossing pools", "climbing one of the trees of the place, getting wet in the wetland water, looking at birds of the place". This is combined with the value of exploration in nature beyond the recreational, which is reflected in the appreciation of the log: "I enjoyed being able to draw and connect deeply with the environment", "smelling the different herbs and learning to recognize them", "seeing the curiosity that the experience aroused in my son. He was delighted", "an entertaining way to connect respectfully with nature".

[18]The option of this activity was not part of the alternatives, which is described later in a free manner.

[19]This may have been affected by the fact that the participants did not keep their logs, and therefore having the collection of herbs and plants as learning over time was not a possible condition.

[20]Introduced species known as "mosquito fish", which has spread in some agricultural and irrigation ponds.

The condition of well-being in relation to the experience carried out is highlighted in statements that had to be selected within a spectrum of greater or lesser degree of agreement. To a large extent, people declared to "totally agree" or "agree" with the following statements: "I feel happy and I enjoy when I am in a river or wetland" (95%), "Visiting this place linked to the riverside gave me more well-being than others located in the city center" (95%), "The proximity to urban rivers and wetlands is very important for my well-being"(91%), "I would like to spend more time in the place I visited if I could" (93%), "I feel identified with the nature of this riverside place" (87%); and finally, "The place seems easily accessible if I wanted to return on my own" (53%), which leaves a significant challenge for access to these places.

Finally, the survey included a section related to the perception of value acquired through the experience, in which it asked "Regarding the experience carried out, mention the main aspect with which you identify", in front of 4 possible alternatives. A majority percentage chose "the experience of the Bioroute allowed me to better perceive the values of the Mapocho River" (62%) and not insignificant percentage declared "the tour of the Bioroute radically changed my perception of the Mapocho River" (29%). The least chosen option was "touring the Bioroute was just a good experience to interact with nature" (9%) and with no selection appears "The tour experience did not motivate me" (0%). This is of great interest for the implementation of accessible infrastructure in wetlands and their surroundings, as it indicates that enabling meaningful experiences in these environments can shape behaviors favorable for a better understanding and care of these ecosystems.

3 Conclusions

The work presented based on the Mapocho Wetlands Bioroutes as a proposal for revaluation based on a method of empirically conducted itineraries, hand in hand with organizations and communities, contributes to significantly evidencing the rich interaction and multiple benefits that people perceive in their approach to wetlands and how this is perceived as a relevant factor in terms of their own well-being. From the proposed qualitative method, it is possible to verify the value that people assign to direct contact with these ecosystems and how that value increases when they have access to experiences that enable a qualified and sensitive approach. The richness in the records is a living expression of what people perceive in contact with these natural environments and eloquent evidence of the wide range of possibilities that this interaction offers. The experience carried out therefore constitutes a contribution to the potential planning, design and implementation of access infrastructure and tours in these environments—such as permanent Bioroutes—conceived as an experience that allows a broad sensory record, and not only the prevalence of sight or viewpoints. The experience with other sensory dimensions—such as the sound and olfactory landscape –are equally or more valuable for people, and, based on what is expressed in the records, they play a central role in generating states of well-being: listening to the sound of water, birds, the wind rustling the trees, are recorded as a source of relaxation and calm. Similarly, the affective memory that smells generate is evident in the records, opening a field of exploration in the design of itineraries or tours to approach these environments. At the same time, far from being associated with a recreational place, people perceive these riparian wetland environments as reservoirs of attributes and opportunities to 'connect' in a more sensitive way with nature and at the same time value being able to learn from these ecosystems as an essential aspect in the relationship with the environment.

References

Brito, R. (2019). "Río Mapocho Aguas Abajo: Paisaje [Alterado] en Proceso. Reclamación de Archipiélagos extractivos en Maipú-Padre Hurtado". Thesis for the degrees of Architect and Master in Landscape Architecture, Pontificia Universidad Católica de Chile.

Brunelli, M. (2019). "From guided activities, to the interpretation of nature in the United States (1872-1920). The origins of the professional practice of heritage interpretation: between protection and education." *Interpretation Bulletin* (40). https://boletin.interpretaciondelpatrimonio.com/index.php/boletin/article/view/390/388

De Groot, R.S., Stuip, M.A.M., Finlayson, C.M. and Davidson, N. (2006). Valuation of wetlands: Guidelines for valuing the benefits derived from the services of wetland ecosystems, Ramsar Technical Report No. 3/No. 27 of the CBD Technical Series. Secretariat of the Ramsar Convention, Gland, Switzerland, and Secretariat of the Convention on Biological Diversity, Montreal, Canada.

De Urresti, A. (2019). Urban Wetlands History of a pioneering and citizen environmental protection law. Senate of Chile. Vice Presidency of the Senate, Senate Extension Center Editions.

Millennium Ecosystem Assessment. (2005). Ecosystems and Human Well-being: Wetlands and Water Synthesis. Washington, DC: World Resources Institute.

Evans, J. & Jones, P. (2011). The walking interview: Methodology, mobility and place. *Applied Geography, 31* (2), 849–858. https://www.researchgate.net/publication/251509902_The_walking_interview_methodology_mobility_and_place_Applied_Geography_312_849-858

Iturriaga, S. (2023). Mapocho Aguas Abajo. Atlas Visual para la Revalorización del Patrimonio y Paisaje ribereño. Santiago de Chile: Ediciones ARQ.

Law 21.202 on urban wetlands (January 2, 2020). Library of the National Congress of Chile. 2023.

Yurac, A. (2018). "Drawing method for the experience of the present moment". Thesis to opt for the Degree of Industrial Designer, Santiago de Chile: Diego Portales University.

Index

C. Rojas Quezada (ed.), *Urban Wetlands in Latin America*, Sustainable Development Goals Series,
https://doi.org/10.1007/978-3-031-69590-2

SPRINGER NATURE

GPSR Compliance

The European Union's (EU) General Product Safety Regulation (GPSR) is a set of rules that requires consumer products to be safe and our obligations to ensure this.

If you have any concerns about our products, you can contact us on ProductSafety@springernature.com

In case Publisher is established outside the EU, the EU authorized representative is:

Springer Nature Customer Service Center GmbH
Europaplatz 3
69115 Heidelberg, Germany

The manufacturer's authorised representative in the EU is Springer Nature Customer Service Centre GmbH, Europaplatz 3, 69115 Heidelberg, Germany. If you have any concerns regarding our products, please contact ProductSafety@springernature.com

Printed and bound by CPI Group (UK) Ltd, Croydon, CR0 4YY
01/10/2026
02225966-0020